D1134712

Measurement Theory and Practice in Kinesiology

Terry M. Wood, PhD

Weimo Zhu, PhD

Editors

Human Kinetics

Library of Congress Cataloging-in-Publication Data

Measurement theory and practice in kinesiology / Terry M. Wood, Weimo
   Zhu, editors.
      p. ; cm.
   Outgrowth of: Measurement concepts in physical education and exercise
   science. c1989
   Includes bibliographical references and index.
   ISBN 0-7360-4503-1 (hard cover)
   1. Physical fitness--Testing.   2. Physical education and training--Evalua-
   tion.   I. Wood, Terry M., 1949-   . II. Zhu, Weimo, 1955-   .
   III. Measurement concepts in physical education and exercise science.
   [DNLM: 1. Physical Fitness--physiology.   2. Exertion--physiology.
   3. Kinesiology, Applied.   4. Mathematical Computing.   5. Physical Education and Training.   QT 255 M48585 2006]
   GV436.M4264 2006
   613.7--dc22

                                                        2005028906
ISBN-10: 0-7360-4503-1
ISBN-13: 978-0-7360-4503-2

The Web addresses cited in this text were current as of September 22, 2005, unless otherwise noted.

Acquisitions Editor: Loarn D. Robertson, PhD; **Managing Editor:** Jeff King; **Copyedi-
tor:** Patsy Fortney; **Proofreader:** Joanna Hatzopoulos Portman; **Indexer:** Patsy Fortney;
Permission Manager: Dalene Reeder; **Graphic Designer:** Fred Starbird; **Graphic
Artist:** Kathleen Boudreau-Fuoss; **Photo Manager:** Sarah Ritz; **Cover Designer:** Fred
Starbird; **Photographer (cover):** Miyoung Lee; **Art Manager:** Kelly Hendren; **Printer:**
Sheridan Books

Printed in the United States of America   10   9   8   7   6   5   4   3   2   1

Human Kinetics
Web site: www.HumanKinetics.com

United States: Human Kinetics
P.O. Box 5076
Champaign, IL 61825-5076
800-747-4457
e-mail: humank@hkusa.com

Canada: Human Kinetics
475 Devonshire Road Unit 100
Windsor, ON N8Y 2L5
800-465-7301 (in Canada only)
e-mail: orders@hkcanada.com

Europe: Human Kinetics
107 Bradford Road, Stanningley
Leeds LS28 6AT, United Kingdom
+44 (0) 113 255 5665
e-mail: hk@hkeurope.com

Australia: Human Kinetics
57A Price Avenue, Lower Mitcham, South
Australia 5062
08 8277 1555
e-mail: liaw@hkaustralia.com

New Zealand: Human Kinetics
Division of Sports Distributors NZ Ltd.
P.O. Box 300 226 Albany,
North Shore City, Auckland
0064 9 448 1207
e-mail: info@humankinetics.co.nz

To Margaret "Jo" Safrit

Extraordinary scholar, compassionate mentor,
valued colleague, and friend.
Your contributions to measurement
in physical education and exercise science
serve as the foundation for this text
and for a new generation of measurement specialists.

Contents

List of Contributors

Chapter 1—
Terry M. Wood, PhD, Oregon State University, Corvallis, OR, USA
Chapter 2—
David A. Rowe, PhD, East Carolina University, Greenville, NC, USA
Matthew T. Mahar, EdD, East Carolina University, Greenville, NC, USA
Chapter 3—
Ted A. Baumgartner, PhD, University of Georgia, Athens, GA, USA
Chapter 4—
Weimo Zhu, PhD, University of Illinois at Urbana-Champaign, Urbana, IL, USA
Chapter 5—
Patricia Patterson, PhD, San Diego State University, San Diego, CA, USA
Chapter 6—
Weimo Zhu, PhD, University of Illinois at Urbana-Champaign, Urbana, IL, USA
Chapter 7—
Allan S. Cohen, PhD, University of Georgia, Athens, Georgia, USA
Chapter 8—
Richard C. Gershon, PhD, Center for Outcomes, Research and Education, Evanston, IL, USA
Betty A. Bergstrom, PhD, Promissor, Inc., Evanston, IL, USA
Chapter 9—
Fuzhong Li, PhD, Oregon Research Institute, Eugene, OR, USA
Peter Harmer, PhD, Willamette University, Salem, OR, USA
Chapter 10—
Ilhyeok Park, PhD, Seoul National University, Seoul, South Korea
Robert W. Schutz, PhD, University of British Columbia, Vancouver, British Columbia, Canada
Chapter 11—
Weimo Zhu, PhD, University of Illinois at Urbana-Champaign, Urbana, IL, USA
Anré Venter, PhD , University of Notre Dame, Notre Dame, Indiana, USA
Chapter 12—
Terry M. Wood, PhD, Oregon State University, Corvallis, OR, USA
Chapter 13—
Michael J. LaMonte, PhD, MPH, The Cooper Institute, Division of Epidemiology, Dallas, TX, USA
Barbara E. Ainsworth, PhD, MPH, San Diego State University, San Diego, CA , USA
Jared P. Reis, MC, San Diego State University, San Diego, CA, USA
Chapter 14—
Richard A. Washburn, PhD, University of Kansas, Lawrence, KS, USA
Rod K. Dishman, PhD, University of Georgia, Athens, GA, USA
Gregory Heath, Cardiovascular Health, Atlanta, GA, USA
Chapter 15—
Marilyn A. Looney, PED, Northern Illinois University, DeKalb, IL, USA
Chapter 16—
Andrew S. Jackson, PED, University of Houston, Houston, TX, USA

Preface

In 1989 Human Kinetics published *Measurement Concepts in Physical Education and Exercise Science*, the first concise reference of current measurement theory and practice for graduate students and scholars in physical education and exercise science. Over the past 17 years much has changed in the theory and practice of measurement. Advances in item response theory, structural equation modeling, and statistics coupled with the increasing sophistication of computer technology and programming have brought a wide array of advanced test construction and modeling methodologies to the desktops of researchers and practitioners. This book brings graduate students and working professionals interested in the study of physical activity up-to-date with recent advances in measurement theory and practice. Like its precursor, this book is aimed at the nonspecialist in measurement and can be used by students in graduate-level introductory measurement courses or as a reference by researchers and clinicians in the allied health professions.

The authors for this text were chosen for their nationally recognized expertise and their ability to bring difficult concepts to life for readers with minimal grounding in measurement or statistics. Whenever possible, emphasis has been placed on conceptual understanding and practical application, rather than on mathematical interpretation and complex formulas.

This book is composed of 16 chapters organized into four parts. Part I, Measurement Basics, consists of four chapters aimed at bridging the material from the 1989 text to this one and establishing a foundation for the remainder of the book. Chapters on validity, reliability and error of measurement, and item response theory combine an overview of measurement concepts with a brief introduction to current issues. Part II focuses on current issues in measurement ranging from ethical issues, test fairness, and test equating to computerized adaptive testing. The primary thrust of part II is the use of item response theory in tackling some of the most difficult issues in measurement practice.

Advanced statistical techniques useful in the measurement process provide the focus for part III. Structural equation modeling has become a popular method for validating constructs, assessing measurement error, and testing theoretical models in several areas related to the study of physical activity, most notably exercise and sport psychology. Additional topics in part III include new methods for analyzing repeated measures, particularly in the context of longitudinal analyses, and methods for analyzing very large and very small data sets.

Part IV, Measurement in Practice, examines measurement issues in selected topics related to physical activity. Chapter 12 highlights assessment issues generating from the current educational reform movement in physical education and is a must-read for teacher educators, preservice teachers, and inservice teachers. Readers interested in the issues surrounding accurate and logistically feasible measurement of physical activity in large-scale epidemiological studies will appreciate the treatment of these topics in chapters 13 and 14. Using tests for clinical decision making is the focus of chapter 15. Topics examined include the sensitivity, specificity, and predictive value of diagnostic tests in clinical settings. Part IV ends with a discussion of the challenging issues and recent advances in preemployment physical testing.

To facilitate reading this book, key words are set in boldface type and defined in a glossary at the end of the book.

PART **I**

Measurement Basics

CHAPTER 1

Introduction

Terry M. Wood

Measurement has become an integral part of human existence. I have joked to students in my measurement classes that we are measured "from womb to tomb" in the sense that measures are taken on humans prior to birth (e.g., sonograms) and throughout one's life (e.g., measures of physical growth and development, intellectual capacity, academic ability, job performance, attitudes, opinions, and so on). Then, at the very end of life, we are measured for the box!

In exercise science and other health professions, measurement is at the very core of research and professional practice. Physicians, physical therapists, physical fitness trainers, and athletic trainers use tests to diagnose, prescribe, and monitor patients and clients; physical education teachers measure and evaluate students in their classes to monitor progress toward educational objectives. In the research setting, tests are administered to participants to examine the veracity of hypotheses or to describe the current status of individuals regarding certain variables.

In all of these endeavors, a fundamental question is: What do the numbers resulting from testing mean? Other ways to ask this question are as follows:

- How certain can we be that the numbers we obtain from administering a test are reflective of the characteristic we are attempting to measure?
- Can the test scores be relevantly interpreted for other populations, purposes, or settings?
- Do the test scores consistently reflect the characteristic we are measuring (i.e., how much **measurement error** exists)?
- What is the mathematical relationship between the numbers we obtain from testing and the true ability of examinees on the underlying characteristic being measured?

Clearly, without adequate answers to these questions, the numbers obtained from testing patients, clients, students, or research participants are suspect. The consequences of interpreting scores with little meaning can range from disgruntled students, to incorrect conclusions regarding research hypotheses, and to physical harm to patients or clients. It is little wonder that measurement and test construction courses are required in many undergraduate and graduate programs in the health professions.

EVOLUTION OF MEASUREMENT THEORY AND PRACTICE

Traditionally, measurement theory and practice in exercise science has relied on adapting measurement theory from other academic disciplines such as psychology and education—disciplines that rely heavily on paper-and-pencil test instruments. For example, over the past 30 years **classical test theory** (Lord & Novick, 1968), developed for paper-and-pencil test instruments in education and psychology, has been used as the theoretical basis of measurement theory in exercise science, whereas the *Standards for Educational and Psychological Testing* (American Educational Research Association, American Psychological Association, & National Council on Measurement in Education, 1999), first published in 1954, has served as the guideline for testing practice. As several authors point out in this book, measurement in the motor and clinical domains presents unique issues that are not addressed in the cognitive or affective literature. Moreover, some of the measurement procedures advocated in the educational and psychological literature do not readily transfer to motor performance and clinical measures.

The measurement process (i.e., the process that examines the meaningfulness of test scores) under classical test theory is an indirect process that requires a rather large degree of inference (Suen, 1990). For example, imagine that we want to measure the characteristic "attitude toward physical activity" or ATPA. We develop 20 questions answered on a true–false scale and sum the total number of *true* responses to obtain a score that is supposed to reflect a person's ATPA. A person's total score on the ATPA scale is known as the **observed score**. From a measurement perspective, two crucial questions must be addressed:

1. Given that we can imagine that the person taking the ATPA scale has a hypothetical "true" value for whatever characteristic this scale is measuring, how precisely does the observed score match the true score? This concept is know as **test reliability** and is described in detail in chapter 3 by Ted Baumgartner. If the observed score does not precisely measure the person's true score, then the measurement device has a degree of measurement error. If the measurement error is zero, the observed score measures the true score exactly.

2. How well does a **true score** reflect the characteristic (or construct) that it is supposed to measure? Note that an observed score can be reliable in that it precisely measures some hypothetical true score, yet it may lack validity if the true score does not reflect the intended characteristic. For example, our ATPA scores might reflect the true scores of those who complete the scale, yet the construct actually being measured might be something other than ATPA (e.g., opinions of physical activity or response distortion). In other words, the ATPA scale might reliably measure the wrong characteristic! The

process of determining the degree to which true scores reflect the appropriate construct is known as **test validity** and is treated in chapter 2 by David Rowe and Matthew Mahar. In addition, ethical considerations intertwined with the examination of validity evidence are discussed in chapter 5 by Patricia Patterson.

In practice, the traditional sequence of events in constructing a test involves delineating the construct to be measured, developing a measurement instrument with an appropriate measurement scale (e.g., true–false, Likert scale, semantic differential), and conducting a number of separate studies to provide evidence for the validity and reliability of the instrument using samples of subjects from the intended population. If the validity and reliability studies are successful, then we *infer* that the observed scores measure what they are supposed to measure with sufficient precision.

A major drawback of the traditional measurement process is that the evidence for validity and reliability of observed scores is specific to the population, setting, and purposes for which the validity and reliability studies were conducted. Thus, if the measurement instrument is to be used with a different population, for example, new validity and reliability studies may need to be conducted. Furthermore, this traditional approach provides psychometric evidence about the total test score for a population and makes no formal attempt to link observed scores with the true underlying ability of individual respondents (Spray, 1989). For example, using the traditional indirect approach, we can only infer that observed scores on the ATPA scale accurately and precisely reflect a respondent's true value of ATPA. We have no direct evidence for linking a person's observed score with his or her true underlying level of ATPA.

These issues, among others, led to the development of **item response theory** (IRT) (see chapter 4 by Weimo Zhu) in the educational and psychological literature. Historically, the development of IRT began in the 1930s and 1940s, but the mathematical complexity of the theory delayed practical application until the advent of cheap, powerful computer hardware in the 1980s (Suen, 1990). Since the 1980s, IRT has caused a revolution in educational and psychological measurement as witnessed by the voluminous space devoted to the topic in measurement journals.

IRT mathematically models the relationship between characteristics of questions or items (e.g., item difficulty) and examinee ability on the underlying trait being measured. From a practical perspective, if a student correctly answers a difficult question on a final exam in a measurement class, what is our estimate of his or her underlying ability on the trait being measured (i.e., knowledge of measurement concepts)? Similarly, what is our estimate of that person's ability if he or she fails to answer an easy question? The ability to link a test taker's response on an item to his or her ability on the latent trait being measured has created new measurement opportunities such as equating scores across measurement instruments (see chapter 6 by Weimo Zhu), detecting differential item functioning (DIF) (see chapter 7 by

Allan Cohen), and using computerized adaptive testing (CAT) (see chapter 8 by Richard Gershon and Betty Bergstrom).

It is important to note that IRT augments but does not replace the traditional notion of test validity. The process of test validation outlined in chapter 2 plus the ethical considerations described in chapter 5 are crucial in constructing a test because we must first provide evidence that the items making up a measurement instrument adequately reflect the appropriate underlying or latent trait. Readers are urged to read McDonald (1999) for an in-depth treatment of the relationship between item response theory and classical test theory.

The advent of relatively cheap and powerful computers has led to the widespread availability of sophisticated statistical techniques aimed at modeling relationships among variables. These new techniques have augmented the measurement toolbox, particularly in the examination of test validity. Arguably the greatest impact in the measurement domain has been made by **structural equation modeling** (SEM), discussed in chapter 9 by Fuzhong Li and Peter Harmer. SEM has great utility in the process of **construct validation** via (a) examining the effectiveness of test items in measuring constructs (the measurement model) and (b) testing hypotheses regarding the theoretical relationships among constructs and observable events (the structural equation model). In addition, SEM can be employed in the measurement of change in longitudinal studies via **latent curve modeling** as outlined in chapter 10 by Ilhyeok Park and Robert Schutz. Chapter 10 also introduces **hierarchical linear modeling,** an extension of linear regression analysis, as a potent method for analyzing change in repeated measures data.

Chapter 11 by Weimo Zhu and Anre Venter tackles the difficult issues surrounding the analysis of very small and very large data sets. The ready accessibility of large public databases through Internet access has engendered a relatively new area of research into how to effectively "mine" these extensive depositories of data. At the other end of the spectrum, many data sets employed in exercise science research are relatively small and contribute to the lack of sufficient statistical power when analyzing data. Strategies for dealing with both types of data are presented in chapter 11.

Chapters 1 through 11 present an overview of contemporary measurement issues and methods. Some of the methods, such as those described in chapter 2 (test validity), chapter 3 (test reliability), and chapter 9 (SEM), are common in physical activity research. In contrast, methods such as IRT (chapter 4), DIF (chapter 7), CAT (chapter 8), hierarchical linear modeling and latent curve modeling (chapter 10), and data mining (chapter 11) have yet to gain a secure foothold as a result of their relatively recent introduction in the literature, the lack of sufficient exploration of their utility in physical activity research, and the mathematical and interpretational complexity of the methods. It is clear, however, that the future of measurement in physical activity research will follow the trend toward mathematical modeling and linking test scores to the underlying characteristics being measured that is so evident in other fields of inquiry.

MEASUREMENT IN PRACTICE

Chapters 12 through 16 describe and outline current measurement issues and practice in five areas of professional practice. The current round of educational reform and its reliance on content standards and benchmarks, along with increased emphasis on accountability for student learning, has elevated measurement practice in physical education. Chapter 12, by Terry Wood, summarizes the significant changes in the assessment of student learning in physical education and the accompanying measurement issues that need to be addressed. Chapter 13, by Michael LaMonte, Barbara Ainsworth, and Jared Reis, deals with the challenging issues in measuring physical activity. Chapter 14, by Richard Washburn, Rod Dishman, and Gregory Heath, discusses using epidemiologic methods to link lack of physical activity to disease or injury. Professionals working in the clinical setting will enjoy chapter 15 by Marilyn Looney, which focuses on issues associated with interpreting and evaluating the effectiveness of diagnostic tests. In the anchor position, chapter 16 by Andrew Jackson outlines many of the legal and measurement issues involved in preemployment physical testing.

SUMMARY

Safrit (1989) noted that although the history of measurement practice in exercise science can be traced back to the latter part of the 1800s when the primary interest was in measuring the physical dimensions (e.g., height, weight, girth) of the human body, "measurement as a formal area of study [in exercise science] has never been clearly defined" and "there is no label to represent the study of measurement in motor behavior" (p. 4). This is in contrast to the vibrant growth of measurement theory and practice in diverse fields such as psychology (psychometrics), education (edumetrics), biology (biometrics), and economics (econometrics), which boast their own academic units in universities, their own professional organizations, and a plethora of research publications dedicated to the area of study (e.g., *Psychometrika, Educational and Psychological Measurement, Applied Psychological Measurement, Applied Measurement in Education*).

There has been some progress in the advancement of measurement theory and practice in exercise science over the past 17 years. For example, the inaugural issue of the journal *Measurement in Physical Education and Exercise Science* in 2000 marked a milestone in the dissemination of theoretical and practical issues related to measurement in exercise science, and a kinesmetrics doctoral program has been created at the University of Illinois at Urbana-Champaign (www.kines.uiuc.edu/labwebpages/Kinesmetrics). Yet, Safrit's observations remain true today (see King, 1984; Looney, 1996; and Wood, 1989, for in-depth discussions concerning the plight of measurement as an area of study in exercise science). It is within this context that the authors of this text have chosen measurement topics from both the traditional measurement perspective and the more contemporary mathematical modeling approach in the hope that cutting-edge methodology will open new measurement vistas for researchers and practitioners in physical activity.

CHAPTER 2

Validity

David A. Rowe and Matthew T. Mahar

Validity is widely recognized as the most important concept in measurement. This is illustrated by the following statements describing validity from some of the leading authorities in measurement:

- " . . . the most fundamental consideration in developing and evaluating tests" (American Psychological Association, American Educational Research Association, & National Council on Measurement in Education, 1999, p. 9)

- " . . . pre-eminent among the various **psychometric** concepts" (Angoff, 1988, p. 19)

- " . . . the most critical step in test development and use" (Benson, 1998, p. 10)

Measurement texts by leading measurement specialists in exercise science, such as Baumgartner, Jackson, Mahar, and Rowe (2003); Morrow, Jackson, Disch, and Mood (2000); and Safrit and Wood (1989, 1995) contain similar statements.

The meaning of validity has changed considerably over the past 50 years. Perhaps the most radical and fundamental changes occurred during the 1980s, as described by Angoff (1988), Benson (1995), Cronbach (1989), Messick (1988), and Shepard (1993). In exercise science, we have typically followed the conceptualization of validity contained in the psychological measurement (psychometrics) literature, especially the various versions of the *Test Standards,*[1] published jointly by the American Psychological Association, the American Educational Research Association, and the National Council on Measurement in Education (1954, 1966, 1974, 1985, 1999), and seminal articles about validity written by two leading psychometricians, Lee Cronbach (1971, 1988, 1989; Cronbach & Meehl, 1955) and Samuel Messick (1975, 1980, 1981a, 1981b, 1988, 1989, 1994, 1995a, 1995b, 2000).

Describing in detail these changes in psychometric theory is not the main purpose of this chapter for two reasons. First, we could not possibly improve on the descriptions in those sources that we have cited, and to attempt to do so would be redundant. Recognizing, however, that some nonmeasurement specialists in exercise science may be unfamiliar with the psychometrics literature, we summarize here the major developments leading to the current state of thinking about validity among psychometricians:

- Validity and validation are now focused on the intended use and interpretation of test scores, rather than on viewing validity as a property of the test itself. Thus we now talk in terms of supporting the appropriateness of an intended interpretation of scores, instead of demonstrating that a test is valid.

- Prior to the 1985 *Test Standards*, the predominant paradigm delineated different, independent types of validity (e.g., content validity, criterion validity, construct validity; sometimes termed the "Trinitarian doctrine"). Psychometricians now view validity as a unitary concept supported by accumulating different types of validity evidence (e.g., content-related evidence).

- There has been a move away from "weak" programs of validation (involving perhaps a single study or a single piece of evidence, or the haphazard accumulation of correlations among variables) toward "strong" programs. This idea was first posited in a landmark article by Cronbach and Meehl (1955), although it has taken almost a half century to become accepted as the standard for practice in measurement research. Strong programs of validation involve, among other things, the use of methods of scientific inquiry to test hypotheses and the willingness to investigate rival hypotheses to explain patterns of variation in test scores.

- Construct validity is now viewed as being at the core of the validation process, or the "unifying force" (Messick, 1989, p. 92). Because interpretation of test scores intrinsically requires us to understand the construct we are measuring, all types of validity evidence, which formerly might have been viewed as being separate from construct validity, are now viewed in terms of how they help us to evaluate construct validity.

- Consideration of the potential ethical consequences of intended test use, first recognized explicitly by Messick (1980), is now part of the validation process (see chapter 5 by Pat Patterson). This change came about largely because of legal challenges to various educational and job-related tests that were demonstrated to be biased against certain groups within society, such as minorities, women, and people with disabilities.

- Because validity evidence pertains to specific uses of test scores (e.g., for specific decisions, with specific populations, in specific environments), and because there is a potentially limitless set of potential uses, the validation process is now recognized as being a "never-ending process" (Shepard, 1993, p. 407).

The second, and perhaps most important, reason we do not present in greater detail the current thinking about validity in the discipline of psychometrics is because *we are not in the discipline of psychometrics, but of* **kinesmetrics.** (We attribute the coining of this new term to its originator, Dr. Weimo Zhu.) Exercise science measurement specialists have followed psychological measurement theory for good reasons. The measurement of psychological constructs poses severe challenges (mostly because of the intangibility of the constructs involved), and to meet these challenges, psychometricians have applied painstaking rigor to the development of validity theory. Judging from the quality of many measurement studies appearing in journals such as *Research Quarterly for Exercise and Sport* and *Measurement in Physical Education and Exercise Science*, many of our own measurement specialists have now caught up with the developments in the psychometric world, answering Spray's (1989) call for us to "become familiar with . . . advances and techniques in psychometric theory for consideration of their implementation in shared behavior domains" (p. 230). However, we cannot escape the fact that, with the exceptions of exercise psychology and sport sociology, many of our subdisciplines have little in common with psychology and other behavioral sciences. It is time for exercise science measurement specialists to answer Wood's (1990) challenge to develop measurement theory and applications specific to exercise science.

Just as there are very good reasons for becoming familiar with the measurement theories of psychometricians, there are also very good reasons that these theories may not transfer readily to exercise science. Many differences exist between exercise science and psychology. Some of the major differences are as follows:

1. We have a greater variety of types of constructs in the subdisciplines of exercise science. Many nonpsychological constructs in subdisciplines such as exercise physiology (e.g., lipid metabolism and muscle fiber type) and biomechanics (e.g., ground reaction force and movement kinematics) are unique to exercise science.

2. Many subdisciplines of exercise science involve principles of the hard sciences that are governed by exact, well-defined metrics. Units of measurement are predefined (e.g., force in N·m; concentration in mmol; VO_2 in ml), leading to a more exact expectation that scores will not only reflect relative order on a scale, but also accurately reflect the precise amount of whatever is being measured. Sallis and Saelens (2000) differentiated between the concepts of relative validity (the degree to which scores from an instrument correctly place people in their true order on the underlying construct) and absolute validity (the degree of exactness with which the scores reflect the person's true level on the construct).

3. In many of our subdisciplines, we are able to obtain scores via highly objective procedures and with far less measurement error than in the behavioral sciences.

4. Many of our instruments are "passive" (i.e., do not require the active participation of the testee) and therefore remove the person being tested as a source of measurement error. Examples are measures of blood lipid and hormone levels and samples of muscle fibers and blood cells.

5. Whereas much of the psychometric literature focuses on paper-and-pencil instruments, the types of measurement procedures used in exercise science differ greatly from those types of instruments. In some of our subdisciplines we rarely use paper-and-pencil instruments other than for collecting demographic information.

In this chapter we present a paradigm for the process of construct validation in exercise science. The paradigm consists of accumulating evidence at three levels, or stages. The preliminary stage is the **definitional stage,** involving investigation of prior theory and empirical evidence to present a description of the nature of the construct. The intermediate level is the **confirmatory stage,** in which we collect evidence that will either confirm or disconfirm our description of the construct. The highest level is the **theory-testing stage.** Here we test theories of how the construct of interest fits into the broader context of our field of study, including considering how it is related to other constructs and identifying the determinants and outcomes of a person's status on the construct.

Figure 2.1 illustrates three important features, or characteristics, of this three-stage paradigm:

1. Validity research should build on prior evidence in a hierarchical fashion. Research at either of the higher stages should build on specific related evidence from the immediately preceding stage. Therefore, theory-testing research should build on directly relevant evidence at the confirmatory level. This confirmatory evidence should, in turn, have followed from directly relevant evidence at the definitional level. Straight arrows pointing upward within the pyramid illustrate this forward progression.

2. The iterative nature of the model incorporates statements by psychometricians that the process is ongoing. Although the theory-testing stage represents the highest level of validation, it does not denote the end of the process. We can only get closer to a perfect understanding of the construct; we can never achieve perfection and declare our validity research "finished."

3. Some overlap exists between the validity evidence gathered at the different stages, as represented by the shaded areas in figure 2.1. For example, later in this chapter we present the known-difference method as a way to confirm our definition of the construct (i.e., the confirmatory stage of validity research). However, a theoretical rationale should drive the known-difference study design; thus, we are also testing a small part of our theoretical nomological network (i.e., the theory-testing stage of validity research). Similarly, a structural equation model (see chapter 9 by Fuzhong Li and Peter Harmer) may be used to test theories about how the construct of interest is associated with other constructs (i.e., the theory-testing stage). However, within these results we may reconfirm via the measurement model the internal structure of the construct derived from earlier evidence at the confirmatory level.

Throughout the chapter we present examples and scenarios that illustrate relevance to the constructs and subdisciplines of exercise science. We refer to the psychometric literature to show how our paradigm interfaces with current psychometric theory. We strongly recommend that readers who are interested in understanding the meaning of construct validation derived from psychometrics follow up by reading some of the articles cited.

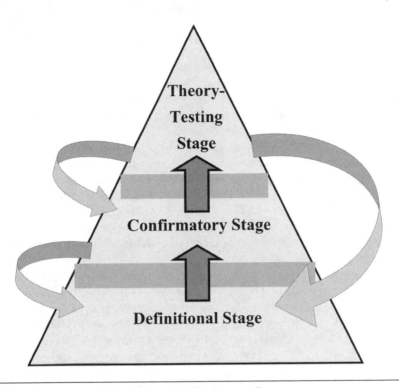

Figure 2.1   Illustration of the three-stage validation paradigm.

DEFINITIONAL STAGE

Obtaining definitional evidence of validity is an appropriate first step in the construct validation process. At the definitional stage, we define the construct of interest (theoretical domain) and propose tests to measure the construct (operational domain).

Historically, the term *content validity* has been used with written knowledge tests to represent definitional evidence of validity. Recognizing that psychomotor tests are different from written knowledge tests, Safrit (1981) described a procedure termed "logical validity" for tests of motor skill. For both of these procedures, expert judgments are used. In the written knowledge test context, experts judge whether items on a test represent all of the important content areas. In the motor skill test context, experts judge whether the movements needed to complete the skill test are representative of the movements required to perform well at that skill in an applied skill performance situation (e.g., a game or competition).

Note that both of these procedures focus on the test and not on the use or interpretation of the test scores. We stated earlier that the inferences made from test scores are validated and not the test itself. Therein lies a major concern with the use of the term *content validity*. Guion (1977) proposed that what is really meant by content validity is content representativeness.

Because of the differences between psychometrics and kinesmetrics, we feel that the term *definitional evidence of validity* more appropriately represents what should take place at this stage. The test-centered focus in this stage of validation is different from the focus in the following stages, but confirmatory and theory-testing evidence cannot be adequately examined unless the construct of interest is appropriately defined. For example, the definitional question What is soccer skill? must be examined prior to developing a test to assess the construct of interest (i.e., soccer skill). Soccer skill can be defined in various ways. For example, kicking accuracy, ball control and dribbling, and receiving ability (i.e., trapping the ball) are all essential components of good performance in soccer. In addition, tests of soccer ability can have various response formats, including kicking at a target, dribbling around moving or stationary targets, and receiving a pass from another person or a ball rebounding from a wall.

Developers of soccer skill tests must first determine the dimensions of soccer skill they are trying to measure. They must then develop tests that assess the dimensions that were defined in this stage of validation. Valid assessment of any motor skill or ability should start with defining the dimensions to be assessed. Development of the test (i.e., the operational domain) follows the definition of the most important dimensions of the skill or ability. We emphasize here that the construction of the theoretical domain drives the development of the operational domain.

Several problems can occur during the definitional evidence stage when the theoretical and operational domains are specified (Messick, 1989). One potential problem is construct underrepresentation, which occurs when the operational domain is too narrowly defined and the test does not adequately measure the construct of interest. For example, if only one test of soccer skill (say, a kicking accuracy test) is used to assess the construct of soccer skill, then the test may provide an incomplete assessment of a person's soccer skill. A person who can kick accurately but has poor ball control, poor dribbling skills, and trouble receiving the ball may be misclassified as highly skilled if only one test is used.

Construct irrelevancy is another potential problem that can occur during the definitional stage. Construct irrelevancy occurs when the test measures something other than what was intended. For example, if a soccer dribbling test is set up in such a way that participants are required to dribble around cones spaced far apart at regular intervals, then "good" performance on that test may be a function of running speed rather than a function of soccer dribbling skill. Interpretation of the results of this test as a measure of soccer dribbling skill would lack validity.

A final consideration related to the definitional stage concerns terminology. Kelley (1927) recognized two fallacies related to terminology. First, it is a fallacy to assume that two tests with the same name measure the same construct. He termed this the "jingle fallacy." The "jangle fallacy" occurs when one assumes that two tests with different names measure different constructs. Researchers need to consider the specific dimensions they want to assess and make sure the tests they choose measure these dimensions.

In summary, the definitional stage is a starting point for construct validation. The construct of interest should be well defined, and the operational domain should be specified to match this definition. The next sections of this chapter describe methods to obtain validity evidence after definitional evidence has been considered.

CONFIRMATORY STAGE

In the confirmatory stage of construct validation, we gather data to evaluate our definition of the construct. Based on previous work in the area, the researcher hypothesizes interrelationships among variables. Various methods are then used to test these hypothesized relationships. Five methods for testing these relationships are described next in this section: (a) regression techniques to obtain criterion-related evidence, (b) the multitrait–multimethod matrix to examine convergent and discriminant evidence, (c) factor analysis, (d) known-difference evidence, and (e) item response theory.

Criterion-Related Evidence

Criterion-related evidence of validity is demonstrated by examining the relationship between scores on a test and scores on a criterion measure of the construct of interest. Selection of the criterion measure is very important because the interpretation depends on how well the test scores correlate with the criterion measure. Thus, the criterion measure should be the most accurate measure of the construct. For example, in criterion-related validity studies that focus on estimation of aerobic capacity, measured $\dot{V}O_2$max, is the recognized gold standard and is thus used as the criterion measure.

Two types of criterion-related designs are typically used and provide **concurrent evidence of validity** or **predictive evidence of validity**. The designs differ based on the intended use of the data. Concurrent designs are used when one desires to use the test scores as a surrogate measure for the criterion. We often want to substitute the surrogate measure for the criterion when the test is more feasible to administer than the criterion measure. For example, those interested in assessing aerobic capacity may administer a field test (such as the PACER) instead of using measured $\dot{V}O_2$max (the criterion) because the PACER is less expensive and more feasible to administer. Predictive designs are used when one needs to make decisions about future behavior or future outcomes (e.g., performance or health outcomes).

Concurrent validity studies have been widely used to provide criterion-related validity evidence for measures of physical activity, fitness, and skill performance. Concurrent validity studies involve choosing a test for which we already have strong construct validity evidence to serve as the criterion measure, developing a surrogate measure of the construct (e.g., a field test), and administering both the criterion measure and surrogate measure to a large sample from the population of interest.

A Pearson correlation between the surrogate measure and the criterion measure is calculated. Most concurrent validity studies, especially with tests designed to estimate aerobic capacity or body composition, will provide results from simple or multiple regression analysis (e.g., Cureton, Sloniger, O'Bannon, Black, & McCormack, 1995; Leger, Mercier, Gadoury, & Lambert, 1988; Slaughter et al., 1988). The coefficient of determination (R^2) provides an indication of the variability in the criterion measure that can be predicted from the surrogate measure and can help researchers determine the extent to which the surrogate measure and the criterion measure are assessing the same construct.

The accuracy of the regression equation that is developed to estimate the criterion measure is examined with simple and multiple correlations, squared correlations, and the standard error of estimate (SEE). The SEE presents the accuracy of the regression equation in the units of measure of the criterion variable. These indexes of accuracy should be compared with values published in similar studies.

To test the accuracy of a regression equation, researchers should cross-validate it on a different sample from the one on which it was developed. **Cross-validation** should be conducted before the regression equation is widely recommended for general use. Cross-validation will help determine the stability of the regression equation and whether any systematic bias (i.e., overestimation or underestimation) in prediction exists. The average error (AE) or average difference between the criterion score (Y) and the predicted value of the criterion (Y') using the regression equation is calculated as

$$AE = \sqrt{\frac{\sum (Y - Y')^2}{N}} \tag{2.1}$$

where N is the sample size.

The assumption of independence of observations should be considered when interpreting the results of regression analyses. Combining multiple observations from the same participants (e.g., perhaps the same participants were measured at three different speeds on the treadmill) into one regression analysis violates the assumption of independence of observations. Consequences of violating this assumption are that the resulting correlation between the test and the criterion may be misleading and that the standard errors probably are misleading. Rather than combining multiple observations from the same people into a regression design, it is more appropriate to randomly select one data point for each person to include in the analysis. This solution requires that an adequately large sample be used for validation studies.

One use of predictive designs is when one is interested in predicting the criterion measure some time in the future from the surrogate measure, such as when success in college is predicted from Scholastic Assessment Test scores. Similar analyses are often used for predictive designs as are used for studies that examine concurrent evidence of validity. The difference between these designs is a function of when the criterion is measured. Alternatively, in some predictive designs we are not necessarily trying to predict scores on a criterion measure. Researchers, for example, may be trying to predict health outcomes such as premature death, myocardial infarction, or falling (e.g., Blair et al., 1989; Myers et al., 2002). In these types of designs, we are assessing the validity of scores on some test as predictors of a later outcome or performance. Different statistical analyses (e.g., logistic regression, survival analysis) are sometimes used in such cases.

In summary, to demonstrate concurrent evidence of validity, we demonstrate that the surrogate measure correlates well with and accurately predicts the criterion. If scores on the surrogate measure correlate highly with the criterion, then the confidence we already have in the criterion measure can, to some extent, be conferred on the surrogate measure. Predictive evidence of validity is used to examine whether the test under investigation is a reliable predictor of some future event.

Multitrait–Multimethod Matrix

Campbell and Fiske (1959) introduced the **multitrait–multimethod (MTMM) matrix** as a method to evaluate convergent and discriminant evidence of validity. **Convergent evidence of validity** is demonstrated when different measures of the same construct are moderately to highly correlated. **Discriminant evidence of validity** is obtained when measures of different constructs do not correlate as highly as measures of the same construct.

A person's score on any measurement is influenced by a number of factors. The factor that we hope has the greatest influence on the obtained score is the person's true score on the construct being measured. For example, if we are assessing aerobic capacity, we hope that differences in scores from the test result from differences in the true level of aerobic capacity. Factors other than a person's true score also can affect the score obtained from a test. For example, the method used for the assessment can be another systematic influence on the obtained score. Certain people tend to score more highly than others on tests of physical performance regardless of the construct that is being measured. In addition, the lower the reliability of a test, the more the scores are influenced by measurement error (see chapter 3 by Ted Baumgartner). The MTMM matrix provides a method to systematically evaluate the relative influence of the construct and method of measurement on test scores.

To construct an MTMM matrix, more than one **trait** must be measured with more than one method. A trait is a relatively stable characteristic of a person, such as muscular strength or extroversion. The simplest example would involve measurement of two different traits, each with two different methods. A two (traits)-by-two (methods) MTMM matrix will result in a four (rows)-by-four (columns) matrix containing correlations among the measures.

The issue of what constitutes a different method is important to consider. Researchers should be guided by the intended use of the test results when determining different methods of measurement. Although the MTMM method has rarely been employed in exercise science research, several examples from the psychology and educational measurement literature can lend some guidance. Different methods in prior research have included subjective versus objective ratings, multiple choice versus incomplete sentences, paper and pencil versus computer administered, questionnaire A versus questionnaire B, and verbal versus visual stimuli.

A simple two (traits)-by-two (methods) MTMM matrix is presented in table 2.1. In this design two traits were measured each with two different methods. Resulting correlations represent reliability coefficients computed separately and inserted along the diagonal, and bivariate correlations between various traits and methods. Appropriate reliability evidence should be guided by the intended use of the data. That is, the intended use of the data will help the researchers decide whether estimates of internal consistency or stability reliability are most appropriate for their design (see chapter 3 by Ted Baumgartner). Ideally, these reliability estimates should be obtained from the same sample that generated the bivariate correlations in the MTMM matrix.

Convergent validity coefficients (denoted CV in the table) represent correlations between the same trait assessed with different methods (i.e., monotrait–

Table 2.1   A Simple Multitrait—Multimethod Correlation Matrix

	Method 1		Method 2	
	Trait A	Trait B	Trait A	Trait B
Method 1				
Trait A	$R_{xx'}$			
Trait B	DV1	$R_{xx'}$		
Method 2				
Trait A	CV	DV2	$R_{xx'}$	
Trait B	DV2	CV	DV1	$R_{xx'}$

$R_{xx'}$ = reliability coefficient; CV = convergent validity coefficient (monotrait, heteromethod); DV1 = divergent validity coefficient (heterotrait, monomethod); DV2 = divergent validity coefficient (heterotrait, heteromethod).

Evaluation criteria:
1. Reliability coefficients should be high.
2. Convergent validity coefficients (CV) should be significantly higher than zero, and clinically, or practically, high.
3. Convergent validity coefficients (CV) should be higher than heterotrait, monomethod discriminant coefficients (DV1).
4. Convergent validity coefficients (CV) should be higher than heterotrait, heteromethod discriminant coefficients (DV2).
5. Within each correlation matrix, the pattern of correlations should be similar (this is more evident in larger matrices from designs with several traits and several methods).

heteromethod coefficients). Discriminant validity coefficients represent correlations between different traits measured with the same method (i.e., DV1 or heterotrait–monomethod coefficients) or correlations between different traits measured with different methods (i.e., DV2 or heterotrait–heteromethod coefficients).

Campbell and Fiske (1959) recommended a series of steps to guide interpretation of the MTMM matrix results. First, the convergent validity coefficients should be sufficiently high to warrant further investigation. Second, the convergent validity coefficients should be higher than the discriminant validity coefficients (DV2) in the same row and column of the same correlation block (only one correlation block is available in a two-by-two design). Third, the convergent validity coefficients should be higher than the discriminant validity coefficients (DV1) that have the similar measurement method. If the data do not fit this pattern, this indicates that a substantial part of the variance in test scores is attributable to the method used to measure the construct. As a result, we would lose confidence in the scores as indicators of the construct. Fourth, the pattern of heterotrait correlations in each monomethod block (i.e., DV1) should be similar.

Factor Analysis

Internal evidence of the construct is evaluated with **factor analysis** procedures. Internal evidence refers to the dimensionality of the construct. Loevinger (1957) also described internal evidence as *structural validity*. In research that examines questionnaire development, traditional factor analysis designs are used to examine whether individual items from a single subscale appear to measure the same dimension of the construct. For example, Rudisill, Mahar, and Meaney (1993) used factor analysis to determine that a set of items measured the unidimensional construct of motor-skill perceived competence.

Jackson and Coleman (1976) used factor analysis procedures to examine the construct of aerobic capacity. Performance scores from multiple running tests (e.g., 440 yd, 600 yd, 9 min run, 12 min run) were examined to determine which of the tests were assessing the same construct. Interpretation of the results of the factor analysis allowed the researchers to conclude that distance run tests needed to be at least 1/2 mi in distance to assess the construct of aerobic capacity. Shorter running tests measure a different construct.

Confirmatory factor analysis procedures (see chapter 9 by Fuzhong Li and Peter Harmer) can be used to test hypothesized dimensional structures (e.g., that several measures of upper body strength are measuring the same construct). Poor model fit indicates that the various measures of upper body strength are measuring different dimensions and possibly that they should not be used interchangeably.

Known-Difference Evidence

Two slightly different designs can be used to demonstrate **known-difference evidence** of validity. Cronbach and Meehl (1955) originally termed this approach the *known-groups method*, whereby two or more populations that are hypothesized to differ on a construct are compared. Another approach is to conduct an experiment in which an intervention is hypothesized to change the levels of the construct of interest and to compare preintervention and postintervention values. In the first approach, statistically significant and meaningful mean differences among groups would provide known-difference evidence supporting validity. In the latter approach, known-difference evidence supporting validity is obtained if the measurement instrument is able to detect a meaningful change in the hypothesized construct. Conversely, if results do not support the hypothesized differences or changes, then confidence in the test scores as measures of the construct would decrease.

Known-difference evidence studies are well suited for measures of physical skills and physical activity, in which it may be relatively easy to identify groups known to differ or to hypothesize changes consequent to an intervention. Selections of the different populations used to compare the experimental interventions should be guided by the intended use of the data to answer important questions. For example, if a skills test were intended for use with higher-level athletes, a statistically significant mean difference between mean scores of professional and high school athletes would be inappropriate validity evidence for this purpose. It would be more appropriate to investigate whether a mean difference exists between professional and semiprofessional athletes.

Item Response Theory

Item response theory (IRT) is an advanced statistical technique that was developed in the psychometric literature as an attempt to obtain more information from responses to test items (see chapter 4 by Weimo Zhu). In classical test theory, information about a person's true score and about test parameters are inferred from scores on the total test. In IRT, each item and each person's response to each item are treated as sources of information. That is, a person's true score and characteristics are estimated from individual items and responses to those individual items, not just from total test scores.

An assumption of IRT is that a single **latent trait** underlies each person's performance on each test item (unidimensionality). Thus, the observed score on a test is a function of this latent trait. In addition, the error in an observed test score is a function of this latent trait. This means that the impact of error on observed test scores may vary with different levels of ability. IRT provides researchers with the ability to measure the construct of interest with any set of test items that measure that unidimensional trait (i.e., the researcher is not limited to a particular test). This allows for the possibility of measuring the construct by administering different items to different people.

These characteristics of IRT allow for several applications that are not easily accomplished with classical test theory, including calibration of large pools of test items (known as item banking), computerized adaptive testing (i.e., tailoring tests to individuals based on their latent ability that underlies performance; see chapter 8 by Richard Gershon), minimizing misclassification errors in mastery testing, assessing change in ability over time, and test equating (i.e., placing two or more tests on a common scale; see chapter 6 by Weimo Zhu). Safrit, Zhu, Costa, and Zhang (1992) used IRT to estimate the difficulty level of various sit-up tests, and Zhu (1998) provided a description of the use of IRT in test equating. Safrit, Cohen, and Costa (1989) presented a tutorial, which described several potential benefits of the use of IRT.

In summary, we gather data to confirm (or disconfirm) hypotheses about the nature of a construct and how well tests measure that construct in the confirmatory stage of construct validation. We described five different methods for this purpose and recommend that evidence from several of these types of designs be examined. Extensive research demonstrating different types of evidence of validity can provide us with confidence regarding the dimensionality of the construct and confidence that our intended interpretations of data from the test are appropriate.

THEORY-TESTING STAGE

The theory-testing stage of construct validation involves using the scientific method to test theories about the construct of interest. Messick (1989) was the first to recognize that theory testing should be a major focus of a strong program of construct validation. Testing the many theories that can surround a construct can be a daunting task. Researchers, however, should recognize that establishment of validity is a continuous process (Shepard, 1993). Cronbach (1989) proposed that validation should be a "community process" that requires "widespread support in the relevant community" (p. 164). This community approach encourages the application of different perspectives to the process of validation.

The **nomological network** and **structural equation modeling (SEM)** can be used to organize knowledge learned from previous studies and to develop coherent theories about constructs. The nomological network is a hypothesized network of interrelationships between the construct of interest and other constructs (Cronbach & Meehl, 1955), whereas SEM is a statistical technique used to examine links in the nomological net. At the early stages of theory development, the network is limited. As the body of research evidence increases, the network may be modified and will develop in size and complexity.

An example of a theoretical model in exercise science, presented in figure 2.2, was proposed in a review article by Rejeski and Focht (2002). These researchers extended previously supported theoretical explanations for determinants of functional limitation in older adults. Their model incorporates all of the characteristics of the nomological network described by Cronbach and Meehl (1955) for theory-driven construct validation. In addition, the model demonstrates the recommendation of Cronbach and Meehl that we start with simple models containing a small number of constructs and build on this as we accumulate a body of evidence. Rejeski and Focht's model contains constructs of pathology (abnormal disruptions of the body's biological processes), impairment (abnormalities in the anatomy, physiology, or emotions), physical symptoms (e.g., pain), self-efficacy (belief in one's ability to accomplish specific functional tasks), and functional limitations (actual impairments in the ability to perform functional tasks). Within the model diagram, each construct is represented by a circle. The arrowhead lines indicate the direction of influence of one construct on another. Some lines indicate direct effects. For example, the line from active pathology to physical symptoms indicates that pathologies lead directly to symptoms (a logical and theoretically defensible idea). Other lines indicate indirect, or mediating, effects. For example, the line leading from impairments to self-efficacy and subsequently to functional limitation suggests that impairments indirectly lead to functional limitation via their effect on self-efficacy.

To test this model using SEM techniques, we would use a cross-sectional design conducted with older adults (the population for whom the model was suggested). Every person would be measured on every construct in the model, using instruments for which we have acceptable previous evidence of validity (without this, we would be operating at the theory-testing stage prematurely). We would then generate a large matrix containing all of the covariances (a covariance is simply an unstandardized correlation) between each pair of measures. The covariance matrix would then be used as input to a SEM data analysis. Description of the complex procedures involved in performing and interpreting structural equation modeling can be found in several books (Bollen, 1989; Bollen & Long, 1993; Hoyle, 1995; Schumacker & Lomax, 1996). Several excellent articles that provide an overview of this technique have been published in the journal *Sociological Methods and Research*, and a relatively new journal, *Structural Equation Modeling*, is devoted to methods and applications of the procedure. In addition, Fuzhong Li and Peter

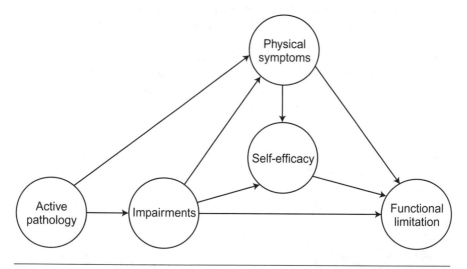

Figure 2.2 A model diagram for the construct of functional limitation in older adults.

Adapted, by permission, from W.J. Rejecki and B.C. Focht, 2002. "Aging and physical disability: On integrating group and individual counseling with the promotion of physical activity." *Exercise and Sport Sciences Reviews*, 30(4):166-170.

Harmer present a comprehensive description of SEM and its application in exercise science in chapter 9 of this text.

Major potential outcomes of SEM are as follows:

1. Summary statistical indexes are used to evaluate the overall fit of the proposed model to explain the data set.
2. Statistical indexes allow comparison of two or more competing models.
3. Parameter estimates are used to draw inferences about the size and direction of relationships among different constructs in the model.
4. Factor loadings allow determination of how well the instruments are measuring the intended constructs within the model.
5. Modification indexes can lead, via empirical evidence, to modifications of the model.

Clearly, structural equation modeling provides powerful techniques to improve the ability to test hypothesized models. New software packages, such as AMOS (Smallwaters Corporation, Chicago), are making the procedure easier to use; however, with increased ease of use comes the increased risk that those with an inadequate understanding of the procedure will use it inappropriately.

In the theory-testing stage, the nomological network is proposed based on previous research and is then tested with an appropriate statistical technique. Expertise and experience are required to correctly use and interpret these advanced statistical techniques, emphasizing the importance of Cronbach's (1989) recommendations regarding the importance of collaboration within the community of researchers when developing a strong program of construct validation.

REPORTING VALIDITY

The validation process may involve any of a multitude of different research designs and statistical analyses. It is impossible within the space available to list prescriptive requirements of what should be reported for each of the possible designs and analyses. Researchers should consult statistical texts and statistical experts to ensure that their analyses and results are reported correctly and completely. It is evident, however, that few researchers report sufficient information about validity evidence to enable the reader to judge the quality of their research (Whittington, 1998). Recommendations for information that should be documented when presenting measurement studies are contained within pages 9 through 70 of the *Test Standards* (American Psychological Association, American Educational Research Association, & National Council on Measurement in Education, 1999). Six important considerations when reporting validity results are as follows:

1. *Specify the situation(s) to which the validity evidence should be applied.* Validity is context specific. Too often, a statement that "previous validity

evidence has been obtained for this instrument" is taken at the global level, implying that no further information is required for uses of the instrument. Validity studies should include specific conclusion statements regarding appropriate uses and interpretations, specific contexts, and specified populations. Future users of the instrument in different contexts, with different populations, or for different purposes would then be under no illusion that this previous evidence supports their intended use.

2. *Use language that is appropriate to our current understanding of construct validity.* The 1985 *Test Standards* introduced the idea that validity should no longer be viewed as a property of a test, but rather as an evaluation of the "the appropriateness, meaningfulness, and usefulness of the specific inferences made from test scores" (American Psychological Association, American Educational Research Association, & National Council on Measurement in Education, 1999, p. 9). Today, it is still common to find in the exercise science literature statements such as "the test is valid," "the factorial validity of the test was supported," or "the validity of the test has been proven previously." Statements such as these, especially when read by those unfamiliar with measurement theory, serve only to perpetuate outdated ideas of validity. We should instead use phrases such as, "the validity of using percent fat standards for cheerleading team selection was not supported by the evidence," or "we conclude from the confirmatory factor analysis results that competition anxiety comprises separate cognitive and somatic components." Similarly, phrases such as "the study is designed to investigate the content validity of . . ." should be reworded to "the study is designed to evaluate content-related *evidence* for . . .", in recognition of the current unified perspective that there are different types of validity evidence, supporting construct validity, rather than different types of validity (Wood, 1989).

3. *Justify the validation level of the current study.* Researchers should present information, or cite relevant studies, to demonstrate that the current study is at the appropriate point in the line of inquiry. Evidence of reliability of the instrument in the intended measurement context is a prerequisite to *all* validity studies. An example of other appropriate evidence is exploratory factor analysis, or strong theory, to justify conducting a confirmatory factor analysis study. Similarly, a study in which researchers test a multiple-construct theoretical model should contain evidence supporting the measurement models of the model components (individual constructs). On a related note, suggestions for future research should be specific to logical next steps in the line of inquiry, and related to a similar intended use or interpretation of a test, rather than general suggestions such as conducting the same study in other populations.

4. *Evaluate the research findings against the criterion of the current state of knowledge.* Too often we use convenient standard criteria for evaluating

correlations between constructs, or attach too much importance to statistical significance. For example, a multiple R of .70 between a set of predictors and a gold standard criterion is only "acceptable" if it improves on prior research in the area, and a predictor that "adds significantly" to the explained variance in a criterion is only useful if the percentage of added variance, or the reduction in prediction error, is meaningful. Similarly, because of reliability concerns with instruments used in behavioral aspects of exercise science, an exercise psychologist might be quite satisfied with a moderate ($r = .50$) monotrait–heteromethod correlation. In biomechanics, in which objective instruments are more reliable, a monotrait–heteromethod correlation of $r = .50$ might be considered quite disastrous.

5. *Be honest about limitations.* All researchers know that the reality of data collection is messier than research methods texts suggest. Rarely, however, do we read reports of such in research articles. Researchers should keep a research diary to record any untoward events that occurred during data collection. Readers of research would consequently be able to interpret the results with full knowledge of any information that might necessitate caution. For example, if calibration drift in a skinfold caliper was not discovered until the end of a study, or schoolchildren in a group questionnaire administration were boisterous, researchers should include this in a limitations section, and where possible, provide some indication of how this may influence the interpretation of the results.

6. *Set the study in the context of theory.* Although much of our research evidence is data based, we should be wary of letting data be the major driving force of theory. New diagnostic techniques in factor analysis and SEM software make it too easy to drift from theoretically plausible explanations, on the basis of sample-specific nuances in our data. Similarly, in regression analysis it is too easy to let the data determine the selection of predictor variables (witness the number of studies in which stepwise analyses are used). Exploratory studies will be driven to a greater extent by findings in the data. In such situations we should be cautious in interpreting the results, be open to alternative explanations, and try to disconfirm an explanation as soon as possible. We should be mindful of Popper's (1959) declaration that an explanation of findings is strengthened by unsuccessful attempts to disprove it.

SUMMARY

Validity refers to the degree to which research evidence supports the interpretation and use of test scores. Validation is the *process* of collecting evidence and refining the theory surrounding constructs. The validation process is continuous and context specific, comprises many different types of evidence, and is iterative (therefore ongoing). In this chapter we summarized the many changes in validity theory within the psychometric literature and made a case that psychometric theory does not always fit the subdisciplines of exercise science. We presented a framework within which exercise scientists can apply the validation process. The framework assigns the validation process to three hierarchical levels. The definitional stage involves initial attempts to describe the nature of the construct, using theory and prior research evidence. In the confirmatory stage, researchers design studies to confirm or disconfirm the construct definition. At the highest level of construct validation, the theory-testing stage, researchers investigate how the construct of interest fits into the broader context of the research area. Within each stage, we provided examples to illustrate how the three-stage paradigm fits exercise science, and we also explained how the psychometric validity literature fits into the three-stage paradigm.

Psychometricians usually use the term *test* to refer to instruments for which there is a preconceived idea of what constitutes a "correct" response (e.g., a test of writing ability or aptitude for mathematics). However, within the *Test Standards*, the term is used more generally, for convenience, to include instruments intended for measuring characteristics such as opinions and reactions to situations, or for diagnostic purposes. Similarly, within this chapter we use the term *test* to include all instruments used to measure a person's status on a construct. Within exercise science this includes, but is not limited to, paper-and-pencil instruments, mechanical devices, observational tools, samples of work, and trials of physical performance.

CHAPTER 3

Reliability and Error of Measurement

Ted A. Baumgartner

A group of people is measured, and each person has a score. How much trust or faith can be placed in these collected scores? If these people are measured again on some other day or repeatedly during a day, assuming their ability has not changed, will the same scores be obtained? Evidence of the degree to which scores have changed or will not change is reported by using either a correlation coefficient called a **reliability coefficient** or an estimate of measurement error called the **standard error of measurement**.

Data, a set of scores, may be collected using many different methods such as a test, a measuring procedure, an instrument, or a questionnaire. Throughout this chapter, the term **test** will be used to denote the data measurement or data collection method and the term **score** will be used to denote the value obtained for a person as a result of the measurement or data collection.

RELIABILITY CONSIDERATIONS

How a reliability coefficient is calculated depends on many things: (a) the type of data (categorical, ordinal, interval), (b) the type of test standard (**norm-referenced, criterion-referenced**), (c) the criterion score (mean score, best score), (d) the type of reliability coefficient (internal consistency, stability), (e) reliability needs, and (f) the type of test. So, one reliability technique or coefficient does not fit all situations. Readers needing to review types of data and types of test standards should consult a measurement book such as Baumgartner, Jackson, Mahar, and Rowe (2003); Morrow, Jackson, Disch, and Mood (2000); or Safrit and Wood (1995). Baumgartner and colleagues (2003) presented a comprehensive discussion of selecting a criterion score. As discussed in any measurement book, **stability reliability** refers to the consistency of test scores across days, whereas **internal consistency reliability** refers to the consistency of test scores within a day. See Baumgartner (2000) for a brief overview of estimating stability reliability and some issues related to estimating reliability.

Reliability is a characteristic of a test score and not a test. People used to say, "the test is reliable," but this is no longer considered correct. For a comprehensive discussion of this issue, see Vacha-Haase (1998) and the April 2000 issue of *Educational and Psychological Measurement*. If the test scores are reliable, the scores are unchanging within a day or between 2 days, which are relatively close together.

Many techniques are used to obtain an indication of the reliability of test scores. Each technique yields an estimate of the test score reliability. You might think of the test scores as having a certain amount of reliability at the population level, and each reliability technique yields an estimate of the population reliability value.

RELIABILITY ESTIMATION FOR NORM-REFERENCED PHYSICAL PERFORMANCE TESTS

Physical performance tests with norm-referenced standards are the most common type of test administered in health and human performance (physical education, exercise science, and so on). Estimating the reliability of the scores from these tests is usually necessary. So, the emphasis in this chapter is on estimating the reliability of norm-referenced physical performance tests.

Reliability Theory

Classical reliability theory is presented to obtain a definition formula for describing a reliability coefficient. Then various calculational formulas are presented based on the definition formula.

In theory the **obtained score** (x), which is recorded as the score for a person, is composed of a **true score** (t) and an **error score** (e). Thus, a single obtained score can be described as $x = t + e$. For example, if for a given person, $x = 32$ and $t = 30$, then $e = 2$ ($32 = 30 + 2$). If the person is measured again the same day or the following day, the true score is still 30, but the error score might be -3.0, so $x = 27$ [$27 = 30 + (-3.0)$]. Seldom, however, are the obtained score and true score of a person equal because measurement error usually is present. For each person in a group there is an obtained score composed of a true score and an error score. The variance (s^2) for the obtained scores, true scores, and error scores is calculated and the variance for the obtained scores is composed of the variance for the true scores and the variance for the error scores, so $s^2_x = s^2_t + s^2_e$. Reliability is defined as the ratio of true score variance and obtained score variance or the proportion of obtained score variance that is explained by the true difference among individuals.

$$\text{Reliability} = s^2_t \,/\, s^2_x \text{ or } (s^2_x - s^2_e) \,/\, s^2_x \qquad (3.1)$$

Notice that if obtained scores are perfectly reliable, then there are no error scores. Thus, the variance for obtained scores equals the variance for true scores, and reliability using equation (3.1) is 1.00. If obtained scores are totally due to error scores, then there are no true scores. Thus, the variance for true scores is zero, and reliability by equation (3.1) is 0.00. So, reliability can vary from a minimum of 0.00 (no reliability) to a maximum of 1.00 (perfect reliability).

Equation (3.1) is theoretically based. It is not possible to determine true scores and true score variance. We can, however, calculate obtained score variance and estimate error score variance to estimate reliability using the second form of equation (3.1). We use the analysis of variance (ANOVA) technique to obtain these variance estimates. The ANOVA technique, commonly used by researchers to test differences among group means, is modified to estimate test score reliability.

Introduction to Estimating Reliability

Before discussing the use of ANOVA in estimating test score reliability, several topics must be discussed. Many different techniques can be used to calculate a reliability coefficient, and each technique yields an estimate of test score reliability. For over 20 years it has been known and presented in numerous sources that the Pearson product–moment **correlation coefficient** or **interclass correlation coefficient** is not an acceptable estimate of test score reliability. In most instances the use of an **intraclass correlation coefficient** based on ANOVA values is more appropriate. For a concise discussion of this issue, see Baumgartner and Jackson (1982) or Safrit (1973). Further, to calculate an estimate of test score reliability, each person must have at least two scores (e.g., two trials within a day, 2 days, two raters). How these two or more scores are used to obtain a test score called the **criterion score** for each person influences what formula is used to estimate test score reliability. Often, more scores for each person than will be collected or used in the future are collected to calculate an estimate of reliability. For example, for each person a score is collected on each of 2 days so each person will have two scores and an estimate of test score reliability can be calculated, but in the future the researcher or teacher or practitioner would like to collect a score for each person on 1 day and have faith that the score is reliable. Using research terms, we may think of reliability being estimated in a pilot study for a measurement situation, which will occur later in a research study. This will become clearer as some common situations are presented in which ANOVA is used to obtain criterion score reliability estimates.

In the following sections, two different ANOVA models are discussed in terms of calculating an estimate of criterion score reliability. The formulas for the calculations in ANOVA are not presented, and it is assumed that the values are obtained by using a computer program package such as SPSS. Equations for calculating a reliability coefficient using the ANOVA values are presented.

One-Way ANOVA

In this ANOVA model each person is considered to be a group, and the multiple measures (days, trials, scorers, raters) for each person are considered the data for the group. For example, if each of 10 people is administered three trials of a 40 yd dash within a day, the data are organized as presented in table 3.1. Looking at table 3.1, we see that person 1 scored 5.0 on the first trial and 5.2 on the third trial. Using the one-way ANOVA program in SPSS, a one-way ANOVA is applied to

Table 3.1   Three 40 Yd Dash Scores for 10 People Arranged in a One-Way ANOVA Model

Person									
1	2	3	4	5	6	7	8	9	10
5.0	4.8	4.5	5.2	4.9	4.7	5.3	4.4	4.8	5.0
4.9	4.9	4.4	5.2	5.0	4.6	5.4	4.3	4.9	4.8
5.2	4.6	4.6	5.1	5.2	4.7	5.3	4.2	4.8	4.8

Table 3.2   One-Way ANOVA Results for the Data in Table 3.1

Source	Sum of Squares	DF	Mean Square	F
Between Groups	2.60	9	.29	29.00
Within Groups	.23	20	.01	
Total	2.83	29		

the data in table 3.1 and the results of this analysis are reported in table 3.2. The mean square (*MS*) values in table 3.2 are used to calculate an intraclass correlation coefficient (*R*) as an estimate of the reliability of the test scores.

Mean Score Is the Criterion Score

One equation for calculating *R* is

$$R = (MS_B - MS_W) / MS_B, \tag{3.2}$$

where MS_B = mean square between groups and MS_W = mean square within groups. Equation (3.2) is used when the criterion score is the mean score for each person. For example, the criterion score for person 1 in table 3.1 is 5.03 (i.e., the mean of 5.0, 4.9, and 5.2). Several equations for calculating *R* will be presented, and each equation is for a particular criterion score. A researcher or practitioner should decide on the criterion score and then select the appropriate equation for calculating *R*. Because the data in table 3.1 are multiple trials of a test within a day, deciding that the criterion score for a person is the mean of the person's trial scores is logical. Using equation (3.2) and the information in table 3.2, the estimated reliability of the mean score over three trials for each person is $R = (.29 - .01) / .29 = .97$. Because the value of *R* can vary from 0.00 to 1.00, an *R* of .97 is quite high. This *R* is an indication of internal consistency reliability because the multiple measures (trials) are all within a day.

We need to understand the **mean square** values in table 3.2 and equation (3.2). Mean square between (MS_B) is an indication of difference among the people in mean score. If all people have the same mean score, MS_B will equal zero. The more people differ in mean score, the larger is MS_B. Mean square within (MS_W) is an indication of difference among the test scores of a person. If in table 3.1 person 1 scored 5.0 on each trial, person 2 scored 4.8 on each trial, and so forth, then MS_W is zero. To the extent that people are not consistently scoring the same from trial to trial, MS_W is greater than zero. Variation in the scores of a person from trial to trial indicates a lack of reliability. Mean square within is an estimate of error score variance in equation (3.1). The difference between MS_B and MS_W is an estimate of true score variance in equation (3.1).

Further, we need to understand what factors influence the value of R in equation (3.2). If MS_B is large and MS_W is small, R is high. If MS_B is small even if MS_W is small, R is not high. When people in the group have many different scores, MS_B is large. When people are consistent in their performance from trial to trial, MS_W is small. To obtain a high R, there must be many different scores possible on a test and people receiving many different scores. So, a low R may be due to all the people receiving about the same score. Sometimes changing the unit of measurement to increase the number of different scores can affect R. If height is measured to the nearest foot, probably all college students will have a score of 5, 6, or 7, and R is low. However, if height is measured to the nearest inch, college students will have many different scores and R will be high. Reliability of test scores for people very homogeneous in ability may be low, whereas for people very heterogeneous in ability, reliability may be high. If a pilot study group of people is considerably more homogeneous or heterogeneous in ability than is typical, the R obtained for the scores of the pilot study group might be suspect as a good estimate of the reliability of the scores for the research study group. Groups composed of a small number of people may be more homogeneous or heterogeneous in ability than is typical.

The R may be low because MS_W is large. Mean square within is large when many people in a group are not consistently scoring the same from trial to trial. If people are not ready to be tested or don't understand how to take a test because of lack of experience with the test, their test scores may vary considerably from trial to trial. Making sure people have experience with a test prior to the day they are tested and giving people sufficient warm-up or practice the day of the test may result in a small MS_W and, thus, a high R. Looney (2000) discussed when an intraclass correlation coefficient is misleading. Spray (1982) showed that autocorrelated errors have an effect on intraclass reliability estimates.

Best Score Is the Criterion Score

Commonly, researchers and practitioners administer multiple trials of a physical performance test within a day and use the best score of each person as the criterion score. The reliability estimate for the best score is calculated by the formula

$$R = [MS_B - MS_W] / [MS_B + (k / k' - 1)(MS_W)], \qquad (3.3)$$

where k = the number of trials administered and the score of these trials analyzed using ANOVA, and k' = the number of trials used in the criterion score. For the data in table 3.1 and using the ANOVA values in table 3.2, the estimated reliability of the best score using equation (3.3) with $k = 3$ and $k' = 1$ is $R = [.29 - .01] / [.29 + (3 / 1 - 1)(.01)] = .90$. This R is high but not as high as the R of .97 obtained with equation (3.2) for a criterion score that is the mean of the trial scores. This R is an indication of internal consistency reliability. Generally, the more scores included in the criterion score, the higher is the R for the criterion score. As we see in this example, using three scores in the criterion score (mean score) resulted in a higher R than using one score as the criterion score (best score).

Other Criterion Scores

The researcher or practitioner can decide to use any criterion score. If for the data in table 3.1 the researcher throws out each person's worst score and the criterion score is the mean of the two remaining scores, the estimated reliability for the criterion score using equation (3.3) with $k = 3$ and $k' = 2$ is $R = [.29 - .01] / [.29 + (3 / 2 - 1)(.01)] = .95$.

Notice that if the researcher throws out the trial 1 score for each person and uses the mean of the trial 2 and trial 3 scores in table 3.1 as the criterion score for each person, using equation (3.3) with $k = 3$ and $k' = 2$, the R is .95, the same as throwing out the worst score for each person. Finally, note that when k and k' are equal, equation (3.3) is the same as equation (3.2). So, equation (3.2) is a special case of equation (3.3), and equation (3.3) can be used in all situations.

It is common for a researcher or practitioner to administer a physical performance test to a group of people and then 1 to 7 days later to administer the test to the group of people a second time to obtain an estimate of stability reliability, which is also called **test–retest reliability**. In this situation the researcher is administering a physical performance test twice so an estimate of stability reliability can be calculated, but in the future the researcher would like to administer the tests on 1 day and have evidence that the test scores are reliable. Thus, the criterion score is a score on 1 day. In this situation each person has a score for each of 2 days. Using equation (3.3) with $k = 2$ (scores on 2 days presently) and $k' = 1$ (scores on 1 day in the future), the estimated reliability is calculated by the formula

$$R = [MS_B - MS_W] / [MS_B + (2 / 1 - 1)(MS_W)] = \\ [MS_B - MS_W] / [MS_B + MS_W]. \qquad (3.4)$$

If the researcher administers the physical performance test on each of 2 or more days and decides that the criterion score is the mean of the multiple-day scores, the reliability of the criterion score can be estimated using equation (3.2).

Conducting a Reliability Study

Reliability of the criterion score should be estimated in a pilot study before testing a large number of people in a research study. An example of estimating reliability using a pilot study group is administering three trials of a physical performance test within a day and determining the reliability of the criterion score desired. The researcher may find that a criterion score using the mean of three trial scores is sufficiently reliable, but he or she may also find that only two trials are needed to obtain a reliable criterion score, or that more than three trials are needed to obtain a reliable criterion score. By manipulating k', a researcher can use equation (3.3) to estimate what R might be if the number of trials were increased or decreased. If the data of a three-trial physical performance test are analyzed using ANOVA, and the estimated reliability of a criterion score that is the mean of the trial scores is low, the researcher can estimate the reliability for a criterion score that is the mean of four trial scores. Using equation (3.3) with $k = 3$ (ANOVA conducted on the data from three trials) and $k' = 4$ (estimating reliability for four trials),

$$R = [MS_B - MS_W] / [MS_B + (3 / 4 - 1)(MS_W)]. \qquad (3.5)$$

Two-Way ANOVA

Using the one-way ANOVA model previously discussed, a researcher can classify the test score only as to the person who received the score with no regard to which score it is for the person. With a two-way ANOVA model, both the person and which score it is for the person are used to classify a test score. A two-way layout of the data of 10 people with four trial scores for each person is presented in table 3.3. Notice that the first trial score of person 1 is 8. This score must be written in row 1 and column 1 of table 3.3 to be classified correctly.

With the one-way ANOVA, the sources of variation are between groups (MS_B) and within groups (MS_W) (see table 3.2). MS_W is greater than zero as a result of people not consistently scoring the same from trial to trial or day to day and is considered an estimate of error score variance. If a group systematically improves in mean score from trial to trial or day to day, this will increase MS_W and lower R

Table 3.3   Data of 10 People With Four Trial Scores Arranged in a Two-Way Layout

Person	**Trial**				Person	**Trial**			
	1	**2**	**3**	**4**		**1**	**2**	**3**	**4**
1	8	7	9	7	6	10	6	8	10
2	10	9	9	12	7	6	6	6	6
3	7	7	7	8	8	9	11	12	8
4	7	10	12	9	9	9	9	8	10
5	10	7	11	11	10	6	6	6	7

Table 3.4   Part of the Output From the Reliability Program in SPSS for the Data in Table 3.3

Analysis of Variance				
Source of Variance	Sum of Squares	DF	Mean Square	F
Between People	82.60	9	9.18	
Within People	57.00	30	1.90	
Between Measures	7.20	3	2.40	1.30
Residual	49.80	27	1.84	
Total	139.60	39	3.58	
Grand Mean	8.40			

(see equations [3.2], [3.3], and [3.4]). If systematic change in the mean score of a group of people from trial to trial or day to day is not expected prior to testing the group of people, this systematic change should be considered measurement error and lower the value of R. However, if this systematic change in the mean score of a group of people from trial to trial or day to day is expected prior to testing the group of people, this systematic change should not be considered measurement error and not lower the value of R. Using a two-way ANOVA, the researcher can treat the effect of systematic change as either measurement error or not.

The reliability program in SPSS was used to analyze the data in table 3.3. Part of the output from the reliability program is a two-way ANOVA summary table that is presented in table 3.4. Using the mean square values in table 3.4, a researcher can calculate R as an estimate of test score reliability.

Mean Score Is the Criterion Score

If the researcher or practitioner decides that the criterion score for each person is the mean of the person's trial or day scores, the equation for calculating an estimate of the criterion score reliability is

$$R = (MS_P - MS_R) / MS_P, \tag{3.6}$$

where MS_P = mean square between people and MS_R = mean square residual in table 3.4.

MS_P is equal to zero if all of the people have the same mean score and is greater than zero if the people differ in mean score. It is used to estimate true score variance. MS_R is equal to zero if all of the people have the same pattern in their scores across trials or days and is greater than zero if all of the people do not follow the same pattern. For example, in table 3.3 the scores of person 1 decreased by 1 from trial 1 to trial 2 and increased by 2 from trial 2 to trial 3, whereas other people followed a different pattern. Thus, in table 3.4, MS_R is greater than zero. Mean square residual is an estimate of the error score variance. Much of the discussion in the one-way ANOVA section concerning interpreting R, factors that affect R, how to obtain a high R, and so forth, apply to an R calculated using a two-way ANOVA model.

General Case

The researcher or practitioner may select any criterion score. The formula for calculating an R as an estimate of the reliability of the criterion score is

$$R = [MS_P - MS_R] / [MS_P + (k / k' - 1)(MS_R)], \qquad (3.7)$$

where k = the number of scores collected and analyzed with ANOVA and k' = the number of scores used in the criterion score. Equation (3.7) is similar to equation (3.3), and generally the discussion of equation (3.3) applies to equation (3.7). Equation (3.7) may be used when the criterion score is the best trial score, the mean trial score, the mean of selected trial scores, a score from 1 day, and so forth.

One-Way Versus Two-Way ANOVA

When the reliability program in SPSS is used to analyze a set of test scores, the sources for both a one-way and two-way ANOVA are listed in the ANOVA summary table (see table 3.4). The between-people source in the two-way ANOVA is the same as the between-groups source in the one-way ANOVA. The within-people source in the two-way summary table is the within-groups one-way ANOVA source. Notice that the within-people source is divided into a between-measures source and a residual source that are two-way ANOVA sources. The between-measures source is greater than zero if the trial means or day means are unequal. The residual source was defined following equation (3.6).

The difference between using a one-way ANOVA model and using a two-way ANOVA model when calculating R can be seen by comparing equations (3.2) and (3.6) for the estimated reliability of a criterion score, which is the mean of the trial or day scores. Because $MS_B = MS_P$, the difference between equations (3.2) and (3.6) is MS_W and MS_R. The residual source, however, is the part of the within-people source that is not due to differences among trial or day means, so it would seem that MS_R will be smaller than MS_W, and, thus, the R using equation (3.6) will be higher than the R using equation (3.2).

The ANOVA model should be selected based on which model the researcher thinks is appropriate and not which model will yield the higher R. This author believes that people should be prepared to take a physical performance test, have prior experience with a physical performance test, and be administered appropriate warm-up or practice before they are tested. Thus, any change in a person's score from trial to trial or day to day represents measurement error (lack of reliability) and a one-way ANOVA model should be used.

Rater Reliability

Rater reliability is also called **objectivity**. It is the degree to which two or more people, raters, scorers, or judges agree on the scores of the people tested. When two or more people are in close agreement on the score of each person in the group tested, high rater reliability exists for the test scores. Baumgartner and colleagues (2003) stated that when determining whether test scores are meaningful, the researcher should first estimate objectivity. If the objectivity is acceptable, reliability is estimated. Finally, if reliability is acceptable, validity is estimated.

Rater reliability is estimated by using any of the previous formulas for calculating an intraclass R. Scores of the raters are used in place of trial or day scores when calculating an ANOVA to obtain mean square (MS) values to use in the formulas. Usually equation (3.3), (3.4), or (3.7) is used because in most cases an estimate of the reliability of the scores of a single rater is desired. Zhu (2001) presented interrater reliability in several situations, with several techniques, and referenced a computer program for calculating interrater reliability.

Computer Programs

Although one-way ANOVA computer programs are readily available, there are problems in using them for estimating reliability. The data entry format for a one-way ANOVA design tends to be different from the data entry format for other ANOVA designs and not very efficient in entering a lot of data. A bigger problem is that each person is considered a group in a one-way ANOVA program, and the program will not handle a large number of groups. Thus, for these reasons it is much better to use a two-way ANOVA computer program. The two-way ANOVA program, however, must allow one score per cell. Some two-way ANOVA programs require multiple scores per cell. The reliability program in the SPSS package of programs is excellent for doing the reliability analysis. The ANOVA summary table provided in the reliability program (see table 3.4) contains the MS values for both a one-way and a two-way ANOVA. If MS values based on a one-way ANOVA are needed to compute an intraclass R and the computer program does not provide them, the values can be obtained using the information in the section titled One-Way Versus Two-Way ANOVA of this chapter. In table 3.4 the MS for the within-people source (MS_W) is obtained by summing the sum of the squares for the between-measures source (SS_{BM}) and the sum of the squares for the residual source (SS_R) and dividing the result by the sum of the degrees of freedom for the between-measures source (DF_{BM}) and the degrees of freedom for the residual source (DF_R).

Predicting Reliability

All else held equal, reliability increases as the length of a test is increased. It is assumed that the additional trials are just as difficult as the original trials and neither mentally nor physically fatiguing. Increasing the number of trials within a day for a physical performance test usually increases the reliability of the test scores. The **Spearman-Brown prophecy formula** is used to predict the reliability of the test scores when the length of the test is increased. The formula is presented in many measurement books. Equations (3.3) and (3.7) have the Spearman-Brown prophecy formula included in them. They can be used to predict the reliability of the test scores when the length of the test is changed.

Reporting Reliability

When reporting the estimated reliability for the test scores, a person should precisely describe how the reliability estimate was calculated. A statement that the reliability of the test scores is .73 is not sufficient.

What to Report

Researchers should indicate whether an intraclass R or some other correlation coefficient is calculated as the reliability coefficient. If an intraclass R is calculated, they should indicate whether it is based on a one-way or two-way ANOVA model and the criterion score used (e.g., mean score, best score). Probably no matter what correlation coefficient they calculated as the reliability coefficient, but particularly if they used an intraclass R, researchers should report whether the reliability coefficient is an internal consistency or stability coefficient. Finally, they should report the value of the reliability coefficient and some indication of the goodness or quality of the reliability coefficient (e.g., acceptable magnitude of the reliability coefficient and confidence limits). Morrow and Jackson (1993) provided a comprehensive list of factors to consider when reporting reliability coefficients.

Quality of a Reliability Coefficient

In some books you may find standards for evaluating the goodness of a reliability coefficient (e.g., .90 is excellent, .80 is above average, and so forth). The problem with standards such as these is that they don't take into consideration the many factors that influence a reliability coefficient. Age is a factor in that the very young and very old tend to be less consistent in their performance than other age groups. The type of test is a factor. Physical performance tests tend to have a defined scoring procedure that is usually easy for both the person being tested and the person scoring the test to follow and tend to have more reliable test scores than other types of tests. Internal consistency reliability coefficients tend to be higher than stability reliability coefficients. The more trials used in the criterion score, the higher tends to be the internal consistency reliability coefficient for the criterion score.

One way to evaluate a reliability coefficient is based on its magnitude. A reliability coefficient of .90 is high because it is close to 1.0 (i.e., perfect reliability). People often want a quality label for a reliability coefficient such as excellent, above average, average, below average, acceptable, or unacceptable. Thus, many people label a reliability coefficient based on the value they expected to get in the research context, the value other people have gotten in similar contexts, and the magnitude of the reliability coefficient needed in the context. In one context, therefore, a reliability coefficient of .80 could be labeled good but in another context labeled excellent. This supports the notion that reliability is not an inherent characteristic of a test, but instead is a characteristic of test scores and how they are used. Generally, with physical performance measures, reliability coefficients of at least .80 but sometimes .70 are labeled acceptable. The "worth" of R can also be viewed from the perspective of confidence limits and statistical significance, as discussed later.

Calculations Related to *R*

It is often desirable to do selected calculations related to an intraclass *R* such as calculating **confidence limits for *R***, calculating the sample size needed to calculate *R*, and conducting significance tests on *R*.

Confidence Limits for R

Many people who have taken statistics courses are familiar with calculating confidence limits for a population mean based on the mean for a sample from the population. A similar procedure can be followed for a population reliability coefficient based on the reliability coefficient for a sample from the population. Confidence limits involve stating a degree of confidence and a lower value and an upper value of the confidence limit. For example, a researcher might state that for *R* = .90 he or she is 95% confident that the interval .86 to .92 spans the population value of *R*. In this case, because the upper and lower values of the confidence limit are similar and the lower value of the confidence limit is high in magnitude, the confidence limits are acceptable. If the 95% confidence limits for *R* = .73 are .65 to .79, probably the confidence limits are not acceptable because there is a large difference between the lower and upper values of the confidence limits and .65 is usually considered unacceptable reliability whereas .79 is often considered acceptable reliability.

McGraw and Wong (1996) presented the sampling distribution for *R* and the procedures for calculating confidence limits for *R*. Confidence limits based on their calculations are an option in the reliability program in SPSS.

Sometimes the correlation coefficient used to estimate reliability is not *R*. If the reliability coefficient is a Pearson product–moment correlation coefficient, the calculation of confidence limits is straightforward. Most statistics books (e.g., Ferguson & Takane, 1989) and others (e.g., Morrow & Jackson, 1993) present the procedures for calculating confidence limits for a Pearson correlation coefficient.

Sample Size

Baumgartner and Chung (2001) used the procedures of McGraw and Wong (1996) to calculate 90% confidence limits for three values of *R* and two different criterion scores with sample sizes of 50 and 100. Baumgartner and Chung used both a one-way and a two-way ANOVA model. They found that the confidence limits are the same for a one-way and a two-way ANOVA model, when all other factors (e.g., R, criterion score, sample size) are equal. Confidence limits for R in selected situations are presented in table 3.5. Confidence limits that are narrow, with the lower value at an acceptable reliability value, are desirable. Note in table 3.5 that as sample size increases, or the value of R increases, or both, the width of the confidence limits decreases. If the minimum acceptable reliability is .70, a sample size of 50 to 100 and a calculated R of .80 or higher tends to result in 90% confidence limits with a lower value of at least .70. So, if R is less than .80, sample size should be greater than 100 to obtain a narrow confidence limit with a sufficiently high lower value. These suggested sample sizes of 50 and 100 are larger than the sample sizes of 15 to 30 recommended and used in the past.

Significance Tests on R

The procedures of McGraw and Wong (1996) can be used to conduct a statistical test of the hypothesis that the intraclass reliability coefficient is a specified

Table 3.5   Ninety-Percent Confidence Limits for *R* in Selected Situations

R	Sample Size	Repeated Measures	Criterion Score	Limits
.70	50	2	Mean Score	.52–.81
.70	50	2	Single Score	.56–.78
.70	100	2	Mean Score	.58–.78
.70	100	2	Single Score	.61–.77
.80	50	2	Mean Score	.68–.88
.80	50	2	Single Score	.70–.87
.80	100	2	Mean Score	.72–.86
.80	100	2	Single Score	.73–.85
.90	50	2	Mean Score	.84–.94
.90	50	2	Single Score	.84–.94
.90	100	2	Mean score	.86–.93
.90	100	2	Single Score	.86–.93

Note: R is from either a one-way or a two-way ANOVA model.

value at the population level. If the minimum acceptable reliability is .70, a researcher can test the hypothesis that the reliability coefficient equals .70 at the population level. This test is one of the options in the reliability program in SPSS. Another statistical test that may be of interest but is not addressed by McGraw and Wong is a test of the hypothesis that two intraclass reliability coefficients are equal at the population level. This could be two intraclass reliability coefficients calculated for the same sample. For example, the reliability for test A and test B for a sample of college students is determined. Alternatively, this could be the reliability for a test that is administered to each of two different samples. For example, the reliability of test A for males and for females is determined. McGraw and Wong showed how to convert an intraclass R to a z-score, whereas Haggard (1958) addressed the distribution of the intraclass R. Alsawalmeh and Feldt (1992) presented hypothesis testing of intraclass reliability coefficients. Feldt (1990) discussed the sampling theory for the intraclass reliability coefficient. A simple and straightforward procedure is to calculate confidence limits for each of the two intraclass reliability coefficients. If the two confidence limits don't overlap, one can conclude that the two reliability coefficients are not equal.

STANDARD ERROR OF MEASUREMENT FOR NORM-REFERENCED PHYSICAL PERFORMANCE TESTS

Researchers and practitioners often want to estimate how much error of measurement to expect in any obtained test score. As presented earlier in the chapter, *obtained score = true score + error score*, and error score is lack of reliability. Thus, these researchers desire an estimate of error score for each person. The reliability procedure of calculating a reliability coefficient is called **relative reliability,** whereas the reliability approach of estimating measurement error is called **absolute reliability.** With relative reliability we are determining whether the position or rank of a person relative to the rest of the group is unchanged over multiple measures such as trials, days, or raters. With absolute reliability we obtain an estimate of the amount each obtained score might be in error.

If each person has multiple (two or more) trial or day scores, a standard deviation for the scores of each person can be calculated. The standard deviation of a person is used as an estimate of the error in a score of the person. However, if each person has only one score, the calculation of a standard deviation for each person is impossible. In this case a single estimate of the average amount of measurement error to expect in the score of any person in a group is calculated using the data of all the people in the group. This single estimate is called the **standard error of measurement** (*SEM*) and is calculated by the formula

$$SEM = s\sqrt{1 - r_{xx}}, \tag{3.8}$$

where SEM = standard error of measurement, s = standard deviation for the group, and r_{xx} = the reliability of the test scores of the group. The value of the SEM can be from zero up to the value of the standard deviation for the group. Obviously, small values of the SEM are desirable. Traub (1994) has a detailed description of the SEM.

The SEM is fairly accurate for some people and very inaccurate for other people. Often the SEM is more accurate for people with scores in the middle of the distribution of scores than it is for people with scores at the ends of the distribution of scores. Darracott (1995) used multiple-trial data to calculate a standard deviation for each person and compared the standard deviations to the SEM for the group. The SEM was often considerably different from the standard deviation for a person.

Atkinson (2003) suggested calculating the SEM by testing each person on two occasions (two trials or 2 days), calculating the difference between each person's two scores (D), calculating the standard deviation (s_D) for the D-scores, and calculating the SEM by the formula

$$SEM = s_D / \sqrt{2}. \tag{3.9}$$

Using equation (3.9) does not require any more data collection than is required to calculate any of the reliability coefficients already presented.

Often 68% confidence limits for the obtained score of a person are calculated for the score of each person by the formula

$$\text{Confidence limits} = x \pm SEM, \tag{3.10}$$

where x = the score of a person and SEM = the standard error of measurement calculated by equation (3.8). Nunnally and Bernstein (1994, pp. 239-240) stated that although the SEM is often used to calculate confidence limits for obtained scores, doing so is incorrect because the confidence interval is not symmetrical around the obtained score. If the confidence limits are narrow enough that from a practical standpoint the lower value and upper value are basically equal, the confidence limits are considered narrow enough to be acceptable and the reliability of the test scores is considered acceptable.

Example: The researcher measures the 1-repetition maximum bench-press strength score of each person in a group. For the data r_{xx} = .84 and s = 8.0.

$$SEM = 8\sqrt{.1 - .84} = 3.2. \tag{3.11}$$

A person in the group has a score of 120. Thus, the 68% confidence limits for the obtained score of this person are $120 \pm 3.2 = 116.8 - 123.2$ (i.e., we are 68% confident that this interval spans the person's true score). Because the researcher considers 116.8 and 123.2 as basically equal, the SEM is acceptable and the data are considered reliable.

RELIABILITY ESTIMATION FOR CRITERION-REFERENCED PHYSICAL PERFORMANCE TESTS

Criterion-referenced tests are commonly dichotomously scored, 0 or 1. If a person meets a **mastery cutoff score,** or has the characteristic, the person is assigned a score of 1; otherwise, he or she is assigned a score of 0. For example, if the mastery cutoff score is 5 for a sit-up test, a person with a test score less than 5 is assigned a score of 0 and a person with a test score of at least 5 is assigned a score of 1. Many different terms can be used with criterion-referenced tests to label the two possible outcomes, such as pass or fail, proficient or nonproficient, and mastery or nonmastery.

In terms of reliability of mastery cutoff scores, the consistency with which each person is assigned to the same mastery classification on two different occasions (trials or days) is of interest. Thus, the data organization is a table with two rows and two columns as shown in table 3.6. The frequency for each cell of the table is determined. For example, the frequency for cell *a* in table 3.6 is the number of people who are assigned a score of 0 (nonmastery) on both occasions. The frequency for each call can be determined by hand calculation or by using a cross-tabulation program in any package of statistical computer programs such as Crosstabs in SPSS. Based on the frequencies obtained, several different estimates of assigned test score reliability can be calculated.

One of the reliability estimates is the **proportion of agreement coefficient** (*Pa*) that is calculated using equation (3.12):

$$Pa = (a + d) / (a + b + c + d), \qquad (3.12)$$

where the format and notation in table 3.6 are used. The value of *Pa* can vary from 0.00 to 1.00. In equation (3.12) the a + d value is the number of people classified the same on both occasions (trials or days). The value of *Pa* is inflated by chance. In theory, if people are assigned a score by chance (flip of a coin where heads = 1 and tails = 0), the value of *Pa* over the long run will be .50. Thus, *Pa* should be much closer to 1.00 than .50 to have good reliability.

Table 3.6   A Two-by-Two Table for Criterion-Referenced Tests Administered Twice

		Day 2			Trial 2	
		0	**1**		**0**	**1**
Day 1	**0**	a	b	**Trial 1** **0**	a	b
	1	c	d	**1**	c	d

Note: a, b, c, and d are frequencies; the number of people with a two-by-two score combination.

Another reliability coefficient commonly calculated is **kappa** (k). It is calculated using equation (3.13):

$$k = (Pa - Pc) / (1 - Pc), \tag{3.13}$$

where Pa = proportion of agreement and Pc = proportion of agreement expected by chance = $[(a + b)(a + c) + (c + d)(b + d)] / (a + b + c + d)^2$ in a table such as table 3.6.

The value of k can range from -1.00 to 1.00, although in the reliability context, values < 0 have no meaning. Fleiss (1981, p. 217) indicated that if there is complete agreement ($Pa = 1.0$), k is equal to 1.0, but the minimum value of k depends on the marginal proportions (row and column proportions in a two-by-two table).

Presented in table 3.7 are the data for a criterion-referenced fitness test administered on 2 days a week apart. Using these data,

$$Pa = (40 + 100) / (40 + 20 + 40 + 100) = .70$$

$$Pc = [(40 + 20)(40 + 40) + (40 + 100)(20 + 100)] / \\ (40 + 20 + 40 + 100)^2 = .54 \tag{3.14}$$

$$k = (.70 - .54) / (1 - .54) = .35.$$

The Pa value, .70, is closer to .50 than to 1.00, indicating high reliability. Also, the k value, .35, is very low, indicating poor reliability.

Sometimes with physical activity assessment more than two classifications are used, such as above average, average, and below average. Fleiss (1981, p. 218) presented a formula for calculating k when there are more than two classifications. The k coefficient is an overall value of k that is the weighted average of the individual k values. The formula for k is still the same as equation (3.13), but the values used in the formula are calculated in a different manner.

The procedure by Fleiss (1981) is best shown with an example. On each of two occasions 250 people are rated on a three-point scale as above average (1), average (2), or below average (3). The data for the 250 people are presented in table 3.8. In the table, 50 people were classified 1 on both occasions. The data in table 3.8 are converted to proportions by dividing each cell frequency by the total number

Table 3.7   Data for a Criterion-Referenced Fitness Test Administered on Two Different Days

		Day 2	
		0	**1**
Day 1	**0**	40	20
	1	40	100

0 = not proficient; 1 = proficient

Table 3.8 Data for 250 People Rated on a Three-Point Scale on Two Occasions

		Occasion 2		
		1	**2**	**3**
Occasion 1	**1**	50	15	25
	2	20	20	20
	3	30	30	40

1 = above average; 2 = average; 3 = below average

Table 3.9 Proportions of Total Number of People for Each Cell in Table 3.8

		Occasion 2			
		1	**2**	**3**	**Total**
Occasion 1	**1**	.20	.06	.10	.36
	2	.08	.08	.08	.24
	3	.12	.12	.16	.40
	Total	.40	.26	.34	1.00

of people (250). The proportions are presented in table 3.9. In table 3.9, .20 (50 / 250) of all the people are classified 1 on both occasions.

Step 1. Calculate Pa:
Pa = proportion of agreement = sum of the diagonal values in table 3.9
= .20 + .08 + .16 = .44

Step 2. Calculate Pc:
Pc = proportion of agreement expected by chance
= sum of the product of corresponding row and column totals in table 3.9
= (total for row 1)(total for column 1) + (total for row 2)(total for column 2) + . . .
= (.36)(.40) + (.24)(.26) + (.40)(.34) = .34

Step 3. Calculate k:
$k = (Pa - Pc) / (1 - Pc) = (.44 - .34) / (1 - .34) = .15$

Fleiss (1981) stated that any k value of .76 or higher indicates excellent agreement (reliability), .40 to .75 indicates fair to good agreement, and .39 or lower is poor agreement beyond chance. Also, he mentioned that k could be interpreted as an intraclass R with quantitative ratings. The intraclass R was discussed earlier in this chapter. Later in this chapter the alpha coefficient that is equivalent to R with quantitative ratings will be discussed. Thus, the alpha coefficient could be used rather than k. See Looney (1989) for an in-depth discussion of criterion-referenced reliability.

RELIABILITY ESTIMATES FOR SCORES FROM KNOWLEDGE TESTS AND QUESTIONNAIRES

Researchers and practitioners in physical education and exercise science use knowledge tests and questionnaires. Knowledge tests have characteristics that are different from physical performance tests. Thus, the reliability techniques used with knowledge test scores tend to be different from the reliability techniques previously presented. Questionnaires may have characteristics similar to knowledge tests. However, if a questionnaire does not have characteristics similar to knowledge tests, then reliability techniques used with knowledge test scores are not appropriate for use with questionnaire scores. Earlier in this chapter it was stressed that the formula used to calculate an intraclass reliability coefficient (R) must be appropriate for the criterion score (e.g., mean score, best score). It is now stressed that the reliability technique used must be appropriate for the data that are influenced by the characteristics of the knowledge test or questionnaire.

Characteristics of a Knowledge Test

This discussion is based on objective knowledge tests (true–false, multiple choice, matching) in which the potential answers are provided for each **item** (question or statement) of the test. Each item has a correct answer. Each item is either answered correctly and the item is scored 1, or answered incorrectly and the item is scored 0. The score of a person for the test is the number of items answered correctly. Usually, all of the items of a knowledge test measure knowledge about a construct or topic and, thus, a total score that is the number of items answered correctly is appropriate. For example, if the knowledge test is designed to measure fitness knowledge, each test item deals with fitness knowledge; a total score reflecting the number of items answered correctly is appropriate; and the higher the total score, the more the person knows about fitness.

Reliability Techniques for Knowledge Tests

Traub (1994) provided a comprehensive discussion of reliability issues and techniques pertaining to knowledge tests. The stability (test–retest) reliability of a knowledge test is seldom, if ever, calculated. It requires administering the knowledge test on two different days. Most people score better the second day because they remember correct answers to items from the first test administration, and learn correct answers by studying or talking to other people. Improvement in test scores from the first day to the second day seems to indicate poor test score reliability. Internal consistency reliability of knowledge test scores, therefore, is usually calculated. An exception to this is when there is an A form and a B form of a test and the two forms are equal in content, difficulty, and so on. In this case the A form is administered on one day, the B form is administered on another day, and stability reliability of the knowledge test scores is calculated. With internal consistency reliability of knowledge test scores, each item of the test is considered to be like a trial when internal consistency reliability of physical performance test scores is estimated with an intraclass R.

An estimate of the internal consistency reliability of knowledge test scores is obtained by using the **Kuder-Richardson formula 20** (r_{20}):

$$r_{20} = [k / (k - 1)][(s^2_x - \Sigma pq) / s^2_x], \tag{3.15}$$

where k = the number of test items, s^2_x = the variance of the test scores $[\Sigma X^2/n - (\Sigma X)^2 / n^2]$, p = the percentage answering an item correctly, $q = 1 - p$, and Σpq is the sum of the pq products for the k items.

Data for a 10-item knowledge test are presented in table 3.10. Analysis of the data in table 3.10 is presented in table 3.11. Using equation (3.15) and the information in tables 3.10 and 3.11, the reliability of the test scores is $r_{20} = [10 / 9][(4.64 - 2.08) / 4.64] = .61$. Because reliability coefficients much closer to 1.00 than 0.00 are desirable, this coefficient of .61 would at best be interpreted as low reliability.

Another estimate of internal consistency reliability of knowledge test scores is obtained by using **Kuder-Richardson formula 21** (r_{21}). Kuder-Richardson formula 20 is tedious to use when calculating by hand with a large number of people, a large number of test items, or both. So, r_{21} is used. Use of r_{21} is based on the assumption

Table 3.10 Data for a 10-Item Knowledge Test

Person	Item										
	1	2	3	4	5	6	7	8	9	10	X
A	1	1	1	1	1	1	1	1	1	1	10
B	1	1	0	1	0	0	1	1	1	0	6
C	0	1	0	1	1	0	0	0	1	1	5
D	1	0	1	0	1	1	0	0	0	0	4
E	1	1	1	1	0	1	0	1	1	1	8

X = total test score; M_x = 6.60; s^2_x = 4.64; 1 = item answered correctly

Table 3.11   Analysis of the Data in Table 3.10 for Calculating Reliability Coefficients

Item	1	2	3	4	5	6	7	8	9	10
p	.8	.8	.6	.8	.6	.6	.4	.6	.8	.6
q	.2	.2	.4	.2	.4	.4	.6	.4	.2	.4
pq	.16	.16	.24	.16	.24	.24	.24	.24	.16	.24

p = percentage answering item correctly; $q = 1 - p$

that all test items are equally difficult. When items differ in difficulty, r_{21} underestimates the reliability coefficient. Thus, r_{21} should be interpreted as the minimum reliability for knowledge test scores. Calculation of r_{21} is by the formula

$$r_{21} = \left[(k)(s_x^2) - (\overline{x})(k - \overline{x}) \right] / \left[(k - 1)(s_x^2) \right], \tag{3.16}$$

where k = the number of test items, s_x^2 = the variance of the test scores, and $\overline{x}$ = the mean of the test scores. Using equation (3.16) and the information in table 3.10, the reliability of the test scores is

$$r_{21} = [(10)(4.64) - (6.60)(10 - 6.60)] /$$
$$[(10 - 1)(4.64)] = .57. \tag{3.17}$$

There isn't much difference in the reliability coefficients obtained by using r_{20} (.61) and r_{21} (.57). All other factors being equal, the larger the standard deviation is, the higher is the reliability coefficient. Thus, high reliability is harder to obtain with homogeneous groups than it is with heterogeneous ones.

A third and common estimate of the internal consistency reliability of knowledge test scores is obtained by using the **alpha coefficient.** Alpha is commonly provided when computer analysis of a knowledge test is provided. Nunnally and Bernstein (1994) stated that with dichotomous items (items scored as either right or wrong) r_{20} is the same as the alpha coefficient. If the items have more than two possible answers (such as answers A, B, C, D, E on a knowledge test or questionnaire) and the items are not scored right (1) or wrong (0), Kuder-Richardson formula 20 cannot be used. When each possible answer for the item is assigned a numerical code, such as A = 1, B = 2, the alpha coefficient can be used. The alpha coefficient may be used with ordinal data (Ferguson & Takane, 1989; Nunnally & Bernstein, 1994). In fact, with ordinal or interval data the alpha coefficient is the same as the intraclass R calculated with equation (3.6). The alpha coefficient is an estimate of the reliability of a criterion score that is the sum or mean of the trial or item scores in 1 day. The alpha coefficient (r_a) is calculated using the following formula:

$$r_a = [k / (k - 1)][(s_x^2 - \Sigma\, s_j^2) / s_x^2], \tag{3.18}$$

where k = the number of trials or items, s^2_x = the variance for the sum or mean of the trial or item scores in 1 day, and $\Sigma\, s^2_j$ = the sum of the variances for the trials or items. Because knowledge test items are usually dichotomously scored, an example using equation (3.18) will not be presented here; however, an example will be presented later in the chapter in the Reliability Techniques for Questionnaires section.

Characteristics of a Questionnaire

Questionnaires are used for collecting information concerning attitudes, beliefs, practices, and so forth. Questionnaire items are usually like knowledge test items in an objective knowledge test because potential answers are provided. The number of potential answers for each questionnaire item is usually the same and varies from two to five. Three examples of attitude scale items follow:

Example 1: Exercising is enjoyable.
  A. Agree   B. Disagree

Example 2: Exercise is enjoyable.
  A. Agree   B. No opinion   C. Disagree

Example 3: Exercising is enjoyable.
  A. Strongly agree   B. Agree   C. No opinion   D. Disagree
  E. Strongly disagree

Before calculating an estimate of reliability for an attitude scale, we need to consider several issues. These issues deal with whether the attitude scale is like a knowledge test (if so, reliability techniques used with knowledge tests can be used with the attitude scale). All the items in a knowledge test measure a single topic or construct, there is a correct answer to each item, and a total score is possible, which is the number of items answered correctly. With some attitude scales all the items measure a single topic, but with some attitude scales this is not true. Also, although some attitude scales have a correct answer for each item, often this is not the case. When all items in an attitude scale measure the same topic, the researcher might decide that there is a socially acceptable answer that is basically like a correct answer, and the total score of a person is the number of socially acceptable answers. If all items measure the same topic and there are two potential answers to each item such as *agree* and *disagree* (see example 1 of the preceding attitude scale examples), it seems that a total score reflecting the number of answers of one type (such as *agree*) is justified. This is equivalent to coding *agree* = 1, *disagree* = 0, and a total score is the sum of the points.

If all items on an attitude scale measure the same topic and there are three potential answers for each item (see example 2 of the preceding attitude scale examples), the responses could be coded *agree* = 3, *no opinion* = 2, and *disagree* = 1. These scores might be considered interval data because in theory the attitude change from *no opinion* to *agree* is the same as that from *no opinion* to *disagree*. With three possible answers, however, you could be concerned whether people with the same total score are equal in attitude. If there are only two items on the attitude scale, there are three different ways a person could have a total score of 4 (1 + 3, 3 + 1, 2 + 2). With four or more items on the attitude scale, the number of different combinations of 1, 2, and 3 all leading to the same total score is quite large.

If there are five potential answers to each item on an attitude scale (see example 3 of the attitude scale examples), usually each potential answer is assigned a numerical code such as *strongly agree* = 5, *agree* = 4, *no opinion* = 3, *disagree* = 2, and *strongly disagree* = 1. These are codes and not measured scores. The only requirement when assigning codes is that each answer is assigned a different code. For example, rather than using a 5-4-3-2-1 coding system, a 9-8-5-3-1 coding system could be used. A coding system results in ordinal and not interval data. Some authors advocate summing the ordinal scores to get a total test score (Nunnally & Bernstein, 1994, p. 13), whereas others believe that only interval scores should be summed to obtain a total score for each person (Zhu, 1996). In theory, the attitude change from *no opinion* to *agree* is the same amount of change as from *no opinion* to *disagree* and the attitude change from *agree* to *strongly agree* is the same amount of change as from *disagree* to *strongly disagree,* but the attitude change from agree to *strongly agree* or *disagree* to *strongly disagree* is more than the attitude change from *no opinion* to *agree* or from *no opinion* to *disagree*. If a researcher believes that the attitude change from *no opinion* to *agree* is a change of one but from *agree* to *strongly agree* is twice as much as the attitude change from *no opinion* to *agree,* the coding of potential answers could reflect this with the codes being 1-3-4-5-7 (*no opinion* = 4).

In most situations researchers have no idea how to code the potential answers to reflect the amount of change in attitude. An issue in regard to coding of possible answers is the interpretation of the total score. For example, if there are two items on an attitude scale, and the 1-3-4-5-7 coding system is used, is 8 = 1 + 7 the same as 8 = 7 + 1 or 3 + 5 or 5 + 3? It seems the more items there are on the attitude scale, the more researchers should question whether two people with the same total score are equal in attitude as reflected by the total score. In general, it seems that the more potential answers to the attitude score items there are and the more items there are, the harder it becomes to consider an attitude scale to be similar to a knowledge test and the less appropriate it is to use the reliability techniques used with knowledge tests on attitude scale data.

Consider two items from a questionnaire designed to collect information concerning exercise practices :

1. How often do you exercise each week?
   A. Every day   B. 5–6 days   C. 2–4 days   D. 1 day   E. Never

2. Where do you exercise?
   A. At a gym   B. At an agency like a YMCA   C. At a park or outdoors
   D. At home   E. I don't exercise

Using any coding system does not seem to justify a total score for these two items because they don't measure the same topic and the potential answers to item 2 have no logical order. Thus, reliability techniques used with knowledge test scores are inappropriate for use with the scores of this questionnaire.

Reliability Techniques for Questionnaires

If the questionnaire is similar in characteristics to a knowledge test in that all the questionnaire items measure the same topic and a total score seems appropriate, any of the reliability techniques used with knowledge test scores are appropriate with questionnaire scores. The alpha coefficient (equation [3.18]), however, is commonly used because it is the same as r_{20} (equation [3.15]) with dichotomous data and can be used with ordinal data. With interval data the alpha coefficient is the same as the intraclass R calculated with equation (3.6). The data in table 3.12 are from an attitude scale with *strongly agree* coded 5 and *strongly disagree* coded 1. A total score that is the sum of the codes for the answers selected is appropriate. A computational example for the alpha coefficient using equation (3.18) and the data in table 3.12 is presented here.

Step 1: Using the data in table 3.12, calculate values need in equation (3.18).
$k = 6$, $s^2x = 18.41$, $s^2_1 = .55$, $s^2_2 = .57$, $s^2_3 = .41$, $s^2_4 = .86$, $s^2_5 = 1.43$,
$s^2_6 = .70$, $\Sigma s^2_j =$ sum of item variances $= 4.52$

Step 2: Calculate the alpha coefficient.
alpha $= [k / (k - 1)][(s^2_x - \Sigma s^2_j) / s^2_x] = [6 / 5][(18.41 - 4.52) / 18.41] = .91$

A reliability coefficient of .91 is considered very good.

Table 3.12   Responses to an Attitude Scale

| | Item | | | | | | |
Person	1	2	3	4	5	6	Total
A	4	4	5	3	4	5	25
B	3	4	4	4	3	4	22
C	2	3	4	2	2	3	16
D	4	5	4	4	5	5	27
E	3	3	3	3	3	3	18
F	4	4	5	4	5	4	26
G	3	4	4	3	5	4	23
H	4	5	4	5	5	5	28

1 = strongly disagree; 2 = disagree; 3 = no opinion; 4 = agree; 5 = strongly agree

If the questionnaire is not similar in characteristics to a knowledge test, reliability techniques used with knowledge test scores should not be used with the questionnaire scores. In this case a researcher can create several pairs of items in the questionnaire, with items in a pair expected to yield the same information. Some example pairs are presented in table 3.13. Paired items should be placed far enough apart in the questionnaire that people responding do not recognize the items as measuring the same information and have forgotten their response to the first item in the pair. Items in pairs should have the potential for different responses between items if reliability is a problem. For example, it would be inappropriate if for both items 11 and 42 a person were asked his or her age because the person is not likely to change his or her response between items. Paired items should be worded differently and may be opposite to each other as pair 2 is in table 3.13.

For each pair a reliability coefficient can be calculated using k (equation [3.13]) or Pa (equation [3.12]) or the alpha coefficient (equation [3.18]). If the reliability

Table 3.13   Examples of Paired Items on a Questionnaire

Pair	Item
1.	5. Exercise is pleasurable.
	A. Agree B. Disagree
	42. Exercise is enjoyable.
	A. Agree B. Disagree
2.	15. Everyone should get some exercise each day.
	A. Agree B. No opinion C. Disagree
	51. Some exercise each day is unnecessary.
	A. Agree B. No opinion C. Disagree
3.	26. How many days a week do you participate in an exercise program for at least 30 minutes a day?
	A. 7 days B. 5–6 days C. 3–4 days D. 1–2 days E. 0 days
	55. The number of days a week I exercise for at least 30 minutes in an exercise program is
	A. 0 days B. 1–2 days C. 3–4 days D. 5–6 days E. 7 days

coefficients for all or most of the pairs of items are acceptable, it is reasonable to conclude that the scores from the entire questionnaire are reliable.

OTHER RELIABILITY TOPICS

We cannot present every detail on a topic in just one chapter of a book. The material and techniques presented here are those of use to many people. Hopkins (2000) presents a variety of statistical and measurement information in his Web-based book including many of the techniques presented in this chapter. Atkinson (2003) discussed the **coefficient of variation** in reporting reliability, particularly concerning the reliability of instrumentation. Generalizability theory is not presented, but some of the techniques and equations (3.3, 3.7) presented have some application to the topic. Considerable attention is given to the topic in the educational and psychological literature. Morrow (1989) provided an excellent tutorial on generalizability theory with reference to physical education and exercise science. Other references with a chapter on the topic are also available (e.g., Thompson, 2003). Shavelson and Webb (1991) published a primer on generalizability theory, and Brennan (2001) wrote an excellent book on the topic. A few examples of applications of generalizability theory are Morrow, Fridye, and Monaghen (1986); Turner and Bouffard (1998); Tobar, Stegner, and Kane (1999); and Stegner, Tobar, and Kane (1999).

Reliability, measurement error, and issues related to these topics are covered extensively in educational measurement and psychological measurement books. Much of the material presented in these books is based on tests with dichotomously scored items such as knowledge tests. For interested readers, Nunnally and Bernstein (1994) and McDonald (1999) are excellent resources. In addition, research on reliability techniques and standard error of measurement are regularly published in journals such as *Educational and Psychological Measurement, Applied Psychological Measurement,* and *Journal of Educational Measurement.* Measurement research with data from physical education and exercise science is published in *Measurement in Physical Education and Exercise Science, Research Quarterly for Exercise and Sport,* and *Medicine and Science in Sports and Exercise.*

SUMMARY

Some estimate of the reliability of test scores is desired in many situations. In many of these situations several different techniques and formulas could be used to obtain an estimate of test score reliability. In this chapter it is stressed that the technique or formula used to estimate test score reliability must be in agreement with the measurement context and the criterion score. Thus, when calculating an estimate of test score reliability for your data, you must be knowledgeable and select the appropriate technique and formula. Likewise, when another person reports a reliability coefficient, you must be knowledgeable enough to determine whether the correct technique and formula were used.

CHAPTER 4

Constructing Tests
Using Item Response Theory

Weimo Zhu

Test construction is the process of developing appropriate measurement devices, from which valid, reliable, and objective inferences can be drawn from test scores. Test, or measurement, theory is the theoretical foundation used to guide the process and practice of test construction. **Test theory** also refers to procedures for estimating the key characteristics of the test or measure, such as validity and reliability. It is dictated by the measurement model, which has a unique set of assumptions and is based on a statistical or mathematical model (Suen, 1990). Many test theories, such as true score theory, strong true score theory, criterion-referenced measurement, and item response theory (Berk, 1980; Crocker & Algina, 1986; Gulliksen, 1950; Lord, 1952, 1980; Lord & Novick, 1968; McDonald, 1999; Rasch, 1960; Wright & Stone, 1979) have been developed since the beginning of the 20th century. Among them, true score theory, also known as classical test theory (CTT), has been used most often in measurement practice and is the most commonly used test theory in kinesiology, a field studying human movement including both physical education and exercise science. CTT, however, is known for several major limitations (e.g., sample dependence in determining item characteristics and an incorrect assumption that variance of errors of measurement is the same for all examinees) (Hambleton & Swaminathan, 1985).

Around the1950s and 1960s, a new test theory called item response theory (IRT) was developed in educational measurement practice. Its relatively slow development was accelerated in the1980s by the growing accessibility of personal computers and the development of application software. Today, IRT is the most dominant theory for test construction in all major testing organizations or agencies such as the Educational Testing Service (ETS) and American College Testing (ACT). After Spray introduced IRT to kinesiology in 1987, many successful applications have been reported (e.g., Looney, 1997; Safrit, Cohen, & Costa, 1989; Zhu, 1996; Zhu & Safrit, 1993).

The purpose of this chapter is to provide an introduction to IRT. After describing a number of commonly used IRT models, major model assumptions and fit statistics are introduced. Key IRT features, advantages, limitations, and applications, including some examples in kinesiology, are described along with commonly used

IRT software. Finally, new developments in IRT and future research directions are addressed.

WHAT IS IRT?

To fully understand IRT, a description of the relationship among a test, test items, examinees, and test scores will be helpful. A test is the device, or the yardstick, that we use to measure what we are interested in capturing (e.g., a scale for body weight and a sit-up test for abdominal strength). In psychological and cognitive testing, a test usually consists of a number of items, which are the basic components of a test. A **test item** is either a statement or question that states in clear, unequivocal terms the psychological or cognitive tasks required of the examinees taking the test. Testing is the interaction between a test (or item) and an examinee, the two most important facets involved in measurement (see figure 4.1). For a tester to make appropriate inferences regarding the abilities of the examinees being measured, a test must be able to provide accurate information about examinees. Compared to test A, tests B (able to measure the examinees at a low level only) and C (able to measure the range of the examinees, but with poor accuracy) clearly are not as effective. A judge or observer may also be involved in testing (e.g., judging a gymnast's performance or observing a child's physical activity in a physical education class). When a judge is involved in the testing, the interaction becomes a three-facet interaction and the judge's severity and consistency become factors that may affect measurement accuracy (see figure 4.1)

In CTT, the facets involved in testing are often set on different scales (e.g., total scores for examinees and proportion correct values for item difficulty) and exam-

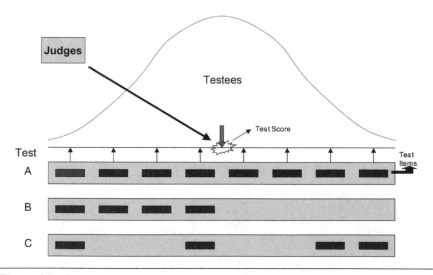

Figure 4.1   Facets involved in testing interaction.

ined separately. In contrast, in IRT, the facets involved in the testing interaction are examined simultaneously. The top of figure 4.2 shows an example using two facets, examinee and item. If an examinee's ability (θ_3) is higher than the difficulty (b) of a test item, the probability, or chance, that the examinee will successfully complete the item should be higher. In contrast, if the examinee's ability (θ_1) is lower than the **item difficulty,** the probability of success should be lower. Finally, when the examinee's ability (θ_2) is the same as the item's difficulty, the probability of success should be half-and-half. Now, let's model this two-faceted relationship mathematically. We first determine the difference between an examinee's ability (θ_j) and the difficulty of a test item (b_i), which can, theoretically, range from minus infinity to plus infinity.

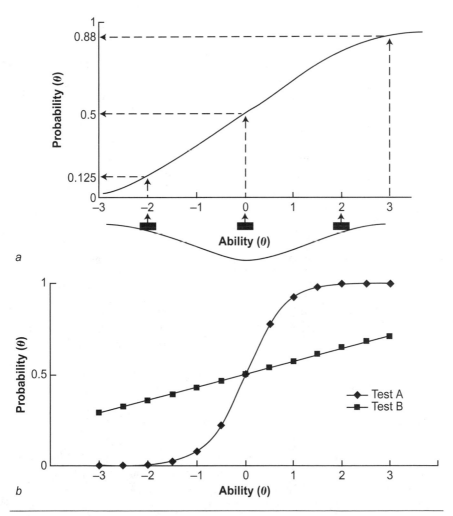

Figure 4.2   (a) ICC and difficulty parameter; (b) ICC and discrimination parameter.

By taking this difference $(\theta_j - b_i)$ as an exponent of the natural constant $e = 2.71828$, then $e^{(\theta_j - bi)}$ falls in a range between zero and plus infinity. We can then bring $e^{(\theta_j - bi)}$ into an interval between zero and one by forming the following ratio:

$$P_i(\theta) = \frac{e^{(\theta_j - b_i)}}{1 + e^{(\theta_j - b_i)}}. \tag{4.1}$$

This equation expressed literally is

Chance of a person to get an item right =

$$\frac{e\left(\text{Ability of Person j} - \text{Difficulty of Item i}\right)}{1 + e\left(\text{Ability of Person j} - \text{Difficulty of Item i}\right)}.$$

Through equation (4.1), the relationship between examinees and an item can be summarized into an S-shaped curve (figure 4.2a). The curve is called the **item characteristic curve** (ICC), also known as the trace line. The ICC is a regression function that relates the probability of success on an item to the examinee's ability measured by the test. Note that although the ICC is a monotonically increasing function, the relationship described is not linear because change in ability at both the lower and upper ability range is not linear with the change of probability in scoring an item correctly. To summarize, IRT is a set of statistical functions that model the relationship between facets involved in testing. The models predict the probability of a specific item response as a function of the latent ability variable and item characteristics (e.g., item difficulty).

COMMONLY USED IRT MODELS

In the IRT literature, equation (4.1) is known as a one-parameter logistic, or **Rasch model.** The "one-parameter" in the model name refers to the item characteristic known as item difficulty, which is defined as the point on the ability scale at which the probability of a correct response to the item is .5. In other words, an examinee's probability of answering an item correctly depends on the location, or difficulty, of an item. The more difficult an item (toward the right side of the scale), the lower the probability of answering the item correctly. For example, if examinee 1's ability $(\theta_3,$ top of figure 4.2) is 2.0, his or her chance of answering correctly an item with a difficulty of zero is .88. The probability decreases to .73 if the difficulty of the item is 1.0.

In addition to the location of an item, the shape of an ICC could have an impact on the probability of an examinee answering the item correctly. The steeper the ICC, the greater the sensitivity of an item to change in the examinee's abilities. Figure 4.2b summarizes two ICCs with the same difficulty, but with different steepness. ICC-A has a steeper slope than ICC-B, which means that in a certain range of ability, a slight change in ability will lead to a larger difference in the score of

Item A than the score of Item B. This steepness feature is called "discrimination" in IRT literature. A **discrimination parameter** can be added to the Rasch model to describe this feature mathematically:

$$P_i(\theta) = \frac{e^{a_i(\theta_j - b_i)}}{1 + e^{a_i(\theta_j - b_i)}},$$
(4.2)

where a_i is the discrimination parameter of the model. To make the logistic function as close as possible to the normal ogive function, a scaling factor ($D = 1.7$) is sometimes added in front of the a_i parameter. This practice is only associated with certain IRT software (e.g., LOGIST; Wingersky, Barton, & Lord, 1982).

In cognitive testing, an examinee may guess the correct answer when he or she is not sure of the correct answer to a multiple-choice or true–false question. Guessing behavior is usually modeled on the lower asymptote of the ICC tail—the higher the tail, the more the guessing. Guessing behavior can be mathematically modeled by adding another parameter c to equation (4.2):

$$P_i(\theta) = c_i + (1 - c_i)\frac{e^{a_i(\theta_j - b_i)}}{1 + e^{a_i(\theta_j - b_i)}}.$$
(4.3)

As a result, the model now becomes a three-parameter model. It should be noted that because other factors may be involved in guessing (e.g., attractive alternatives in a multiple-choice item), some researchers have proposed not to call the c-parameter the guessing parameter. The three logistic models presented previously are the most commonly used models for cognitive tests consisting of dichotomously scored items.

In practice, especially in affective and survey research, the data are often not scored dichotomously. In fact, most of the data in kinesiology are scored into three or more categories, or scored continuously. Fortunately, multiple-category, or polytomous, models have been developed to meet these needs. For example, a number of polytomous models have been developed from the Rasch model, such as the rating scale (Andrich, 1978) and the partial credit models (Wright & Masters, 1982). A more detailed treatment of the polytomous models can be found in van der Linden and Hambleton (1997). When more than two facets (e.g., raters or other subgroup characteristics) are involved in testing, multiple-facet models are required (Linacre, 1989). Finally, if more than one dimension has been included in the measurement, multidimensional models are available. Multidimensional models are discussed later in this chapter.

Selection of a particular model for IRT application often depends on users' training, experience, and beliefs, and the data available. A difference in preference sometimes leads to a hot debate concerning which model should be employed. A balanced view is that no single model will fit every situation and, if there is no additional significant measurement advantage, a low-cost model (e.g., requiring small sample size) that can provide a simple and practical interpretation of the testing results should be chosen.

ASSUMPTIONS RELATED TO IRT

Compared to CTT, IRT is known as a strong-assumption theory. Therefore, although IRT has a number of advantages over CTT, these advantages cannot be realized if the IRT model's assumptions are not met. Two major assumptions related to IRT are unidimensionality and local independence.

Unidimensionality

Except for a few new multidimensional models, most IRT models assume that all testing items in the instrument measure a single latent trait or ability (e.g., gross-motor function). In practice, this assumption of **unidimensionality** may not be met perfectly because more than one trait, such as motivation and familiarity, may be involved in the performance of a motor task. The common practice, therefore, is to consider the assumption adequately met as long as a dominant factor or component determines the performance of the task (Hambleton, Swaminathan, & Rogers, 1991). When more than one factor or component clearly influences performance, the assumption may be violated and multidimensional IRT models are needed.

Local Independence

Local independence means that an examinee's responses to the test items are statistically independent (i.e., the probability of examinee responses to an item will depend only on this examinee's ability, not the connection among the items). In practice, an examinee's responses to several test items are likely related when the items share common traits (e.g., measure the same domain of knowledge). This seems contradictory to the local independence assumption. However, after these common traits are removed or held constant, the items should become uncorrelated. Thus, local independence should be called more accurately "conditional independence" (Hambleton et al., 1991). Whether or not data meet the assumptions should be examined before interpreting the results of an IRT analysis.

ESTIMATION OF ITEM AND ABILITY PARAMETERS

One of the major tasks in applying IRT is estimating item and ability parameters. Depending on the IRT model employed, item parameters such as difficulty, discrimination, and guessing represent characteristics of an item. Ability parameters of the examinees also have to be estimated because the purpose of developing tests is to determine the latent abilities of the examinees tested and differences among them. In practice, parameter estimation is accomplished via IRT software. To better understand and interpret the results of an IRT analysis, however, some basic knowledge of parameter estimation is helpful.

Item Estimation

Let us start by estimating item parameters while assuming we already know the abilities of the examinees involved. Say we have a large group of examinees and they responded to N items in a test. As expected, there will be a large variability in examinees' abilities (figure 4.3a). If we divide the examinees into J groups along the ability scale so that all the examinees within a group have a similar ability θ_j, there will be m_j examinees within group j, where $j = 1, 2, 3, \ldots J$. Within an ability score, r_j examinees answered a given item correctly. The higher the group's ability, the more correct answers the examinees get. Therefore, at a specific ability level of θ_j, the observed proportion of correct response is Proportion $(\theta_j) = r_j/m_j$, which is an estimation of the probability of correct response at that ability level. If we repeat the same computation at each level of θ_j and plot proportion (θ_j) along the ability scale, the result is an S-shaped band, as illustrated in figure 4.3. Note that the band is the summary of raw data, and our task is, using the information available, to mathematically estimate a best-fit curve for this band. In fact, any of the three logistic models described earlier can be used to estimate this curve.

One may quickly realize that this curve fitting is like our common practice of fitting a linear regression line. There is a key difference in the IRT curve fitting, however. In contrast to regression, in which the regressor variable (also known as the independent variable) is observable, the regressor variable (i.e., the ability parameters) in IRT is unobservable. In fact, it is the latent nature of the ability parameters that makes the estimation in IRT much more complex (Hambleton et al., 1991). The actual estimation procedures (e.g., maximum likelihood [ML] estimation) usually require complex and labor-intensive computations, which must be performed for every item in a test. Only a brief description of the estimation procedure will be described in the following ability estimation section, but more details can be found in a thorough description by Baker and Kim (2004).

Because of the similarity between regression and IRT parameter estimation, one of the useful features in regression, called **group invariance,** is shared also by IRT parameter estimation. In regression, group invariance means that the estimated intercept and slope of a linear regression will remain stable even if different samples are employed in the estimation processes. In IRT, this means that parameters of items (difficulty, discrimination, or guessing) will not be affected by different samples. Figure 4.3b illustrates the group invariance feature in item parameter estimation. Although there are significant differences in ability in both groups involved in the estimations, the estimated difficulty and discrimination parameters are almost the same. This is because both groups' data were fitting a section of an underlying ICC, one on the low tail and the other on the high tail of the curve. It should be pointed out that the invariance feature also applies to

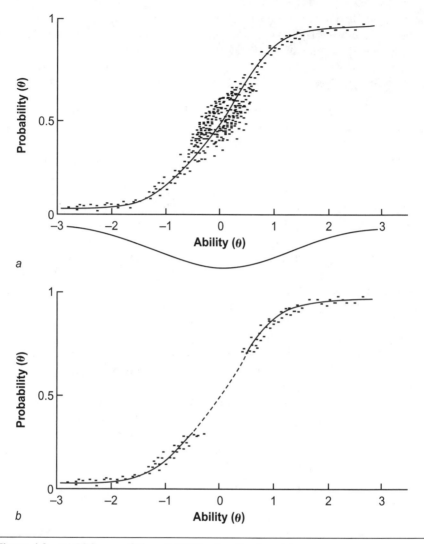

Figure 4.3   (a) ICC curve fitting; (b) ICC estimation invariance.

ability estimation, described next (i.e., examinees' ability estimates are invariant from items employed in the estimation).

Ability Estimation

Similar to the discussion of item parameter estimation, understanding ability estimation becomes easier if item parameters are assumed to be known. Once again, we will use the ML procedure to illustrate the basic concepts of ability estimation.

The ML procedure is an iterative process that starts with an arbitrarily selected value for the ability of the examinee and the known item parameters. Using the IRT model selected, the probabilities of correct response to each item by the examinee are estimated. Assuming the arbitrarily selected ability parameter is not the true parameter, an adjustment to the ability estimate is needed to obtain improvements in the agreements of the computed probabilities with the examinee's item response vector. This "adjust/examine" process is repeated until the adjustment becomes so small that the change in the estimated ability is negligible. The iterative process is repeated for each examinee individually until everyone's ability estimate is determined.

An example of ability estimation is presented in table 4.1. Table 4.1 is based on a test consisting of three items with difficulty, discrimination, and the examinee's responses to the items as follows (more details can be found in Baker, 1985):

	Difficulty	**Discrimination**	**Response**
1	−1	0.8	1
2	0	1	0
3	1	0.6	1

Ability estimation is computed using the following equation:

$$\hat{\theta}_{s+1} = \hat{\theta}_s + \frac{\sum\limits_{i=1}^{N} a_i \left[u_i - P_i\left(\hat{\theta}_s\right) \right]}{\sum\limits_{i=1}^{N} a_i^2 P_i\left(\hat{\theta}_s\right) Q_i\left(\hat{\theta}_s\right)}, \tag{4.4}$$

where $\hat{\theta}_s$ is the estimated ability of an examinee within iteration s; a_i is the discrimination parameter of item i, $i = 1, 2, \ldots N$; u_i is the response made by the examinee to item i (1 = correct and 0 = incorrect); $P_i\left(\hat{\theta}_s\right)$ is the probability of correct response to item i, which is estimated by the two-parameter model introduced earlier (equation [4.2]); and $Q_i\left(\hat{\theta}_s\right)$ is the probability of an incorrect response to item i, which equals $1 - P_i\left(\hat{\theta}_s\right)$. The estimation starts by selecting an arbitrary ability value $\hat{\theta}_s = 1$. At the end of the first iteration, the ability adjustment equaled −.7889. Because it was a rather large value, it was added to the originally selected $\hat{\theta}_s$ and a new $\hat{\theta}_s$ (.211103) was created, which is used for the next iteration. At the end of the third iteration, the ability adjustment value became so small (.000846) that the iteration was stopped and the finial estimated ability value for this examinee became .309737 (.000846 + .308045) (see table 4.1 for more details).

Standard Error of Estimation

Conceptually, the ability in IRT is a latent trait that can only be measured indirectly or estimated. How much can we trust the estimated ability level? Similar to using reliability and standard error of measurement (SEM) to determine the quality of a measure or score in CTT (see chapter 3 by Ted Baumgartner), we can also compute

Table 4.1 Iteration Changes in Ability Estimation

b	a	U	P	Q	a(u – P)	a*a²(PQ)
–1	0.8	1	0.832018	0.167982	0.134385	0.089449
0	1	0	0.731059	0.268941	–0.73106	0.196612
1	0.6	1	0.5	0.5	0.3	0.09
					–0.29667	0.376061
		Adjustment =	–0.7889			
	Ability + adjustment =		0.211103			

b	a	U	P	Q	a(u – P)	a*a²(PQ)
–1	1	1	0.773293	0.226707	0.226707	0.175311
0	1.2	0	0.567682	0.432318	–0.68122	0.353404
1	0.8	1	0.350145	0.649855	0.519884	0.145628
					0.065372	0.674342
		Adjustment =	0.096942			
	Ability + adjustment =		0.308045			

b	a	U	P	Q	a(u – P)	a*a²(PQ)
–1	1	1	0.789846	0.210154	0.210154	0.165989
0	1.2	0	0.595994	0.404006	–0.71519	0.346731
1	0.8	1	0.368002	0.631998	0.505599	0.148849
					0.00056	0.661569
		Adjustment =	0.000846			
	Ability + adjustment =		0.308891			
	Final ability		0.309737			

a precision index called the **standard error** (SE) to determine the accuracy of the ability estimation:

$$SE\left(\hat{\theta}\right) = \frac{1}{\sqrt{\sum_{i=1}^{N} a_i^2 P\left(\hat{\theta}\right) Q\left(\hat{\theta}\right)}}. \tag{4.5}$$

Like SEM in CTT, $SE\left(\hat{\theta}\right)$ is a measure of the variability of the values of $\left(\hat{\theta}\right)$ around the examinee's latent parameter value θ (i.e., if we use the same test to measure the examinee many times, most of the examinee ability estimates will be close to the examinee's true ability value with a few very low and high variations). Unlike SEM, however, $SE\left(\hat{\theta}\right)$ is a local precision index that is associated with a specific ability value $\left(\theta\right)$. By examining equation (4.5) more carefully, one can see that the term under the square root sign is the denominator of equation (4.4). Thus, $SE\left(\hat{\theta}\right)$ is a corresponding product of the estimated ability. For example, $SE\left(\hat{\theta}\right)$ for the previous example is 1.51156 (1 / .661569). This is rather large, most likely because only three items were employed in the estimation. The concept that measurement precision is determined locally is a very useful feature of IRT and will be described more thoroughly later.

Joint Estimation and an EM Algorithm

When either ability or item parameters are known, the estimation of the other class of parameters is relatively easy, and the procedure is often called conditional estimation. In practice, however, this is often not the case, especially at the beginning process of calibrating a set of items when both ability and item parameters are unknown. When both ability and item parameters are unknown, joint estimation and an expectation maximization (EM) algorithm must be employed. Basically, a paradigm is to be established in the estimation so that both "artificial data" and the parameter estimates can be obtained iteratively (see Baker & Kim, 2004, for more details).

Estimation Procedures

In addition to the ML procedure introduced earlier, many other advanced estimation procedures are employed in IRT software, such as marginal maximum likelihood estimation, joint maximum likelihood estimation, and Bayesian estimation. A simple and practical introduction to these procedures can be found in Suen (1990) and Baker and Kim (2004).

ADDRESSING MODEL–DATA FIT

Item and ability estimates may not be able to be used to accurately evaluate items and examinees if there is not a good fit between the IRT model selected and the test data. Generally, the question of whether a model fits the data should be addressed in three general steps:

1. The assumptions of the model should be examined to determine how well the data satisfy them.
2. The features of the model (e.g., invariance) should be examined.

3. The goodness-of-fit should be conducted to determine the degree of fit between predicted and observable outcomes.

Hambleton and Swaminathan (1985) provided an extensive review of this topic. A brief summary follows.

Inspecting Model Assumptions

Although IRT is characterized by its strong assumptions, these assumptions will rarely be completely met by test data (Lord, 1980; Lord & Novick, 1968). Evidence has shown that the models are robust to moderate departures from their assumptions (Hambleton & Cook, 1983; Wainer & Wright, 1980). The extent of this robustness, however, is not well understood because (a) the assumptions can be violated in different ways and (b) the seriousness of the violations depends on the nature of the examinee sample and the related application. Commonly inspected assumptions are unidimensionality, local independence, equal discrimination indexes, minimal guessing, and nonspeeded test administration. Only the first two will be discussed here because of their importance. For unidimensionality, a comprehensive review is provided by Hattie (1984), in which 88 different indexes for assessing dimensionality are examined. Most of the methods in earlier literature were found unsatisfactory, whereas the methods that were developed by nonlinear factor analysis and the analysis of residuals were recommended.

For local independence, four common statistical approaches—the odds ratio test, the proportion of correct under independence, the Mantel-Haenszel chi-square test, and the Goodman-Kruskal gamma test—are commonly used (Rosenbaum, 1984). Both the odds ratio and the proportion of correct under independence are developed on the same algebraic base, and both are used only when an item or a subset of items is scored dichotomously. The Mantel-Haenszel procedure is similar to a chi-square test for independence (Mantel & Haenszel, 1959). Finally, the Goodman-Kruskal gamma test is an index of relationships for discrete ordinal data arranged in a bivariate frequency table (Roscoe, 1975).

Inspecting Expected Model Features

Invariance, or estimation that is sample or item free, is one of the most important features of IRT. When item parameters in IRT are not dependent on the ability level of the examinees responding to the item, the group is invariant in item estimation (Baker, 1985) or invariance of item parameter estimates can be claimed (Hambleton, 1989). To test the feature of group invariance, item parameter estimates from different subgroups of the population must be compared. If the estimates are invariant, the correlation plot of the estimates should be linear with an amount of scatter that reflects only sample size error (Hambleton, 1989). The common way to test group invariance is to compare item characteristic curves (ICC) from different estimations. First, the item parameters need to be estimated from different subgroups; then the ICC can be developed from these parameters. Because ability scales are often arbitrary in an IRT analysis, the ICCs must be matched to the same scale (e.g., using a linear transformation), before comparing (see chapter 6 by Weimo Zhu).

Inspecting Overall Model–Data Fit

Perhaps the majority of procedures developed for model–data fit are used to determine the overall fit of model, item, and examinee. These tests can be classified into two general categories: chi-square related and residuals based. Although chi-square-related procedures have been extensively studied and widely applied to test data, the effect of sample size on the statistical model fit has a serious flaw. It has been noted that these statistical fit tests should not be completely trusted (Hambleton, 1989) because "when sample sizes are large, nearly all departures between a model and a data set (even those where the practical significance of the differences is minimal) will lead to rejection of the null hypothesis of model–data fit" (Hambleton & Murray, 1983, p. 72).

Residual analysis combined with graphical analysis is considered a good alternative to chi-square-based tests (Ludlow, 1986; Murray & Hambleton, 1983). Two indexes, raw residuals (RR) and standardized residuals (SR), are often used in residual analysis. The indexes reflect the degree of misfit between the test data and the expected item performance based on the chosen model. The major difference between raw and standardized residuals is that raw residuals are simpler to calculate and easier to interpret, whereas standardized residuals take into account the associated sampling error. The experimental comparison of these two indexes (Murray & Hambleton, 1983) has shown that both provide very useful fit information, but the latter presents a more accurate picture of model–data fit.

Computer Simulation for Model–Data Fit Testing

The development of computer simulation software has provided measurement specialists with a useful new tool to study model–data fit and help determine the effect of different factors such as sample size, test length, and item difficulty on the model–data fit. The value of employing a computer simulation analysis, often called a Monte Carlo study, is that the true or underlying parameters are known and can be compared with those estimated by an IRT model. Two indexes, correlation coefficients and the root mean square error (RMSE), are most often used in this procedure. The correlation coefficient measures the intercorrelations between the known item or ability parameters and the estimated parameters, but it is not sensitive to systematic error. RMSE, in contrast, is used to detect the absolute deviation between known and estimated parameters. RMSE is computed by averaging the squared difference between known and estimated parameters and then taking the square root. The accuracy of parameter estimation is sometimes described as recovery of item parameters, in the sense of recovery of true parameters by estimation. Research applying a computer simulation model includes the accuracy of parameter estimation (Lord, 1975; Swaminathan & Gifford, 1979), effect of sample size (Hambleton & Cook, 1983), selecting IRT model (Yen, 1981), and so on. Caution should be taken, however, when applying the findings of a computer simulation because the simulated data may not completely resemble the data collected in applied settings.

SOME UNIQUE FEATURES AND ADVANTAGES OF IRT

Besides the invariance feature of IRT and related advantages (i.e., items and abilities are not dependent on samples or items employed), several additional features make IRT unique and useful. These are common metric scale, item and testing information, and "no" reliability.

Common Metric Scale

In contrast to the different scales for item properties (e.g., p-value for item difficulty) and examinee response (e.g., raw or standardized scores) in CTT-based test construction, IRT employs a common scale, called the ability scale (Hambleton et al., 1991), for both item characteristics and examinee ability. Because ability is a latent trait and cannot be measured directly, an arbitrary metric scale must be set up so that the relationship between items and examinees can be placed along the scale accordingly. This setup is accomplished through the estimation procedures described earlier. In theory, because of the nature of invariance, this process should not be affected by item characteristics, or examinees' ability, or both. In practice, because the scale has to be set up on either the mean of items or examinees, there will be a scale difference when different items or groups of examinee are employed. This difference, however, can be adjusted by using well-developed statistical methods (see chapter 6 by Weimo Zhu).

The ability scale can also be considered as a log-odds scale. The unit of a log-odds scale is called "logit," the contraction of "log-odds unit." The log-odds scale is a scale with ratio scale properties. **Logits** can be thought of as probabilities—the probability that an examinee with a given ability will successfully complete an item. For example, a difference of .693 logits between two examinees corresponds to a doubling of the odds ($e^{(.693)} = 2.0$) for one's success than another's, whereas a difference of approximately 1.4 corresponds to a quadrupling of the odds. The log-odds scale, therefore, has the following important features: (a) It is a "linear" model because all facets (e.g., examinee's ability and item difficulty, in this case) can be represented as fixed positions along one straight line and (b) logits are of equal intervals, therefore additive. In measurement, additive measures are also called *linear measures*. For more information on the log-odds scale and logits, as well as the metric scale under more complex IRT models, see Wright (1977) and Hambleton and Swaminathan (1985, pp. 57-61).

Item and Test Information

Other special features of IRT are item and test information functions. *Information* in general is defined as to "know something about a particular object or topic" (Baker, 1985, p. 81). Testing thus can be considered a process of obtaining information about examinees. In statistics, defined by Sir R.A. Fisher, information (I) is the reciprocal of the precision with which a parameter can be estimated (Baker, 1985). That is, $I = 1 / \sigma^2$, where σ^2 is the variance of the parameter estimators, a precision measure of the variability of the estimates around the value of the parameter.

In IRT, information related to an item is called the **item information function** and is defined specifically for each IRT model. For example, the item information function of the two-parameter model is defined as

$$I_i(\theta) = a_i^2 P_i(\theta) Q_i(\theta). \tag{4.6}$$

Because $a = 1$ in the Rasch model, the item information function in the Rasch model becomes

$$I_i(\theta) = P_i(\theta) Q_i(\theta). \tag{4.7}$$

The **test information function,** $I_T(\theta)$, can be computed by summing all item information functions in a test at a particular ability level as follows:

$$I_T(\theta) = \sum_{i=1}^{N} I_i(\theta), \tag{4.8}$$

where N is the number of items in the test. According to equation (4.6), the test information function of the two-parameter logistic model can be expressed as follows:

$$I_T(\theta)_{\text{for two-parameter model}} = \sum_{i=1}^{N} a_i^2 P(\hat{\theta}) Q(\hat{\theta}). \tag{4.9}$$

One will immediately recognize that the test information function is the same term within the square root of the standard error equation (see equation [4.5]). Thus, the relationship between the standard error $SE\,\hat{\theta}$ and test information function can be expressed as

$$SE(\hat{\theta}) = \frac{1}{\sqrt{I_T(\theta)}}. \tag{4.10}$$

To illustrate, assume that there are two tests with five items each. The difficulties of items (bs) are as follows:

	Item 1	Item 2	Item 3	Item 4	Item 5
Test A	−1	−.05	0	0.05	1
Test B	2	2.2	2.4	2.6	2.8

Using equations (4.7) and (4.8), item and test information functions were computed and summarized in table 4.2. As one may expect, because of the distributions of items (i.e., more variability in item difficulties in test A and more difficult items in test B), test A has a large range of test information with a focus at the center, and test B, a more difficult test, can provide more information at the high end of the ability scale (see also figure 4.3).

Test Relative Efficiency

One of the useful applications of the test information function is to compare the efficiency of tests or subtests for evaluation or selection purposes (Lord, 1977, 1980). **Relative efficiency** (RE) can be computed by comparing the information function of one test with another test at a particular ability level as follows:

$$RE(\theta) = \frac{I_A(\theta)}{I_B(\theta)}. \tag{4.11}$$

Using again the example in table 4.2, $RE(-3, -2, -1, 0, 1, 2, 3) = 11.45271, 9.107661, 5.799434, 2.858506, 1.182109, .4999945,$ and $.2410053$, respectively (see also figure 4.4b). With RE and the information from a targeted examinee population, a more appropriate test can be selected.

Global "Reliability" Is No Longer a Concern

Because IRT is able to provide precise information of a measure at a specific ability level, test reliability used in traditional CTT test construction (see chapter 3 by Ted Baumgartner) is no longer a concern. As introduced earlier, SEM in IRT is associated with an ability level, and there is no guarantee that a test can be reliable over the whole range of the scale. In fact, because of the development of local precision measurement in IRT, there is a call for a change in treatment of "reliability" as follows: (a) The reliability should be interpreted under the framework of the measurement theory employed and (b) test developers should try to provide local precision information even if a test is constructed under CTT (e.g., using conditional standard error coefficient) (American Educational Research Association, American Psychological Association, & National Council on Measurement in Education, 1999).

Finally, IRT and its unique features have made many other measurement practices easier (e.g., examining item bias and differential item functioning [see chapter 7 by Allan Cohen] and conducting test equating [see chapter 6 by Weimo Zhu]).

Table 4.2 An Example of Item Information "$I_i(\theta)$" and Test Information "$I_T(\theta)$" Function

	$I_i(\theta)$					
	Item 1	Item 2	Item 3	Item 4	Item 5	$I_T(\theta)$
Test A						
−3	0.104994	0.070104	0.045177	0.028453	0.017663	0.266390
−2	0.196612	0.149146	0.104994	0.070104	0.045177	0.566032
−1	0.250000	0.235004	0.196612	0.149146	0.104994	0.935756
0	0.196612	0.235004	0.250000	0.235004	0.196612	1.113231
1	0.104994	0.149146	0.196612	0.238651	0.250000	0.939403
2	0.045177	0.070104	0.104994	0.173343	0.196612	0.590229
3	0.017663	0.028453	0.04517	0.075347	0.104994	0.271633
Test B						
−3	0.006648	0.005456	0.004476	0.003671	0.003009	0.023260
−2	0.017663	0.014556	0.011981	0.009853	0.008096	0.062149
−1	0.045177	0.037632	0.031252	0.025890	0.021402	0.161353
0	0.104994	0.089800	0.076255	0.064358	0.054038	0.389445
1	0.196612	0.177894	0.158685	0.139764	0.121729	0.794684
2	0.250000	0.247517	0.240261	0.228784	0.213910	1.180471
3	0.196612	0.213910	0.228784	0.240261	0.247517	1.127083

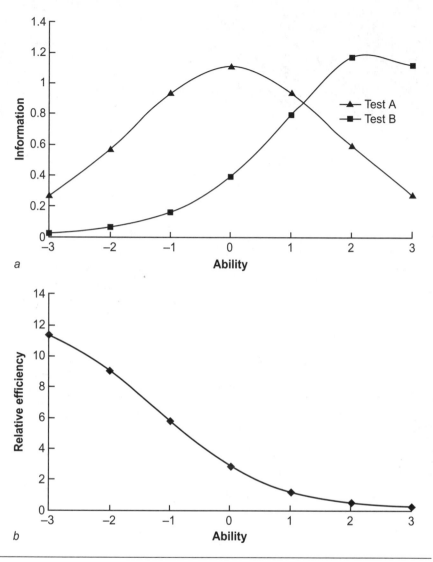

Figure 4.4 (a) Item and test information function; (b) relative efficiency of tests A and B.

ITEM BANK AND IRT-BASED TEST CONSTRUCTION

As a result of the advantage of IRT features, test construction has become more convenient and efficient. Test construction based on a calibrated item bank is a major successful IRT application in modern testing construction. An **item bank** is a collection of items or questions organized and cataloged to take into account the content of each item, as well as other measurement characteristics (e.g., validity, reliability, and difficulty; Umar, 1997). More important, all of the items in the bank share a common scale after IRT calibration.

Several measurement advantages can be achieved with an item bank. First, because all of the items are set on the same scale, the lack of score equivalence among different tests is automatically eliminated, and scores can be directly compared with each other. Second, a stable scale can be developed even if new versions or items are added to the bank later (i.e., the mean and standard deviation of the scale are consistent across different times even if new items are used). Again, a stable scale is essential to use a test across occasions and to communicate the results. Third, because the characteristics of the items are already known, constructing new tests for different purposes of testing becomes much easier. For example, if a researcher is interested in constructing a test to screen a group's ability, items that cover a broad range of the ability can be selected (see the top of figure 4.5). In contrast, if the interest is to construct a test to determine whether an examinee is qualified for a professional certification (e.g., mastery vs. nonmastery), only items around a theoretical cutoff point will be selected. As a result, a shorter test, but one with peaked information at the cutoff point, can be constructed (see the middle of figure 4.5) and a more accurate classification can be achieved.

With a well-developed item bank, a new and tailored way to administer a survey using computers, known as **computerized adaptive testing** (CAT; Wainer, 1990), becomes possible. CAT, originally developed as an American military personnel screening practice, is very similar to a high-jump competition, in which competitors choose their jumping heights depending on their perceived abilities. In CAT, a question from the middle of an ability range that an item bank has defined is administered first to the examinee. If the item is answered correctly, the next question asked is more difficult; if it is incorrectly answered, the next question is easier. This continues until the examinee's proficiency is established to within a predetermined level of accuracy. Because of IRT invariance, examinees' abilities can be estimated even if they respond to different items within CAT. In fact, a more accurate estimation is expected because items selected are closer to an examinee's ability. Because responses have been entered into a computer during the test administration, the final estimation of an examinee's ability can be reported as soon as the test is finished (see the bottom of figure 4.5). More information about CAT can be found in chapter 8 by Richard Gershon.

KINESIOLOGY APPLICATIONS

Spray (1987) called for research to determine whether IRT can offer some improvement over more traditional CTT measurement techniques used in kinesiology. Based on the relatively limited subsequent research efforts, two conclusions can be drawn addressing Spray's challenge. First, when the data are dichotomous or polytomous, all existing IRT models can be applied (e.g., Safrit, Zhu, Costa, & Zhang, 1992; Zhu, 1996; Zhu & Kurz, 1994) and associate measurement advantages

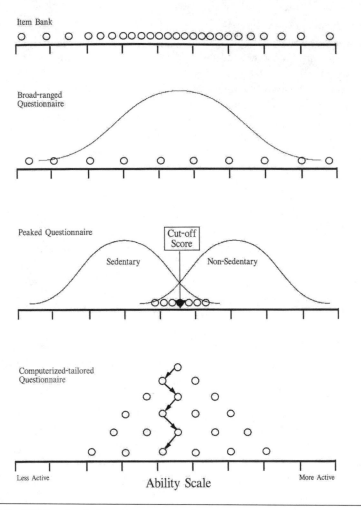

Figure 4.5   IRT item bank–based test construction.

can be taken (e.g., Zhu & Cole, 1996; Zhu, Timm, & Ainsworth, 2001). In fact, some measurement specialists in kinesiology have been active in exploring IRT's advantages in solving critical measurement problems (e.g., in determining optimal categorization [Zhu, 2002; Zhu, Updyke, & Lewandowski, 1997] and detecting judge bias in competitive sports such as figure skating [Looney, 1997]). Because of its relative small sample size requirement and ease of interpreting results, almost all IRT applications in kinesiology used the Rasch model or its extensions. Considering that most of the data in the field are continuous, there is still a significant gap in exploring the continuous IRT model. Except for a few attempts (e.g., Spray, 1990), little progress in developing a continuous IRT model for kinesiology applications has occurred. The second conclusion drawn is that, from an applied perspective, exercise scientists have not taken advantage of IRT measurement properties. In fact, the IRT method has been basically ignored in the field except for a few applications (e.g., the "FitSmart" fitness knowledge test by Zhu, Safrit, and Cohen, 1999). Although it was originally hoped (Wood, 1987) that subdisciplines such as sport psychology, which typically collects polytomous data, would take up the method first, it failed to materialize even though (a) IRT could bring significant measurement advantages to measurement practice (Embretson & Reise, 2000) and (b) there was a call for the IRT application made within sport psychology (Tenenbaum & Fogarty, 1998). Lack of progress may be due to three major reasons described by Wood (1987)—lack of training in the field, the complex nature of IRT, and the nature of psychomotor measurement. To take full advantage of IRT, kinesiology must start making efforts to address these issues.

SOFTWARE

Similar to other advanced measurement techniques, computer software is necessary for IRT applications. Software varies depending on the chosen models, estimation procedures, and fit statistics employed. Operating systems and prices also vary. Characteristics of several commonly used IRT software packages are summarized in table 4.3. Interested readers are referred to the software Web sites for more specifics.

Table 4.3   Summary of Commonly Used IRT Software

Software	Dimen- sionality	Data Format	Source
BILOG	Uni	D	Scientific Software International, Inc.
			7383 N. Lincoln Ave., Suite 100
MULTILOG	Uni	D and P	Lincolnwood, IL 60712-1747, USA
			Tel: 800-247-6113
RARSCALE	Uni	D and P	Fax: 847-675-2140
			www.ssicentral.com
ConQuest	Multi	D and P	ACER Press
			347 Camberwell Road
			Private Bag 55
			CAMBERWELL
			Victoria 3124 Australia
			Tel: 800-338-402
			Fax: + 61-3-9835-7499
			E-mail: sales@acer.edu.au
			www.acer.edu.au/publications

(continued)

Table 4.3 *(continued)*

Software	Dimen-sionality	Data Format	Source
FACETS	Uni	D and P	Winsteps, P.O. Box 811322
			Chicago, IL 60681-1322, USA
Winsteps	Uni	D and P	Tel. & Fax: (312) 264-2352
			www.winsteps.com/facets.htm
MASI-IRT (free software)	Uni	C	Judith A. Spray
			c/o ACT
			P.O. Box 168
			Iowa City, IA 52243, USA
MSP (nonparametric)	Uni	D and P	Assessment Systems Corporation
			2233 University Ave., Suite 200
			St. Paul, MN 55114, USA
PARELLA (nonomonotone items)	Uni	D	Tel: 651-647-9220
			Fax: 651-647-0412
			www.assess.com/
RUMMFOLD (unfolding response)	Uni	D and P	

Uni = unidimensionality; Multi = multidimensionality; C = continuous; D = dichotomous; P = polytomous

IRT LIMITATIONS AND FUTURE DIRECTION

Although many advantages have been demonstrated in IRT-based test construction, IRT is not without limitations. Strong assumption requirements, the need for a large sample size, and lack of a continuous model are perhaps the major limitations associated with IRT. As mentioned earlier, there are several critical assumptions required by IRT models. Although violation of an assumption may not result in critical threats to the credibility of the results, one has to be very careful in interpreting findings whenever a violation occurs. Understanding the data and taking steps to prevent the violation early is perhaps the best approach in dealing with this limitation. The need for large sample sizes for IRT is another limitation. In general, sample sizes of at least 200 and 20 items are needed for the Rasch model. Sample sizes of $N = 500$ and 30 items and $N = 1,000$ and 60 items, respectively, are required for two- and three-parameter models (Suen, 1990). For large-scale testing, the requirement is not a problem. It may become a major threat, however, when applying IRT in laboratory research, in which a sample of 30 or more could be problematic as a result of cost and logistical concerns. Finally, most IRT models are developed for dichotomous and polytomous data, whereas the majority of motor-performance data are continuous. Although Spray (1990) introduced a set of models that can analyze "multiple-attempt, single-item" data and a few continuous IRT models were reported, there is still a significant shortage in the availability of IRT continuous models and related software. Efforts in exploring and developing new continuous models and related application software are crucial.

Although IRT has been in the literature for over 50 years, new research efforts and development are still very active and can be summarized in three areas: new models, new components, and new applications. Because test performance and a score are often determined by more than one single trait, the unidimensionality assumption of IRT is often a limitation in practice. Efforts, therefore, have been made in developing new multidimensional IRT (MIRT) models. A set of new MIRT models (e.g., McDonald, 2000; van der Linden & Hambleton, 1997) and related software (e.g., ConQuest by Wu, Adams, & Wilson, 1998) have been developed. It is expected that additional new models and software will become available in the very near future. To make the IRT assumptions less restrictive, efforts have also been made to develop a set of nonparametric IRT models (Sijtsma & Molenaar, 2002), although no major applications have been reported.

Another new development is the integration of cognitive, trait, or data characteristics into the modeling. The development of psychometric modeling has been criticized for paying too much attention to the "metric" aspect of a model (i.e., quantitative, statistical, or mathematical aspects of a model) and too little on its "psycho" features or characteristics (Suen, 1990). Significant efforts (e.g., Nichols, Chipman, & Brennan, 1995; van der Linden & Hambleton, 1997) have been made to integrate cognitive components into new IRT model development. Very recently, data characteristics also have been taken into consideration. For example, to analyze the clustered nature of school research data, a set of new multilevel IRT models was developed (e.g., Pastor, 2003). It is expected that these developments will lead to a new area of IRT application in the coming years.

Finally, IRT applications have been moved out of the traditional box of educational, cognitive, or knowledge testing, in which measuring examinees' cognitive abilities is the only interest. Today, IRT has been applied to measurement practices in various areas such as medicine, biology, sociology, and kinesiology. Applications have expanded to address many nontraditional measurement issues such as optimal categorization (Zhu, 2002), artistic judgment (Bezruczko, 2002), and genomic measurement (Markward, 2004). This extension is expected to continue.

SUMMARY

Because of its many advantages, IRT has become a major measurement theory for many fields. New development in both theory and practice makes it a very useful measurement theory for kinesiology. Unfortunately, since IRT's introduction to exercise science nearly 20 years ago, it still has not been widely employed in kinesiology measurement practice. Kinesiology urgently needs greater effort in developing IRT applications and training for both measurement specialists and graduate students.

PART II

Current Issues
in Measurement

CHAPTER **5**

Ethical Issues in Measurement

Patricia Patterson

Measuring characteristics of participants in physical activity environments is fraught with ethical issues. To illustrate, consider the following scenario: Kristin is administering a battery of functional fitness tests to Maria, a 70-year-old Hispanic woman who has limited English-speaking skills. Each test in the battery has excellent validity and reliability evidence, based on concurrent validity coefficients, discriminant evidence for the construct validity of function, and intraclass correlation coefficients for test–retest reliability. Kristin has also been through an extensive training program that included test administration procedures and practice as well as discussions of characteristics of adults 65+ years of age. While administering the tests, Kristin notices reticence and fear on the part of Maria. She also feels uncomfortable with her ability to communicate with her client. Kristin completes the test administration and compares Maria's scores with normative scores based on a sample of several thousand seniors across the country. Given that Maria scored in the 10th percentile on all functional tests, Kristin speaks with her about her high risk for disability.

This scenario raises a number of questions. For example, what impact might language difficulty have on test scores? How might differences in cultural attitudes regarding the importance of physical activity affect scores? Could fear of falling or low levels of self-efficacy for performing physical tests play a role? What were the characteristics of the normative sample used by Kristin for score interpretation? Could any of these issues affect the validity of the interpretation of these scores? And finally, what are the consequences of the feedback that Kristin gave to her client? These questions involve issues of ethics in measurement, the topic of this chapter.

In general, **ethics** is considered to be a set of moral ideas and principles that guides our conduct (Etzel, Yura, & Perna, 1998). Much has been written about ethical issues within the context of research settings (e.g., Bird & Housman, 1996; George, 1997; Kroll, 1993; Lynn & Virnig, 1995; Roberts, 1993; Safrit, 1993; Thomas & Nelson, 2001; Zelaznik, 1993). Topics include (a) informed consent and treatment of human or animal subjects, (b) plagiarism, (c) fabrication of data, (d) publication issues, (e) authorship, and (f) appropriate behavior in professional relationships.

Considerably less has been written about ethics in measurement. Cates (1999) discussed ethics in terms of principles that guide the decision-making process of testing, and it is this definition that will provide the framework for examining ethical issues in the context of assessment. That is, this chapter will explore ethical issues involved with test construction, test administration, test use, and test interpretation, which are cornerstones of the assessment process. A brief history of the changes in ethical and technical standards adopted by the American Psychological Association (APA) will be described because they inform the changing face of validity (see chapter 2 by Matt Mahar and David Rowe), which now includes a consideration of the consequential or ethical nature of test use and interpretation. Examples in exercise science will be used to illustrate common ethical issues in measurement.

ETHICAL STANDARDS

The APA has published ethical standards for psychologists since 1953, with revisions occurring in 1963, 1977, and 1981. The original code, with 100 principles, divided testing issues into 18 standards in six subcategories: (a) test user qualifications, (b) the sponsoring psychologist's role in testing, (c) the role and qualifications of the test publisher, (d) the readiness of a test for release, (e) test descriptions in manuals, and (f) confidentiality issues regarding test materials (Haney & Madaus, 1991). Ethical considerations of test construction or application were covered briefly. The 1963 revision, which was published by APA in 1967, decreased the ethical testing standards to three rather broad principles regarding test security, interpretation, and publication, whereas the 1977 and 1981 revisions distilled these standards to a single principle centered on the qualifications of the test user and confidentiality issues of tests and test results.

TECHNICAL STANDARDS

Technical Recommendations for Psychological Tests and Diagnostic Techniques, coauthored by the APA, the American Education Research Association (AERA), and the National Council on Measurement Used in Education (NCMUE) (1954) were developed because "a test manual should carry information sufficient to enable any qualified user to make sound judgments regarding the usefulness and interpretations of the test" (p. 2). The document contained over 160 standards in the following six categories: information dissemination, interpretation, validity, reliability, administration and scoring, and scales and norms (Haney & Madaus, 1991). Even in this early document, there is a hint of the importance of ethics: "Almost any test can be useful for some functions and in some situations. But even the best test can have damaging consequences if used inappropriately" (APA, AERA, & NCMUE, 1954, p. 7, as cited in Haney & Madaus, 1991). The emphasis on the importance of considering the *consequences* of testing clearly binds technical standards to ethical standards.

Subsequent revision of the standards (APA, AERA, & NCMUE, 1966) focused on changes in the conceptualization of validity. Although the 1954 version listed four kinds of validity (content, concurrent, predictive, construct), validity was viewed as having tripartite categories of content, criterion-related (in which concurrent and predictive validity were subsumed), and construct validity in 1966.

The 1970s brought additional changes to the technical standards (APA, AERA, & NCMUE, 1974). As a result of litigation associated with personnel testing, the validity section was predominantly a discussion of criterion-related validity, with over 60% of the validity standards emphasizing this area (Haney & Madaus, 1991). Moreover, a standard concerned with the importance of investigating test bias was an important addition, again arising from the preoccupation with employment testing. The other salient change was the addition of a section pertaining to test use. For the first time direction was given to test users. For example, Standard J1 stated, "A test score should be interpreted as an estimate of performance under a given set of circumstances. It should not be interpreted as some absolute characteristic of the examinee or as something permanent and generalizable to all other circumstances" (APA, AERA, & NCMUE, 1974, p. 68, as cited in Haney & Madaus, 1991). The standard further stated, "A test user should consider the total context of testing in interpreting an obtained score before making any decisions [including the decision to accept the score]" (p. 68). Again, this standard emphasizes the importance of considering the context of testing and the potential for variables such as behavior problems, physical disabilities, language problems, and ethnic or cultural factors to affect score interpretation.

In 1985 the standards were dramatically revised, including changing the title to *Standards for Educational and Psychological Testing* (AERA, APA, & National Council on Measurement in Education [NCME], 1985). The change from the word *tests* to *testing* emphasizes the dynamic nature of testing to include consideration of alternative contexts and applications in regard to test use and interpretation, rather than a simple view of the technical aspects of a test. Of most importance for the discussion of ethics in this chapter is the major change in how the standards treated validity. The validity section begins: "Validity is the most important consideration in test evaluation. The concept refers to the appropriateness, meaningfulness, and usefulness of the specific inferences made from test scores. Test validation is the process of accumulating evidence to support such inferences" (p. 9). Furthermore, "A variety of inferences may be made from scores produced by a given test, and there are many ways of accumulating evidence to support any particular inference. Validity, however, is a unitary concept" (p. 9).

The 1999 edition of the *Standards for Educational and Psychological Testing* (AERA, APA, & NCME, 1999) emphasized two important points that are relevant to the discussion of ethics. First, **consequential validity** was defined as occurring when group differences in test scores occur because of test bias or measuring additional skills that are not a part of the construct, not when differences in test scores result from true differences among groups. And second, the interrelational nature of testing was emphasized by stating the importance of directing attention toward examining how differences in characteristics of test administrators and test takers may affect scores. If an interaction of these characteristics does affect scores, test fairness has been compromised, which results in less confidence in the validity of the score interpretation.

CONSTRUCT VALIDITY

Many measurement specialists, (e.g., Messick, 1975, 1989b; Pedhazur & Schmelkin, 1991; Shepard, 1993), have argued that **construct validity** is the "unitary concept" for validity. For example, Messick stated that "construct validation is the process of marshaling evidence in the form of theoretically relevant empirical relations to support the inference that an observed response consistency has a particular meaning" (Messick, 1975, p. 955). Messick concluded that one must examine the potential consequences of score interpretation and use. Thus, one does not validate a test, but rather the interpretation of data that are generated from a test and the attendant consequences of test use. This view makes it clear that validity is a matter of degree and not an "all or none" proposition (Messick, 1989b); it is a continuing process of validating both interpretive and action inferences.

Validation of interpretative inferences requires examining a construct in multiple ways, similar to Campbell and Fiske's (1959) multitrait–multimethod approach. This approach allows one to ensure that two methods of measuring the same trait relate more highly than measures of different traits using the same method. Thus, one proposes and tests both hypotheses and counterhypotheses. Validation of action inferences requires examining score meaning, value implications, and action outcomes. As Messick (1989b) described, one is appraising "the relevance and utility of the test scores for particular applied purposes and the social consequences of using the scores for applied decision making" (p. 13).

Because interpretation and use of test scores will be influenced by the values held by people and the larger society, construct validity processes are placed squarely in the realm of ethics. To ignore this fact would be tantamount to saying that having technical evidence that the test is a good measure of the construct it is supposed to measure justifies its use in any setting, regardless of the consequences. Clearly, values pervade all aspects of the measurement process (e.g., Cole & Moss, 1989; Lynn & Virnig, 1995; Messick, 1975, 1980, 1981); thus, weighing the actual and potential consequences of test use must become an intrinsic part of the validation process. The next sections of the chapter will take a closer look at the consequential, or ethical, nature of measurement using examples from exercise science.

EXAMPLE 1: ETHICS AND THE MEASUREMENT OF PHYSICAL ACTIVITY (PA)

Given the abundance of research indicating that there is an inverse relationship between PA and risk for hypokinetic diseases, we decide that it is important to measure this construct; that is, we *value* knowledge about physical activity patterns and their relationship to public health concerns. Thus, our intent is to develop a self-report measure of PA that can be used in large epidemiological studies (see chapter 13 by Michael LaMonte, Barbara Ainsworth, and Jared Reis in this text for further information about assessing physical activity).

We review the literature to assist with defining physical activity and select the definition by Casperson, Powell, and Christenson (1985) that states that PA is any bodily movement produced by skeletal muscles that results in energy expenditure. Based on this definition, activity categories are generated that include activities of daily living, occupational activities, and sports and leisure activities, with a list of associated activities that "belong" in each of these categories. Content-related evidence is gathered by having judges rate the items for relevance, representativeness, clarity, and specificity (Patterson, 2000). Factor analytic techniques are employed to examine the structure of the construct (Goodwin, 1999), and convergent and discriminant evidence (Campbell & Fiske, 1959) are analyzed to ensure that scores from the test yield meaningful interpretations.

The previous descriptions of the steps taken in test development are generally viewed as technical issues of construct validation. However, these steps also involve value judgments. Decisions are made about which question to study, what design to use, how the construct is measured, who will be measured, and how the test will be used, all of which are viewed from the perspective of the test developer (Lynn & Virnig, 1995).

The implications of these value-laden decisions during test development significantly affect test use and interpretation. For example, what are the consequences of defining physical activity in another way? What changes would need to be made if one were interested in using the scores from the self-report questionnaire for estimation of caloric expenditure or to classify people into activity levels for an exercise program? These questions require weighing actual and potential consequences of using the self-report PA in a particular way—no easy task. Churchman (1971) suggested using Kantian and Hegelian inquiry processes to explore these questions. Kantian inquiry pits the proposed test use against alternative uses, whereas Hegelian inquiry pits the proposed test use against antithetical uses. These strategies should "draw attention to vulnerabilities in the proposed use and expose its tacit value assumptions to open debate" (Messick, 1980, p. 1020). Moreover, although high scores on the self-report of PA may indicate that a person is physically active, low scores do not necessarily mean that a person is physically inactive. The low scores may be the result of a number of other reasons, including lack of understanding, language and cultural differences, lack of motivation to answer truthfully, and inadequate construct definition, to name a few. The important point here is that there is no easy answer, and perhaps no right or wrong answers, to these issues. If we "confront our assumptions and ideologies, then their influence upon our interpretations may be taken into account, or at least our responsibility for them will no longer be as easily disclaimed on the grounds of innocence" (Messick, 1975, p. 965).

That is, we cannot eliminate biases or values; but if we identify them, we can track their role in data collection, analysis, and interpretation.

Self-report PA scores may also be used as one of several outcome variables of an intervention study. We often judge the efficacy of an intervention program by statistically analyzing the changes in the outcome variables, as well as evaluating the associated costs of the intervention in relation to the magnitude of change in the outcome variables. If important differences occur as a result of a reasonably cost-effective program, one might conclude that the intervention was a success. However, this view overlooks two important points. First, to judge the value of an outcome, one must examine the nature of the processes that produced the outcomes (Messick, 1975). Dewey (1939) described this means–ends examination as a two-way street. That is, evaluation of the means (intervention) takes place in terms of the ends (outcomes), but also the ends must be evaluated in terms of the means that produced them. Second, an intervention study will likely yield several outcomes in addition to the intended ones. One must take both intended and unintended outcomes into consideration when judging the worth of the program. For example, in the case of a PA intervention, an unintended outcome might be that the program will serve to widen the gap between the public health status of those with adequate time, money, and access to the program as compared to those who do not have this access. This difference in value perspective means that the data from the intervention will be filtered in very different ways and result in a conclusion that the intervention is not a success.

This example illustrates the intrusion of values in research settings and how each value premise can result in an incomplete or selective understanding of the construct, or both. It further underscores the importance of viewing test use and test interpretation from multiple perspectives. As Kaplan (1964) said, "Fortunately, science does not demand that bias be eliminated but only that our judgment take it into account" (p. 376). Although this task may seem daunting, it can be guided by linking the two points previously described. If one has an "understanding of the nature and meaning of the processes operating to produce effects," there is a "rational basis for inferring likely effects" (Messick, 1975, p. 963).

Another common issue associated with the measurement of PA is the use of the scores at the individual level, rather than at the group level. Let's go back to the scenario described at the beginning of the chapter. Kristen gave feedback to Maria by comparing her scores on the functional fitness tests with age- and gender-matched scores in norm tables. This seems to be a reasonable and appropriate thing to do. There are, however, unintended implications associated with this action. It is well known that the variability associated with functional fitness scores in seniors is quite large. It is also well known that heterogeneity in a sample can inflate reliability coefficients. These two factors play an important role in how one should view the reliability of measures and subsequently the interpretation of the scores. For example, the reliability coefficients for the functional fitness tests reported by Rikli and Jones (1999a, 1999b) are very high, which might lead one to provide feedback with confidence to seniors. Yet, without considering the root mean square error (RMSE) associated with the reliability of the test, the consequences of this feedback could be misleading (Looney, 2000).

Imagine that the intraclass reliability is .90 for a test that measures the number of times (i.e., repetitions) a person can stand and return to sitting in 30 s. Further, the RMSE associated with the test is ± 2.5 repetitions. You consult norm tables for your client and tell her that her score is in the 50th percentile, about average for seniors her age and gender. However, the difference between the 25th and 75th percentile for these scores is five repetitions. Thus, the real meaning of this score is very different from the feedback just provided. In this case, the consequences of interpreting this score can mean that you may provide information to a client that underestimates her true ability, causing undue concern about function, or overestimates her true ability, causing false confidence about function. Each consequence can have serious implications and underscores the importance of evaluating these ethical measurement issues.

EXAMPLE 2: ETHICS AND THE MEASUREMENT OF EFFECTIVE COACHING

A graduate student in sport psychology was interested in developing a prediction equation for coaching effectiveness. The student selected win–loss record as the criterion measure for coaching effectiveness. The predictor variables included discrepancy scores obtained from the Leadership Scale for Sports (LSS) (Chelladurai & Saleh, 1980), team cohesion scores from the Group Environment Questionnaire (Carron, Widmeyer, & Brawley, 1985), and a number of demographic characteristics of the coach. One hundred fifty female college basketball teams completed the two tests; their coaches provided the demographic information and win–loss records. Multiple regression indicated that three of the five subscales of the LSS as well as two subscales for team cohesion and years' experience coaching were significant predictors, accounting for 81% of the variance in win–loss records. The student concluded that the derived equation not only could be used to predict win–loss record but also would be useful in personnel selection.

This example raises a number of questions. What are the consequences, intended or unintended, of using win–loss record as an indicator of effective coaching? How might response styles in answering the two paper-and-pencil tests affect interpretation? What role could test anxiety or concern that a player's coach will somehow learn how she answered the questions have on the usefulness of the interpretations to be made from these scores? What are the consequences of this approach if used for job selection of coaches?

Let's take a closer look at the appropriateness of using win–loss records in job selection. A number of value judgments are implicit in the previous scenario. First, the student decided that an outcome-based criterion (win–loss record) was the best measure of coaching effectiveness, a view not entirely antithetical to what we read in the sport pages. Predictor variables were selected based on their assumed relevance to coaching effectiveness and sound psychometric properties. The design employed by the student was quite typical and involved examining the correlation between the predictors and criterion scores, with high correlations ($r > |.80|$) indicating that the predictors could be used with reasonable confidence. However, this view of criterion-related evidence is rather narrow and concerned with the predictor–criterion relationship to the exclusion of other evidence.

Messick (1989a) eloquently discussed the problems inherent in this narrow view of evidence. For example, the selection of win–loss record as the criterion must be evaluated, just like all other measures. What if it is deficient in capturing the essence of the construct of coaching effectiveness? For example, what if player satisfaction should be included in this construct? This potential for **construct underrepresentation** supports the importance of viewing this question from a construct validity approach, rather than from the narrow view of criterion-related evidence evaluated in isolation. Messick (1989a) wrote, "construct validity is based on an integration of any evidence that bears on the interpretation or meaning of the test scores" (p.7). This includes an examination of expected relationships as a basis for provisional score meaning as well as the consequences of both test use and interpretation. Messick (1989a) went on to say that "using test scores that 'work' in practice without some understanding of what they mean is like using a drug that works without knowing its properties and reactions. You get some immediate relief, to be sure, but you had better ascertain and monitor the side effects" (p. 8). Unless one can demonstrate that the observed relationships are meaningful in terms of the nomological network associated with the construct of coaching effectiveness, evidence of high correlations does not provide a rationale that justifies either the relevance or use of the scores for personnel selection.

Use of test scores for selection or ability grouping is also linked with issues of **test bias**, or test fairness. But clarifying the meaning of test bias is complicated. For example, if using a test for selection of a coach is viewed as something positive, then this test use is easily accepted. In contrast, if these actions are seen as limiting the opportunities of deserving coaches, then the same decisions are viewed negatively. Consensus on test bias is difficult, if not impossible, because of the diversity of values and beliefs (Cole & Moss, 1989). Its complexity requires examining the issue from multiple perspectives and the testing of rival hypotheses. This exploration provides evidence about bias.

What is bias? Cole and Moss (1989) discussed bias in terms of the inferences to be made from test scores. Bias is present when the meanings and implications of the test scores are different for a particular subgroup than are the meanings and implications for the rest of the test takers. Cole and Moss stated, "bias is differential validity of a given interpretation of a test score for any definable, relevant subgroup of test takers" (p. 205).

Although this definition of bias appears rather straightforward, the process of determining whether bias exists is complex. Bias might be involved in any or all of the following: test selection and use, test content and format, test administration and scoring, and the internal structure of the test. In the case of test administration, it is generally agreed that one should standardize administrative factors such as environmental conditions, test-taking directions, and time. However, this standardization does not ensure protection against bias. For example, if the gender or race of the test administrator who is providing standardized directions influences test scores differentially, the interpretation of these scores is biased. Bias can also be present in test scoring, particularly when scoring methods are not completely objective. Two examples of this phenomenon are scoring written responses and ratings of performance, both of which may be part of the selection process for a job or other situations. If the rater is influenced, either consciously or unconsciously, by the respondent's age, gender, ethnicity, social class, ability level, and so on, there is a danger of bias. Carlson, Bridgeman, Camp, and Waanders (1983) suggested that bias-free ratings can be particularly difficult to achieve when the rater and examinee do not share the same cultural heritage. Potentially different views of what is meant by competent performance may lead the rater and examinee to value different aspects.

Several statistical methods can be used to examine the internal structure of a test, particularly the invariance of the structure. These include confirmatory factor analysis and structural equation modeling (see chapter 9 by Fuzhong Li and Peter Harmer), in which tests of equality can be used to determine whether the test structure is invariant across age groups, genders, ethnic groups, and so on. Moreover, item characteristic curves can be examined using item response theory (see chapter 4 by Weimo Zhu). Test fairness is described in detail in chapter 7 by Allan Cohen.

EXAMPLE 3: ETHICS IN PERFORMANCE ASSESSMENTS

Testing practices that do not link tasks and contexts to real-world challenges have received criticism. Critics of conventional large-scale testing admonish that standardized testing is based on a simplistic stimulus–response view of learning undertaken in a one-event format (Wiggins, 1993). In fact, Resnick (1990) suggested that the assumptions that knowledge can be decomposed into elements and decontextualized are false. Learning is exhibited only when knowledge can be applied fluently and aptly in both a particular context and diverse contexts, which requires moving away from standardized testing. For example, a history teacher might weigh the pros and cons of two different testing scenarios. One choice might be to construct a test of 100 multiple-choice questions as indicators of students' understanding of U.S. history. A second choice might be to construct a performance assessment that requires students to write a paper and give a 10 min oral presentation to their class regarding whether a particular U.S. history textbook should be adopted for use. The presentations and papers are evaluated on how well students support their views with regard to whether the text is accurate, biased, or simply presents a different viewpoint (Wiggins, 1993). In a physical education setting the decision might be between (a) the relevance of a student being able to write an accurate description of a particular defense in basketball for an exam and (b) the ability of a student to perform appropriately during a game in response to an ever-changing offensive team. These discussions have led to the increasing popularity of performance tests as assessments of student learning (see chapter 12 by Terry Wood).

Historically, physical educators have used performance assessments when rating students' abilities during a game setting. The recent development of national physical education standards (National Association for Sport and Physical Education, 2004) and their associated performance benchmarks are indicative of an even more dramatic shift toward performance assessment. Content standards define what a student should know and be able to do to be a "physically educated person," whereas performance standards describe how well a student must perform.

A performance takes place when we produce something of our own using our knowledge and a variety of skills in response to the tasks and context at hand. This definition makes a clear differentiation between skill drills, which are a common part of physical education curricula, and a performance, which requires a student to demonstrate the ability to make appropriate adjustments in a dynamic game environment. Thus, a performance assessment is authentic when it requires the performance to be faithful to criterion situations; that is, the test is representative of the contexts encountered in the "real world."

Performance assessments "typically permit students substantial latitude in interpreting, responding to, and perhaps designing tasks; they result in fewer independent responses, each of which is complex, reflecting integration of multiple skills and knowledge; and they require expert judgment for evaluation" (Moss, 1992, p. 230). This approach poses unique challenges to measurement specialists and test users. Whereas multiple-choice questions are designed to have a single correct answer, performance assessments allow for variations in response to ensure fidelity to the often organic and "messy" criterion situations of successfully performing in a complex society. Therefore, the traditional measurement issues of validity, reliability, and objectivity become even more complex when using performance assessments.

Despite the popularity of performance assessments, validation efforts to date have primarily focused on ensuring that performance assessment content is in agreement with the content standards being assessed. As Linn (2002) stated, "There is considerable emphasis on developing assessments that are aligned with content standards. Far less attention is paid, however, to accumulating evidence needed to judge the adequacy and appropriateness of interpretations and uses made of assessment results" (p. 234). The concern with alignment is rather like believing that content relevance is sufficient evidence for test validity. The more recent view of a unified concept of validity exemplified by construct validity (e.g., Messick, 1989b; Shepard, 1993) illustrates the inadequacy of content alignment as sufficient evidence for the validity of score interpretation or use of performance assessments.

Messick (1994) emphasized the need for a construct-driven view of performance assessment rather than a task-driven view because in some situations, such as an athletic contest, science fair, or arts competition, there is no wish for replicability or generalizability. The performance *is* the focus, and it is examined for its quality to designate a "winner." However, if the real intent is to infer that a performance is indicative of competence of an underlying construct, then an examination of the construct validity of the interpretations of the performance is necessary. Messick's (1975, 1989b) discussions of score meaning, value implications, and action outcomes are relevant here as well. Within a construct validation framework, the intended and unintended consequences of performance assessments must be evaluated from multiple perspectives. Thus, one must consider the validity of the interpretations from these measures, which include consideration of their consequences on instruction, learning, and equity in the classroom. Linn, Baker, and Dunbar (1991) also raised the importance of examining the intended and unintended consequences of performance assessments. Their discussion of fairness issues centers on ensuring that differences among groups are not a result of a "difference in familiarity, exposure, and motivation on the task of interest" (p. 18).

Using Messick's (1989a) language, these differences could introduce construct-irrelevant variance to the score meaning. In other words, a richly contextualized performance task may not be uniformly good for everyone. Some students may be motivated by the task, whereas others may be alienated or confused. Messick concluded, "If the adverse social consequences are empirically traceable to sources of test invalidity, such as undue reading comprehension requirements in a purported test of knowledge or of reasoning, then the validity of the test use is jeopardized, especially for less proficient readers" (p. 11). Therefore, it is incumbent upon test users to examine both the intended and unintended consequences of test use. Given the diversity of our classrooms, this is no easy task; however, the difficulty of the task does not negate the essential nature of this view. As Cremin (1989) noted, "standards involve much more than determinations of what knowledge is of most worth; they also involve social and cultural differences, and they frequently serve as symbols and surrogates for those differences" (p. 9).

As discussed in early sections of this chapter, an awareness of the intrusion of values in performance assessment is critical. For example, values are inherent in choices made about what should be included and emphasized in standards, which then leads to what should be taught and assessed (Linn, 2002). Therefore, an examination of these choices and weighing the consequences of the inclusion of content a, b, c as opposed to d, e, f is necessary.

Another example of the pervasiveness of values in performance assessments is the use of labels to describe the degree to which a student has or has not met the standard. Common labels include *master/nonmaster, pass/fail, proficient, inadequate, minimal, marginal, excellent, outstanding,* and so forth. The selection of these labels also has consequences because of the values inherent in each label. An example of the power of a label can be seen in the common use of the term

Disabled Student Services on university campuses. If a student who has dyslexia views the label *disabled* in a negative way, he or she may be less inclined to use the services. The consequences of the student's interpretation of the label and his or her subsequent decision to avoid seeking assistance may include lower grades, increased anxiety and stress, lower self-esteem, and lowered expectations of the student by the instructor. It is important, therefore, to consider the potential consequences of our choice of labels on test use and interpretation.

Those involved with administering state and national performance assessments struggle with another value-laden issue: the desire to include as many students as possible in these assessments. This decision has resulted in an increasing number of requests for special testing accommodations. The most common request is extra time, but other requests include oral presentations of tests, individualized administration, and translation of tests into different languages (Linn, 2002). These accommodations raise issues not only for the students for whom the accommodations are intended but also for all students taking the test. For example, what if allowing extra time for nondisabled learners better represents their true ability? Shouldn't they also be given that accommodation? Test administrators must consider the consequences of these decisions in terms of their role in both the validity of score interpretation and test fairness.

Lest it appear that test users are the sole arbiters of ethics in measurement, these issues also reside with the responsibilities of measurement specialists. A description of the technical aspects of a test is not inclusive of the full range of construct validation concerns. Publications should also include a discussion of relevant value contexts and the potential consequences of test use. Consideration of alternative uses of the test and the attendant implications is also essential. Messick (1981) made this point very clear when he wrote, "Therefore, from the standpoint of scientific and professional responsibility in educational and psychological measurement, it seems clear that ethics, like charity, begins at home" (p. 19).

SUMMARY

This chapter describes the genesis of ethics in measurement and its link with construct validation processes. Examples in three areas of exercise science illustrate the pervasive nature of values in measurement and outline important points to consider. First, issues associated with measurement, analysis, and interpretation have often been miscast as technical issues. It is impossible to discuss the validity of our interpretations without taking into account their consequences, and this involves a discussion of ethics. Second, viewing the choices we make in test construction and selection, test administration, test use, and test interpretation from multiple perspectives can assist us in exposing our value premises. Third, multiple perspectives can be evaluated by setting up hypotheses and counterhypotheses, using the Kantian and Hegelian inquiry systems described in this chapter and explained at length by Churchman (1971). An awareness of our own paradigms and how they shape the questions we ask, the methods we use, and the interpretations we make is critical to being open to new and different ways of inquiring. And finally, values and biases can never be completely eliminated; however, we can provide an explicit accounting of the rationale for the choices we make in the methods section of our published works. As Cronbach (1980) stated, "value-free standards of validity is a contradiction in terms, a nostalgic longing for a world that never was" (p. 105). Perhaps readers of this chapter will be challenged to rethink their views of measurement and consider the inextricable link between technical and ethical issues.

CHAPTER 6

Scaling, Equating, and Linking to Make Measures Interpretable

Weimo Zhu

Raw scores on a test (e.g., the number of correct scores) could have little practical meaning even if the test is valid and reliable. This is because raw scores are determined often by the characteristics of a test, such as the length of the test, the difficulty of the items, and the time allowed for the test. Raw scores can be very difficult to interpret without connecting them to a reference system. As an example, if a group of examinees all scored well on a very easy test, the obtained high scores tell us little about the true abilities of the examinees. The interpretation of the raw scores becomes even more difficult when two or more test forms are employed simultaneously because one form may be more difficult, or easier, than another. Finally, interpreting raw scores becomes more challenging when different tests are employed to measure the same or a similar construct. In the measurement of physical activity (see chapter 14 by Richard Washburn, Rod Dishman, and Gregory Heath), for example, more than 40 questionnaires have been employed in practice, and scores from these instruments are rarely comparable. Fortunately, a set of statistical methods—namely, **scaling**, **equating**, and **linking**—is available to help set test scores on a reference system so that they are more interpretable. This chapter provides an introduction to these methods. More detailed descriptions can be found in the provided references.

SCALING

To avoid the impact of unique characteristics involved in a particular test, raw scores are often converted into an entirely different set of values called derived scores or scale scores. A **scale score** is simply a transformed raw test score with different measurement units. To transform a high-jump score measured in centimeters (raw score) to a score measured in inches (scale score) is perhaps the simplest example. Scales used in well-known standardized tests, such as the ACT college entrance test (with a mean of 22 and standard deviation [SD] of 3) and the Graduate Record Examination (GRE) (with a mean of 500 and SD of 100), are other familiar examples. The process of developing a score scale is called "scaling" a test (American Educational Research Association, American Psychological

Association, & National Council on Measurement in Education, 1999). Based on the nature of the scale developed and the process employed, scaling procedures can be generally classified into three categories: developer defined, population based, and criterion referenced. Although these methods will be described separately, they are often used simultaneously in test construction practice.

Developer-Defined Scaling

In **developer-defined scaling**, raw scores are transformed into a scale defined by test developers. The new scale has a new mean and SD. Because this method is not affected by the distribution of examinees involved in scaling, it is also called "non-population-referenced" (Yen, 2002). By removing the impact of multiple forms (e.g., variability in test difficulty), developer-defined scaling makes interpretation of test scores more consistent. For example, a GRE score of 600 this year will be equivalent to a score of 600 next year. The interpretation of this kind of scale score, however, is often left to test users. Using the GRE again as an example, the GRE as a requirement for graduate admission is determined by an institution rather than the test developer. Clearly, the major benefit of developer-defined scaling is to make scores easier to interpret and to enhance the comparability of the scores from different forms.

Although many methods have been developed for scaling, they can be generally classified into linear and nonlinear categories (Petersen, Kolen, & Hoover, 1989). Linear transformations of a raw score (X) to a scale score (S) are the most commonly used methods in practice, and are defined as follows (Petersen et al., 1989):

$$S = AX + B, \tag{6.1}$$

where A is the slope and B is the intercept of the linear transformation. The location of the scale scores can be defined by specifying that a particular raw score X_1 is equivalent to a particular scale score S_1. The spread of the scale scores can be established by specifying a second raw-to-scale score equivalence—say, that X_2 is equivalent to S_2. For equation (6.1), this leads to

$$A = (S_2 - S_1) / (X_2 - X_1) = (S_1 - S_2) / (X_1 - X_2) \tag{6.2}$$

$$B = S_2 - AX_2 = S_1 - AX_1. \tag{6.3}$$

For example, if the passing point for a particular version of a test were set as a raw score of 72, this raw score could be specified to correspond to a scale score of 550, the selected passing point; and a raw maximal score of 100, to a scale maximal score of 800. Then, from equations (6.2) and (6.3): $A = (800 - 550) / (100 - 72) = 8.93$, and $B = 800 - 8.93(100) = -93$. According to equations (6.2) and (6.3), raw scores of 50 and 80, for example, are transformed to scale scores of 8.93(50) + (-93) = 353.5 and 8.93(80) + (-93) = 621.4, respectively. Usually, the scaled scores are rounded to the nearest 10 points (e.g., 353.5 and 621.4 become 350 and

620, respectively). All scaled scores less than 200 are reported as 200, which is the minimum value of the scale.

Although a linear transformation is simple and easy to use, it sometimes does not work well at the upper and lower ends of the scales where linear conversion becomes truncated. A set of nonlinear methods, therefore, has been developed (Petersen et al., 1989). Nonlinear methods are considered more flexible. For example, raw scores can be transformed to almost any prespecified distributional shape for a specific group of examinees. These methods, however, are often more complicated to implement and describe, and require a larger sample size.

Finally, when a test is constructed using item response theory (IRT) (see chapter 4 by Weimo Zhu), IRT-based scaling will automatically be employed. In general, IRT scales are determined by the IRT models employed in the analysis and do not force the scale-score distribution into a particular shape. IRT-based scaling has two advantages: (a) Both item and examinees are set on the same scale, and (b) local precision is known along the constructed scale. Another advantage of IRT-based scaling is the ability to provide statistically optimal item weights that produce the most accurate scores (Yen, 2002). Technically, however, IRT-based scaling is a more complex process. More detailed information on IRT-based scaling can be found in the description of equating later in this chapter.

Population-Based Scaling

In **population-based scaling**, also known as "interindividual comparison" (Lyman, 1971), raw scores are transformed onto a new scale based on their relative position in the sample or population data collected. The transformation typically is based on (a) the mean and *SD* of the data, (b) the rank of a score, or (c) the status (such as age and grade) of those obtaining the same score.

In practice, z- and T-score scales are perhaps the most well-known scales derived from the mean and *SD*. A **z-score scale** transformation is defined as follows:

$$z = \frac{X - M}{SD},\tag{6.4}$$

where X is the raw score, M is the mean, and SD is the standard deviation. In scaling sport performance data in which a lower score represents better performance (e.g., time in 50 m dash), the positions of X and M in equation (6.4) can be reversed so that a smaller score can have a larger percentile rank (Safrit & Wood, 1995). Because the z-score scale has a mean of 0 and an SD of 1, it is sometimes challenging to explain the results to the general public. Therefore, z-scores are often further transformed to a **T-score scale**, which has a mean of 50 and an SD of 10 as follows:

$$T - score = 10z + 50,\tag{6.5}$$

where z is the z-score.

Among the scales based on score rank, the percentile and percentile rank scales are the most popular. The **percentile** and **percentile rank** are indexes reflecting the percentage position of a score in a defined group. A percentile is defined as a score value that specifies the percentage of cases in a distribution of scores falling below that score. For example, the 80th percentile is the score that is better than 80% of the scores in a distribution. A percentile rank is defined as the percentage of cases falling *at or below* a specified score in a distribution. For example, let's assume we have a distribution of scores ranging in magnitude from 10 to 60 and we wanted to know the percentile rank of the raw score 55. If we computed the percentile rank to be 85, we can say that 85% of the scores in the distribution fall at or below a raw score of 55. A detailed computation of percentiles and percentile ranks can be found in Safrit and Wood (1995). In addition, application examples are described in the norm section of this chapter.

For scales based on status, **age-based scales** are the most popular in exercise science. In contrast to other scaling systems, in which raw scores are transformed to a scale system with equal units and to a desired distribution form, age-equivalent scales are intended to convey the meaning of test performance in terms of what is typical for a person at a given age. They are used mainly for those ages at which function or ability increases rapidly with age (Angoff, 1971). The major steps in constructing an age-equivalent scale are summarized below. Because many other factors may affect the performance of an examinee, age-equivalent scales should be used with great caution, especially when there is a low correlation between function or ability and age, when the correlation is low in a particular age range, when there is a larger variation of the function or ability at a particular age, or when chronological age is not the best representation of age (Angoff, 1971). In educational settings, because the function or ability differences are often directly associated with the impact of school curriculum, grade-equivalent scales are often employed. In physical education, however, function or ability may be associated with both age and instruction level.

Often the objective of population-based scaling is to compare a score with scores earned by the members of some defined population(s). This is frequently accomplished by constructing norms. A norm is a reference point (often the mean

---

Steps in Constructing an Age-Equivalent Scale (Angoff, 1971)

Step 1: Administrating the test for scaling

- Representative sampling over a range of ages: Group together when samples fall within six months of a particular birthday.
- Item difficulty: Include easy items even for the lowest level and difficult items even for the highest level.

Step 2: Finding the mean (or median) scores
- Find the mean scores at each age interval.
- Plot the mean scores at the midpoint of the age interval.

Step 3: Drawing a smooth curve through the points
- Minimize the distance from the points to the curve.
- Represent the lawful relationship among the points.
- Accomplish the preceding two objectives simultaneously—actually compromise between them.

Step 4: Assigning the smoothed values
- Assign the smoothed values of the means to the age designation of the group.
- These designations are the age equivalents.

Step 5: Obtaining detailed values
- Obtain year-and-month values by interpolating on the curve.

of scores) against which to compare the worth of scores in a distribution. In measurement practice, *norm* often refers to a norm table in which the relationship between raw scores and scale scores is defined. Norms commonly take the form of a percentile distribution, which makes it possible to determine both the status of a population and an individual's relative position in the population so that an interpretive statement can be made. Table 6.1 illustrates 1 mi walk/run norms derived from the National Children and Youth Fitness Survey Study (Ross & Gilbert, 1985) for girls between ages 10 and 18. Using 12-year-old girls as an example, 50% of girls at that age in the United States could complete the 1 mi walk/run test within 11 min in 1985. In contrast, if a girl can complete the test within 9 min and 52 sec, her percentile rank is 75%. One should be aware that norms are often associated with a sample or population at a particular time, and previously published norms may not be applicable to the current population. Depending on the samples employed and the interest of developers, there are many types of norms (e.g., national, local, age and grade equivalents, norms by age and grade). A more detailed description of norms and technical considerations on developing norms can be found in the classical description by Angoff (1971). Safrit and Wood (1995) also provided some good examples of norms in the field of kinesiology.

Criterion-Referenced Scaling

Another way to transform raw scores into an interpretable scale is to connect the scale score with a well-defined "absolute" domain of behavior. In the case of a

Table 6.1 National Children and Youth Fitness Survey Norms by Age for the 1 Mile Walk/Run—Girls (in min and sec)

Percentile						Age					
	10	11	12	13	14	15	16	17	18		
99	7:55	7:14	7:20	7:08	7:01	6:59	7:03	6:52	6:58		
90	9:09	8:45	8:34	8:27	8:11	8:23	8:28	8:20	8:22		
80	9:56	9:35	9:30	9:13	8:49	9:04	9:06	9:10	9:27		
75	10:09	9:56	9:52	9:30	9:16	9:28	9:25	9:26	9:31		
70	10:27	10:10	10:05	9:48	9:31	9:49	9:41	9:41	9:36		
60	10:51	10:35	10:32	10:22	10:04	10:20	10:15	10:16	10:08		
50	11:14	11:15	10:58	10:52	10:32	10:46	10:34	10:34	10:51		
40	11:54	11:46	11:26	11:22	10:58	11:20	11:08	10:59	11:27		
30	12:27	12:33	12:03	11:55	11:35	11:53	11:49	11:43	11:58		
25	12:52	12:54	12:33	12:17	11:49	12:18	12:10	12:03	12:14		
20	13:12	13:17	12:53	12:43	12:10	12:48	12:32	12:30	12:37		
10	14:20	14:35	14:07	13:45	13:13	14:07	13:42	13:46	15:18		

certification examination, the raw-score scale can be transformed into *pass* or *fail* categories, whereas the raw scores of a performance test may be transformed to a multiple-category scale such as *unsatisfactory, partially proficient, proficient,* and *advanced.* In measurement practice, setting such a scale is known as setting standards and is often described under the framework of constructing criterion-referenced tests. In the past, this scaling process was not commonly described under the topic of scaling. Because setting standards is directly related to the interpretation of a scale score and is often integrated with the previous scaling procedures, a brief description of the topic is included. For more details, interested readers are referred to Cizek (2001) and two excellent chapters on criterion-reference measurement by Safrit (1989) and Looney (1989).

Generally, standard-setting methods can be classified into two categories: item centered or examinee centered. In the **item-centered method**, an expert panel is employed to evaluate each item in a test and associate the items with a criterion behavior (e.g., has minimal ability or knowledge to work in a fitness club). Among the many methods available, the Angoff method is perhaps the most popular (Angoff, 1971):

> This procedure is to ask each judge to state the probability that the "minimally acceptable person" would answer each item correctly. In effect, the judges would think of a number of minimally acceptable persons, instead of only one such person, and would estimate the proportion of minimally acceptable persons who would answer each item correctly. The sum of these probabilities, or proportions, would then represent the minimally acceptable score. (pp. 515)

Table 6.2 illustrates the Angoff method. There are 10 items in the test, and each of the five judges is asked to rate the probability that a minimally acceptable person would answer each item correctly. The first judge rated item 1, 90%; item 2, 85%; and item 10, 65%. The mean of five judges on item 1, 90%; on item 2, 78%; and on item 10, 64%. The total mean rating is 75%. The passing score of this test is therefore 4 (0.75×5). Several modified versions of the Angoff method were subsequently developed as described in Cizek (2001).

Another relatively new item-centered method is the bookmark procedure developed by Lewis, Mitzel, and Green (1996). There are six steps to this procedure:

1. Create ordered item booklets based on item difficulty from either IRT analysis or conventional item analysis, and form three to four small rating groups with five to seven judges each.

2. Place bookmarks in the first rating.

3. Review and discuss each judge's ratings within each group.

4. Make the second ratings and calculate the median of judges' ratings within each group.

Table 6.2   Example of the Angoff Method

Item	Judge 1	2	3	4	5	Mean
1	90	85	95	90	90	90
2	85	80	70	75	80	78
3	80	80	85	85	85	83
4	60	65	70	70	60	65
5	85	80	75	75	80	79
6	80	75	75	80	80	78
7	65	65	70	65	65	66
8	75	70	70	70	70	71
9	80	70	75	75	75	75
10	65	65	65	60	65	64
Mean	76.5	73.5	75	74.5	75	75

5. Present the ratings from the small group to the whole group. The final performance standard is set by taking the median of the small group ratings (third rating).

6. Develop performance descriptors for the standards.

The bookmark procedure is illustrated in figure 6.1. The major advantages of this method include that (a) it is appropriate for setting multiple standards, (b) it can handle multiple item types (e.g., dichotomous and polytomous items), and (c) it simplifies the judgmental task.

In the **examinee-centered method**, standards are set based on the difference of the data distributions among groups with known ability or status (e.g., instructed vs. noninstructed groups). The contrasting group method, the borderline group method, the body of work method, and the generalized examinee-centered method are just a few examples of examinee-centered methods. The major advantage of examinee-centered methods is that the subjectivity involved in judge ratings in the item-centered method is significantly reduced. For more information about the examinee-centered methods and related technical requirements, see Safrit and Wood (1995), Cizek (2001), and chapter 12 by Terry M. Wood in this text.

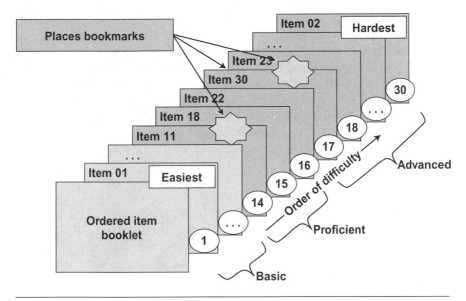

Figure 6.1   Visual presentation of the bookmark procedure.

Reprinted with permission from *Research Quarterly for Exercise and Sport,* Vol. 69, pp. 11-23, copyright 1998 by the American Alliance for Health, Physical Education, Recreation and Dance, 1900 Association Drive, Reston, VA 20191.

EQUATING AND LINKING

Multiple forms of a test are often used to keep a test secure. Different tests or measurement instruments are also used in practice to measure the same underlying construct. For example, several forms of the sit-up test (Safrit, Zhu, Costa, & Zhang, 1992) have been used in practice to measure abdominal strength. Physical activity assessment, as mentioned earlier, is another example. There are more than 40 self-report questionnaires being used to assess physical activity (Kriska & Caspersen, 1997; Montoye, Kemper, Saris, & Washburn, 1996), and new questionnaires are constantly being developed. As a result, few questionnaires are exchangeable.

Developing different forms and tests on the same scale is a very challenging task. **Test equating,** a set of statistical methods developed primarily from the field of educational measurement, is available to address the challenge. Equating has been used to link both different forms of a test and different tests being used to measure a similar construct. More recently, there has been a push to distinguish "test equating" from "linking" (Feuer, Holland, Gree, Bertenthal, & Hemphill, 1999). Test equating should be applied when associating different forms of a test with the same content, whereas linking is applied to associate different tests measuring the same or a similar construct. Recently, Dorans (2004) and Kolen (2004) introduced another term called *concordance* to describe linking different tests. Concordance is defined as "linking scores on assessments that measure similar (but not identical) constructs and in which scores on any of the linked measures are used to make a particular decision" (Kolen, 2004, p. 219). Because most of these terms are used to distinguish linking practices conceptually and have limited impact on the statistical procedures being used, only the terms *equating* and *tests* will be used in this chapter. Readers, however, should be aware of the new changes and be cautious when reporting the results of linking studies. The task of equating, in general, is to establish statistically a conversion relationship among summary scores from two or more test forms or tests. The relationship could be linear or nonlinear depending on the equating method employed.

EQUATING METHODS

Equating methods can be generally classified into two categories according to the testing theory on which they are based: **traditional equating** (Kolen, 1988) and **IRT equating** (Cook & Eignor, 1991).

Traditional Equating

Traditional equating methods are based on classical test theory (CTT) in which "observed scores" are believed to consist of "true scores" and "errors" (Hambleton & Jones, 1993). In traditional equating methods, score correspondence of tests is established by setting the characteristics of the score distributions equal for a specified group of examinees (Kolen, 1988). This means that although the absolute values of the respondents' summary scores may be different from each other, their relative positions in a sample should be approximately the same, as long as both tests take behavioral samples from a similarly defined activity universe. Linear and equipercentile equating are the two most commonly used traditional equating methods.

Linear Equating

In the **linear equating** method, the means and SDs of the two tests for a particular group of examinees are set equal, thus,

$$\frac{X_1 - M_1}{SD_1} = \frac{X_2 - M_2}{SD_2} \tag{6.6}$$

where X_1 and X_2 are summary scores. One may immediately recognize that the elements on each side of equation (6.6) are, indeed, the familiar z-score equation. Linear equating, therefore, can be conceptually considered as the establishment of equivalent z-scores for two different tests. Based on equation (6.1), the conversion constants A and B, known as the *slope* and *intercept* of linear equating, can be derived as follows (Kolen, 1988):

$$X_1 = \frac{SD_1}{SD_2} X_2 = \left[M_1 = \frac{SD_1}{SD_2} M_2 \right] = AX_2 + B \qquad (6.7)$$

therefore,

$$A = \frac{S_1}{S_2} \qquad (6.8)$$

and

$$B = \bar{X}_1 = \frac{S_1}{S_2} \bar{X}_2. \qquad (6.9)$$

If, for example, $\bar{x} = 400$, $S_1 = 150$, $\bar{x}_2 = 35$, and $S_2 = 2$, then $A = 75$ and $B = -2225$. A score of 37 on test 2, according to the conversion based on equation (6.2), is equivalent to a score of 550 on test 1. In this way, test 2 was equated to test 1. However, there is an important assumption in linear equating that the score-distribution shapes of tests 1 and 2 should be the same, or approximately the same, which in practice is often not the case.

Equipercentile Equating

In **equipercentile equating**, score distributions are set to be equal so that the same percentile ranks from different tests are considered to indicate the same level of performance. The first step in equipercentile equating is to determine the percentile ranks for the score distributions on each of the two tests to be equated. Percentile ranks are then plotted against the raw scores for each of the two tests, and so-called "rank-raw score curves" can be constructed. Figure 6.2 illustrates plots of the rank-raw score curves of two hypothetical tests on different scales, in which the relative cumulative frequency distributions (i.e., percentile ranks / 100) were first plotted for each test. As long as the percentile rank-raw score curves are constructed, equivalent scores can be converted from them. For example, to convert a score from test 2 to test 1, simply draw a vertical line up from the scale of test 2 to its own rank-raw score curve. Next, draw a horizontal line to the curve of test 1 and another vertical line down to the scale of test 1 (see also figure 6.2). To reduce sampling error, especially when the sample size is small, these curves are usually smoothed, which is often accomplished using analytical smoothing procedures (Kolen, 1984).

IRT Equating

Item parameters in IRT are independent of the ability level of the respondents answering the item. At the same time, ability is also independent of the performance of other respondents and the items used in tests, if model assumptions are met. This is known as the "invariance" feature of IRT. Because of this feature, the interpretation of item characteristics and respondent ability is consistent in IRT. If parameters in IRT are invariant (i.e., similar estimations should be obtained

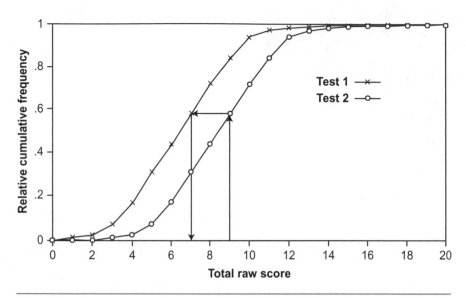

Figure 6.2 Equipercentile equating of two hypothetical tests.

Reprinted with permission form *Research Quarterly for Exercise and Sport,* Vol. 69, pp. 11-23, copyright 1998 by The American Alliance for Health, Physical Education, Recreation and Dance, 1900 Association Drive, Reston, VA 20191.

regardless of which test respondents are to take or which group of respondents will be used to calibrate tests), there seems little need to equate tests. Although the IRT invariance feature is true in theory, it is not completely true in practice. The invariance feature will hold if there is only a single calibration on which a scale is set. Again, this is often not the case in practice because two calibrations are needed when linking one test to another. When there is additional calibration, the IRT invariance feature no longer holds because, in each separate IRT calibration, a score of zero has to be arbitrarily assigned, either to the mean of examinee abilities or to the mean of item parameters. Therefore, to set two tests on the same scale, differences attributable to arbitrary assignments must be adjusted. The process used to adjust the difference is called "equating," which, indeed, is scaling. Thus, in the framework of IRT, scaling (rather than equating) is necessary, but the terms are frequently used interchangeably in the literature (Hambleton, Swaminathan, & Rogers, 1991).

Major steps of IRT equating include data collection, selection of an IRT model, estimation of parameters, determination of scaling constants, and scale transformation. The following paragraphs briefly describe these steps. See Kolen and Brennan (2004) for more information about IRT equating, and Zhu (2001) for an application in motor function assessment.

Data collection starts by selecting a data collection design. Although all three data collection designs described here have been used in IRT equating, the **anchor-**

test design is most commonly used. This is because group abilities involved in the anchor-test design are often different from each other—a situation in which traditional equating methods often do not work well (Cook & Eignor, 1983). After data are collected, the next step is to select an appropriate IRT model. The commonly used IRT models for dichotomous scores are one-, two-, and three-parameter logistic models, although multiple-response-category and multiple-parameter IRT models are now available (see chapter 4 by Weimo Zhu). Item and respondent parameters can be estimated and model-fit statistics calculated via computer software (Hambleton et al., 1991). If the model and data do not fit, a new model is considered or new data are collected. If the model and data fit, the equating can then move to the next step, which places parameter estimates from separate calibrations on a common scale.

Determining scaling constants is the key process when placing parameter estimates on a common scale. Methods for determining scaling constants can generally be classified into four categories: regression, mean and sigma, robust means and sigma, and characteristic curve methods (Hambleton & Swaminathan, 1985). These methods differ from each other in which parameter (e.g., item difficulty or discrimination) or kind of information (e.g., mean and standard deviation) is included in the computations (Hambleton et al., 1991; Stocking & Lord, 1983).

IRT equating is completed after parameters from two separate estimations have been set on the same scale. In practice, however, it may be necessary to further scale ability estimates into other predetermined scales. There are two reasons for this. First, IRT computer programs usually provide a scale called the "logit," which is very similar to z-scores, with zero representing the mean and negative values representing scores below the mean. The general public, however, is often uncomfortable with, and sometimes even confused by, these zero and negative values. Therefore, instrument developers transform ability estimates into scales having only positive values. Second, many assessment programs have established their own scales. For example, the GRE is calibrated on a scale with a mean of 500 and a standard deviation of 100. To report IRT estimates on an established scale, further transformation becomes necessary.

If instrument developers want to transfer the ability estimates to a scale having only positive values, the transformation is straightforward because ability estimates are already set on an interval scale, which means that the scale can be linearly transformed into any other scale. If, however, instrument developers want to transform ability estimates to established scales, a more complex two-step approach is needed: (a) Transfer ability estimates into true scores, and (b) further transfer the true scores into reported scores (Cook & Eignor, 1991; Kolen & Brennan, 2004; Lord, 1980).

Practical Issues in Equating

Equating, like other aspects of instrument development, involves a rather complex process involving many practical decisions. Several key practical issues are reviewed briefly in this chapter, including data collection design, selecting an equating method, equating errors, sample size, and choosing computer programs. For more detailed information, refer to Kolen and Brennan (2004) and Zhu (1998b).

Data Collection Design

Equating starts with data collection. Three commonly used data collection designs are single-group; equivalent-groups; and common-item, nonequivalent-groups designs. In the **single-group design,** two or more tests are administered to the same group of respondents. The advantage of this design is that the measurement error is relatively small. This is because only one group of respondents is involved in the equating process. Therefore, differences among tests are not confounded by differences among groups. Because this design requires respondents to answer many questions, however, fatigue is a major concern. Practice effect (i.e., respondents may change their responses as a result of their responses to previous tests) is another concern. To avoid fatigue and practice effect, the application of some sort of spiraling process should help (e.g., the order of administration of the testing forms can be counterbalanced) (Kolen, 1988; Kolen & Brennan, 2004).

In the **equivalent-groups design,** two tests to be equated are administered to two equivalent groups of respondents. The groups may be chosen randomly, which is why this design is sometimes called the *random-groups design*. The advantage of this design is that it eliminates problems related to the single-group design, such as fatigue and practice effect. Furthermore, minimized administration time means the test can be finished in a single period. The disadvantage of this design, however, is that an unknown degree of error is introduced in the equating process because groups are rarely the same in their ability distributions. To control for sample-related error, larger samples generally are required for this design (Kolen, 1988; Kolen & Brennan, 2004).

Tests are also administered to two different groups of respondents in the **common-item, nonequivalent-groups design** (Kolen, 1988). In contrast to the equivalent-groups design, however, groups can be different from each other in their ability distributions. Furthermore, a set of common, or anchor, items or questions is included in both tests so that differences between the two tests can be adjusted based on common-item statistics. Because the two groups do not have to be equivalent, this design is extremely useful when measuring growth or change in which two groups are known to be nonequivalent, or when it is impossible to administer more than one test per date because of practical concerns. This design is also necessary when developing an item bank, described later. Strong statistical assumptions are required to remove the confounding effects of group and test differences, and, quite often, statistical procedures can provide only limited adjustments (Petersen et al., 1989). These designs are illustrated in figure 6.3.

Practicality is often the major concern in selecting a data collection design. The common-item, nonequivalent-groups design is often more practical because both single-group and equivalent-groups designs require the administration of two tests with little or no time intervening between test administrations, which in practice is often difficult to implement. Many equating procedures may not be applicable

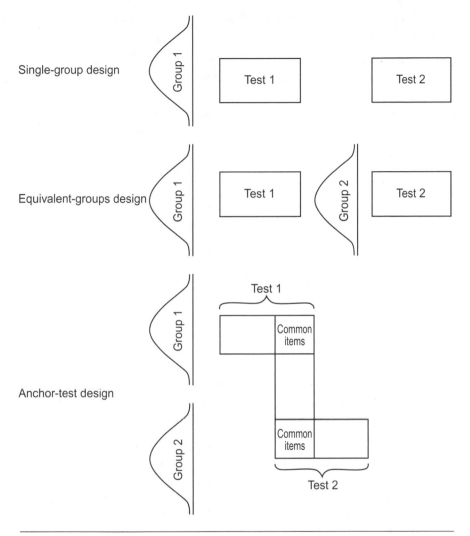

Figure 6.3   Commonly used equating data collection design.

when distributions of common items from two subpopulations are markedly different (Angoff, 1971). In addition, there should be enough common items with representative content to be measured. A rule of thumb is that common items should account for 20 to 25% of the items in either of the tests.

Selecting an Equating Method

Measurement advantages and disadvantages, as well as the tenability of assumptions, accuracy, and practicality (Crocker & Algina, 1986), are the major considerations when choosing an equating procedure. When IRT is used to calibrate a test, IRT equating should be employed because it has the following five advantages over traditional equating (Cook & Eignor, 1983):

1. The quality of equating is improved because better equating at the upper end of the score scale is usually provided, which is extremely important in the case of vertical equating.

2. Greater flexibility is provided by choosing previous versions of a test because all previous versions are calibrated on the same scale.

3. Re-equating becomes easier because if an item is dropped, the shortened test can be easily reconstructed based on the item information from the remaining test items.

4. The impact of sampling error decreases because of the invariance feature of IRT equating.

5. Pre-equating becomes possible when applying IRT equating. Pre-equating prepares raw-to-scale score conversion tables prior to administering a question to accomplish quick score reporting. IRT equating makes pre-equating possible because items in the "new" test have been previously administered, and they will be invariant when applied to new groups.

One of the major limitations of IRT is that it can handle only dichotomous or polytomous data. When the data are continuous, traditional equating methods are required. Compared to linear equating, which assumes that the only differences between the distributions of tests 1 and 2 are the mean and the variance, equipercentile equating makes fewer restricted assumptions. The equipercentile methods have larger equating errors than linear equating. Thus, linear equating is generally preferable if the distributions of z-scores from tests to be equated are approximately the same. In contrast, if the distribution shapes of tests to be equated are not the same, linear equating may not be appropriate and equipercentile equating should be employed. In the past, equipercentile equating was considered more complicated to carry out. With today's computing power, practicality is no longer a constraint in selecting an equating method because commonly used equating computer programs often include several equating methods. The RAGE equating computer program (Zeng, Kolen, & Hanson, 1995), for example, provides equating results by the linear, unsmoothed equipercentile method and several smoothed equipercentile methods.

Equating Errors and Evaluation Criteria

Two kinds of errors, random and systematic, often occur in the process of test equating (Kolen, 1988). Random equating errors occur when samples are used to estimate parameters (e.g., population means, percentile ranks, or item difficulties). Generally, random error is a major concern when the sample size is small. Therefore, increasing the sample size can reduce random error. Employing appropriate data collection designs may also help to reduce random error. Systematic errors occur when assumptions, or conditions, of a particular data collection design or an equating method are violated. In the single-group design, for example, failure to take fatigue and practice effect into consideration may lead to systematic errors. Systematic errors can occur whenever new and old tests differ in content, difficulty, or reliability. Every effort should be made to control and monitor random and systematic errors in the process of test equating.

To control, or at least monitor, equating errors, a number of evaluation criteria and procedures have been proposed (e.g., weak equity [Yen, 1983], standard errors [Angoff, 1971], standardized difference score [Wright, 1968], and cross-validation [Kolen, 1981]). According to an extensive review by Harris and Crouse (1993), none of these criteria or procedures applies to all equating situations. More research is needed in this area. Even so, to ensure the quality of test equating, some sort of evaluation plan is strongly recommended whenever test equating is conducted or a new equating procedure is introduced. Kolen and Brennan (2004) provided a comprehensive discussion of equating including a description of the characteristics of equating situations and related evaluation plans and methods.

Sample Size

Sample size has a direct impact on random equating error. Larger samples generally lead to better equating results. Kolen and Brennan (2004) proposed two rules of thumb for sample size. The first is based on SD units; the second is based on comparisons with identity equating. For traditional equating, a sample size of 400 per test is generally needed for linear equating and slightly over 1,500 for equipercentile equating. For IRT equating, a sample size of 400 is generally needed for the Rasch model, and 1,500 for a three-parameter model (Kolen & Brennan, 2004). When the sample size is small, other equating approaches (e.g., log-linear smoothing [Livingston, 1993] and the collateral information method [Mislevey, Sheehan, & Wingersky, 1993]) may be considered.

Computer Programs

Scale equating, like other areas of test development, takes advantage of modern computers. Computer programs for test equating, including both traditional and IRT equating methods, have been developed along with an excellent test-equating book by Kolen and Brennan (2004). In addition, Baker (1993) has developed a computer program for IRT nominal and graded-response models. These computer programs can be obtained at no cost by contacting the program authors.

KINESIOLOGY APPLICATIONS AND FUTURE RESEARCH DIRECTIONS

Scaling practice is not foreign to kinesiology. The topic of z- and T-score scaling has been covered in almost every undergraduate measurement textbook, and both scaling methods are commonly used in measurement practice. Percentile norms and age-equivalent norms have been reported for many standardized tests (Burton & Miller, 1998; Docherty, 1996; Heyward, 1991; Safrit & Wood, 1995), although some of them were developed many years ago and are now outdated. The *Test of Gross Motor Development* (Ulrich, 2000) provides an excellent example of scaling. Using a large representative national sample ($N = 1,208$), percentiles, standard scores, quotients, and age equivalents were computed and reported along with related norms. *FitSmart,* developed by Zhu, Safrit, and Cohen (1999), is another good example. Based on a large national sample ($N = 4,025$), two forms of the test were calibrated using the Rasch model and the one-parameter IRT model and equated onto the same scale. User norms for both forms were developed using raw scores, and their conversion to standard scores and percentile ranks were reported. More important, criterion-referenced standards were developed for the whole test and subtests, respectively, which makes *FitSmart* a good example of how to combine several different scaling methods into the development of a single test.

In comparison, the application of equating procedures has been rather inactive in kinesiology. Only a few applications (Kang & Zhu, 1998; Shin & Zhu, 1998; Zhu, 1998a, 1998b, 2001; Zhu & Kang, 1998; Zhu et al., 1999) are reported despite the fact that score incomparability is a major problem in measurement practice. This is not surprising because the concept of test equating is still relatively new to kinesiology (see Zhu, 1998a, 1998b, 2001, 2002, and Zhu et al., 1999). Increasing the awareness of the problem in score incomparability and continuing to promote the concept and applications of testing equating is a critical task for measurement specialists in the field.

Lack of appropriate equating procedures and software to handle continuous data with different response formats is another major challenge. In exercise science, test security is rarely a problem. Different tests are the major interest of linking. Tests are often constructed to measure slightly different constructs (e.g., leisure vs. occupational activities in physical activity assessment) with different scoring formats. Existing test equating procedures and computer programs often cannot adequately handle these kinds of tests or response formats. For example, the RAGE equating computer program developed by Zeng and colleagues (1995) is limited to a total of 80 items, or score categories, in each form. An effort to explore new equating procedures (e.g., the Kernel method by Davier, Holland, and Thayer, 2004), modify existing software according to the nature of continuous data in kinesiology, and evaluate the appropriateness of their utility is greatly needed.

SUMMARY

Scaling is the statistical procedure through which raw scores are transformed to scale scores, and the interpretation of the scores becomes easier, more consistent, and meaningful. Commonly used scaling methods include developer-defined, population-based, and criterion-referenced methods. In test construction practice, these methods are often used simultaneously. Test equating is a set of statistical methods to help place two or more tests on the same scale. Many issues are associated with score incomparability in kinesiology. Yet, adequate efforts in training graduate students and practitioners on the concepts and methods of test equating are lacking. There is an urgent need to explore new test equating procedures, as well as a need for software that can address kinesiology's equating needs.

CHAPTER 7

Item Bias and Differential Item Functioning

Allan Cohen

Items or tasks that function differently in different groups represent a threat to the validity of a test (Thissen, Steinberg, & Wainer, 1988). This is so because items that function differently are actually measuring one or more dimensions that are not the primary focus of the test. These items may be unfair if they cause examinees of the same ability in one or more groups to have different probabilities of success. For this reason, detection and study of these items is a critical part of the development of any test.

Most of the work on detection of **differential item functioning** (DIF) has been done in the context of knowledge or achievement testing, so it is primarily in that context that we discuss this problem. In this chapter we focus on the different methods for detection and for understanding the causes of this phenomenon. The chapter begins with a discussion of the differences between **item bias** and DIF. Next, a multidimensional framework is presented within which to examine DIF and then consider some different methods for detecting and understanding the problem.

BIAS AND DIFFERENTIAL ITEM FUNCTIONING

The terms *item bias* and *DIF* are sometimes used interchangeably. An item is identified as a DIF item as a result of a statistical test. This result may also be termed *bias*. Bias is sometimes inferred based on judgmental evidence. Some measurement researchers, in fact, prefer the term *DIF* because it is considered less emotionally charged. In fact, there is general agreement that an item that functions differentially may not be biased, but a biased item must function differentially. *DIF* is a more useful term than *bias* when the focus of the inquiry is on statistical differences in performance among groups on a particular item or task. When the discussion is about differences in performance on nonitem kinds of tasks (such as a basketball shooting task or a sit-up task), however, the term *bias* is clearly less useful. Because the primary focus of this chapter is on statistical methods for detecting differences in item performance, the term *DIF* is used. These methods refer to the problem of detection of differences in performance on tests as a result of nonrandom sources of variation that distort the test results for members of a particular group (or groups). This variation is considered extraneous to the intended construct to be measured and represents a serious threat to the validity of the measurement procedure.

Defining DIF

DIF is said to occur when the probability of a correct response differs for examinees in different groups conditional on ability (Pine, 1977). An item is determined to function differentially when examinees of the same ability but in different groups (e.g., males and females) have different probabilities of success on the item. This definition of DIF is useful for dichotomously scored items, such as when responses are scored as either correct or incorrect.

Measures of physical activity are often scored in more than two categories. This is common with performance measures such as sit-ups, bowling, shooting basketballs, running, and gymnastics. These kinds of tasks are often scored polytomously (i.e., in more than two categories) or continuously. Consequently, DIF needs to be defined more broadly for tasks such as these. Stated more generally, therefore, differential functioning in the item or task occurs when the response function for that item or task differs in different groups (Cohen, Kim, & Baker, 1993). This latter definition of DIF is more useful for performance types of items for which dichotomous scoring may not be either possible or desirable. Even though we may switch from items to tasks in our discussion, in this chapter we will continue to use the term DIF.

DIF Is Conditional

DIF is not used simply to describe differences among groups. Rather, DIF is used to describe differences among groups that are otherwise of the same ability. An extreme example serves to make this point: Males, on average, tend to be stronger than females. If we administer a running test over a 100 m course, the most likely result is that the average speed over this distance for a random sample of males will be less than that for a random sample of females. This represents a gender difference. It is not, however, an indication of DIF. DIF does not occur simply because one group is faster or stronger than another. DIF would occur if the same watch registered a faster time for male runners than for female runners even though both sets of runners actually covered the 100 m course in the same time. Looney, Spray, and Castelli (1996) described DIF related to gender in the context of a basketball shooting task. DIF in a basketball shooting context arises because some shooting tasks are harder for females than for males who are otherwise of the same level of basketball shooting ability. Another common realization of DIF occurs in a testing context in which examinees who have the same raw score (i.e., number correct score) on the test have different probabilities of getting an answer correct.

MULTIDIMENSIONAL FRAMEWORK FOR DIF

Multidimensionality, a fact of life for any test, arises because it is all but impossible to construct test items that measure only a single dimension. People do not arrive at answers to questions or solutions to tasks in the same way. A common example is the measurement of arithmetic ability using word problems. When arithmetic

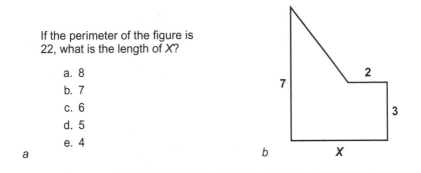

Figure 7.1 (a) Sample algebra word problem; (b) sample symbolic algebra item.

problems are described in words, such as the one in figure 7.1a, examinees will use both verbal ability and quantitative ability to solve the problem. Some types of items, however, are more unidimensional than others. An algebra test composed of items such as that in figure 7.1b requires primarily symbolic manipulation and consequently will usually be highly unidimensional. Even with tests that are highly unidimensional such as a basketball shooting task, however, some additional dimensionality will invariably be present. The purpose of DIF detection is to check whether this additional dimensionality intrudes on the measurement of the ability that is the focus of the test. The common approach to studying DIF detection is to determine whether this intrusion is related to some examinee characteristic of interest such as gender or ethnicity.

Sometimes we do seek to measure more than one dimension with a test item or set of test items. When the intent of the test is to measure both arithmetic ability and verbal ability, then both dimensions are considered primary dimensions. If the focus of the test is on measuring arithmetic ability only, however, and not on measuring verbal ability, then the arithmetic ability is the primary dimension and the verbal ability is a nuisance dimension.

DIF can usefully be characterized as the presence of one or more secondary or nuisance dimensions accompanying the primary ability dimension (or dimensions) that is intended to be measured (Ackerman, 1992). DIF occurs when examinees of equal ability on the primary dimension(s) differ on the nuisance dimension(s). As noted earlier, this is a conditional statement in which examinees of equal ability on the primary dimension(s) are compared.

DIF As a Threat to Validity

It is helpful to think of the multidimensional framework for DIF in the context of test validity. Messick (1989) identified two aspects of consequential validity for test scores: **construct-irrelevant variation** and **construct underrepresentation.** Construct irrelevant variation is theoretically unjustified because it measures a construct that is not part of the test specifications. If running ability is not a primary dimension in the measurement of bowling accuracy, for example, then its presence in the bowling accuracy score would be an example of construct-irrelevant variation. Construct-irrelevant variation, in other words, comprises variation that is extraneous and therefore undesirable for the measurement of the primary construct of interest. Group differences, which appear on a test because of the presence of construct-irrelevant variation, may not be reflective of the same primary construct in both groups. In the context of the previous framework, running ability is a nuisance dimension and would be considered a source of construct-irrelevant variation when measuring bowling accuracy. Construct-irrelevant variation can lead to DIF.

Construct underrepresentation occurs when a test fails to measure the complete construct. This arises when the test provides an inadequate measure of the construct. As an example, if a measure of fitness knowledge has only items that consider cardiovascular fitness, the test does not fully measure the complete domain of fitness knowledge. The presence of construct underrepresentation is related to poor test construction practices but does not necessarily lead to DIF.

DETECTING DIF

Methods for detecting DIF rely on the use of statistical procedures for identifying items or tasks that function differentially. A number of statistical methods have been developed that provide a statistical test for determining whether an item functions differentially. Some of these methods are described here.

A variety of nonstatistical procedures is also available for identifying possible bias in items. These procedures are not considered DIF detection methods. Some of these methods predate the development of statistical methods (e.g., Ziecky, 1993). Others have been developed more recently (Li, 2001; Roussos & Stout, 1996). Nonstatistical methods generally rely on judgments made about characteristics of the items. Some nonstatistical methods are used to design items or tasks so that the likelihood of DIF is minimized. Other nonstatistical methods such as the *ETS Sensitivity Review Process* (Educational Testing Service [ETS], 1987) are used to review items once they have been developed.

The objective of the sensitivity review process used by ETS is to "encourage the use of materials that acknowledge the contributions of women and minority group members" and to "eliminate materials that women or minority group members are likely to find offensive or patronizing" (Linn, 1993, p. 356). This process is quite helpful for detecting characteristics of items that might be causes of some kinds of DIF. Roussos and Stout (1996) provided a more direct connection to the test development process.

Statistical Methods for Detecting DIF

As noted, DIF is a statistical concept, and statistical methods are required to detect it. Statistical methods for detecting DIF are based on expectations that arise from the particular model we use to describe the item response function. Both parametric and nonparametric methods for DIF detection are available. Some of the more common ones are those available for **dichotomously scored items.** Extensions of these methods to **polytomous items** and to groups of items are also available. These are more complicated, but essentially they are extensions of DIF detection methods for dichotomous models. (The interested reader is referred to Chang, Mazzeo, & Roussos, 1996; French & Miller, 1996; Portenza & Dorans, 1995; and Zwick, Thayer, & Mazzeo,1997, for discussions of issues related to polytomous DIF.) In this chapter we focus only on methods for DIF detection with dichotomously scored items.

It is possible for DIF detection to be done in multiple groups; however, to simplify the following discussion, we use the typical DIF detection framework of two groups, a **reference group** and a **focal group**. The focal group is usually the group against which the studied item is suspected of being biased, and the reference group provides the basis for comparison. In a gender DIF detection study, for example, the sample of females is specified as the focal group and the sample of males as the reference group. We will use this reference–focal group distinction to describe the DIF detection methods.

Finally, it is important to note that DIF detection is done between items (or bundles of items) that are on the same score scale. The score scales for tests are not automatically on the same metric. Therefore, the first step in a DIF detection study is to determine the appropriate transformation to express items on a common score scale. This is known as equating. Classical test theory and item response theory (IRT; see chapter 4 by Weimo Zhu) methods for establishing this transformation are described in detail in Kolen and Brennan (2004). For a discussion of test equating in exercise science, see chapter 6 by Weimo Zhu in this text.

Nonparametric DIF Detection Methods

Nonparametric methods use some version of the raw (or number correct) score for conditioning. They are called nonparametric because they do not assume a parametric form for the item response function. We illustrate this type of method using two of the most common statistics, the Mantel-Haenszel D-DIF statistic, and the Simultaneous Item-Bias Test or SIBTEST.

Mantel-Haenszel Statistic

The **Mantel-Haenszel statistic** (MH; Holland & Thayer, 1988) is used to compare the performance of two groups of examinees one item at a time. In the MH procedure examinees are grouped into classes on the basis of a matching variable. Most often, the matching variable is the total score on the test. The MH procedure provides a significance test and also a measure of the effect size. This latter feature is useful because DIF detection alone does not always tell a complete story.

The MH procedure is an observed-score-based procedure that detects DIF by testing a constant odds-ratio hypothesis. The hypothesis is based on a comparison of the frequencies in figure 7.2. The MH statistic is calculated as follows: Assume

	Correct	**Incorrect**	**Total**
Reference group	A_{ij}	B_{ij}	n_{1ij}
Focal group	C_{ij}	D_{ij}	n_{2ij}
Total	m_{1ij}	m_{2ij}	n_{ij}

Figure 7.2   Cells in the table used to calculate the Mantel-Haenszel statistic for an item.

that we have a test score that can take values $j = 1, \ldots, J$. The data for each of the I studied items can be arranged in the I separate $2 \times 2 \times J$ contingency tables. The contingency table for a single item for score group j is shown in figure 7.2. The rows of this table consist of the frequencies of the number of examinees in the reference and the focal groups for score group j. The columns consist of the frequencies of examinees who have given correct and incorrect responses to the studied item for score group j. There are n_{ij} examinees in the data, all in score group j on the matching variable.

The MH measure of DIF is called MH D-DIF and is given as

$$\text{MHD-DIF} = -2.35 \times \log \hat{\alpha}_{MH}, \tag{7.1}$$

where $\hat{\alpha}_{MH}$ is the Mantel-Haenszel (adjusted) odds-ratio estimator given by

$$\hat{\alpha}_{MH} = \frac{\sum_j A_{ij} D_{ij} / n_{ij}}{\sum_j B_{ij} C_{ij} / n_{ij}} \tag{7.2}$$

The quantities A_{ij}, B_{ij}, C_{ij}, and D_{ij} in equation (7.2) are the frequencies in each of the four cells identified in figure 7.2. There is a separate two-by-two table for each of the J score groups defined on the matching variable. A value of zero for $\hat{\alpha}_{MH}$ indicates no DIF in the studied item, a positive value indicates the studied item favors the focal group, and a negative value indicates the studied item favors the reference group. In most cases, the matching variable for the MH statistic consists of the total test score including the studied item. It is also possible to use an external measurement (e.g., an electronic monitoring device for measuring a person's physical activity) as a matching variable. The intent in finding an appropriate matching variable is to make certain it measures the same construct at the items being studied.

Simultaneous Item-Bias Test

The **Simultaneous Item-Bias Test,** or SIBTEST (Shealy & Stout, 1993), provides a test of the following hypothesis that there is no DIF in the studied item against the alternative hypothesis that there is some DIF in the studied item:

$$H_0: \beta_i = 0 \text{ compared to } H_1: \beta_i \neq 0, \tag{7.3}$$

where β_i indicates the amount of DIF for an item. β_i is calculated as the weighted expected score difference on the studied item between examinees in the focal and reference groups who are of the same ability.

Items on the test are divided into two groups: those items that are suspected of functioning differentially and those that are not. Those suspected of measuring both the primary and nuisance dimensions(s) will be the studied items. The items that are not suspected of DIF are determined to be measuring only the primary dimension(s). These non-DIF items normally comprise the matching variable. This means that, for the SIBTEST, the matching variable does not include the studied item.

Examinees are placed into groups based on their scores on the matching variables. The weighted mean difference between the reference and focal groups is determined as

$$\hat{\beta}_i = \sum_{j=1}^{J} p_j d_j \tag{7.4}$$

where p_j is the proportion of examinees in the focal group in group j on the matching variable and $d_j = \overline{M}'_{jR} - \overline{M}'_{jF}$ where $\overline{M}'_{jF}$ is the mean of examinees in the focal group in group j on the matching variable. Note that the means have an apostrophe (') indicating that they are adjusted to account for any differences in ability between the focal and reference groups.

The test statistic for the SIBTEST is

$$SIB = \frac{\hat{\beta}_i}{\hat{\sigma}\left(\hat{\beta}_i\right)}, \tag{7.5}$$

where $\hat{\sigma}\left(\hat{\beta}_i\right)$ is the standard deviation of $\hat{\beta}_i$. SIB in equation (7.5) is distributed normally with a mean of 0 and a standard deviation of 1. This means it can be tested using large sample normal curve probabilities.

Parametric DIF Detection Methods

Parametric methods for DIF detection have been developed using the models from IRT. These methods are called parametric because they are based on statistics that assume parametric forms for the item response functions. Models under IRT are very useful because they are based on strong assumptions about the form of the item response function. One of the most useful of these assumptions for DIF detection is the assumption of item parameter invariance. This assumption states that item parameters will be the same when estimated in different samples from the same population. This assumption can be tested directly under IRT.

Under IRT, the probability of a correct response for an item for the three-parameter logistic model is given as

$$P(\theta) = c + \frac{(1-c)}{1+\exp(-a(\theta-b))}, \tag{7.6}$$

where a is the slope of the item response function (IRF) at the point of inflexion, commonly known as the discrimination parameter; b is the location of the IRF at the point of inflexion, also known as the difficulty parameter; and c is the lower asymptote of the IRF, commonly known as the guessing parameter. An example of an IRF for a three-parameter item is shown in figure 7.3. The parameters of the IRF are $a = 1.0$, $b = 1.0$, and $c = .1$.

In the typical two-group DIF context, we have two IRFs (see figure 7.4). One IRF is estimated for the reference group, and one IRF is estimated for the focal group. Two general approaches to DIF detection under IRT are (a) to compare differences in the parameters of the IRFs estimated in the focal and reference group, and (b) to test the significance of the area between the IRFs estimated in the reference and focal groups. Lord's chi-square and the likelihood ratio test detect DIF by comparing differences in the item parameters. Raju's (1990) two area measures detect DIF by testing the significance of the area between item response functions.

Metric Transformation

The model parameters estimated in the reference and focal groups must be on the same metric for DIF comparisons to be made under IRT. Current algorithms for

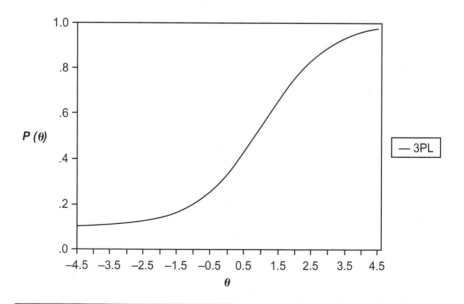

Figure 7.3 Item response curve.

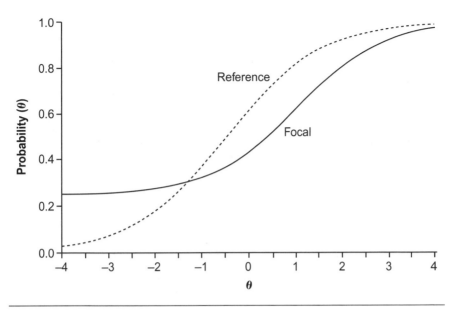

Figure 7.4   Item response functions for reference (R) and focal (F) groups.

estimating IRT model parameters do not place the estimates on the same scale or metric. This means that the model parameters estimated separately for the reference group and the focal group are not on the same scale. Under IRT, the transformation from one scale to another is linear. This transformation is determined differently, depending on how the item parameters are estimated. Two ways to transform the item parameter estimates obtained from the focal group to the metric of the reference group are concurrent calibration and linking.

Concurrent calibration is used when the item parameters can be estimated simultaneously in the two groups. Concurrent calibration is included in IRT programs, such as MULTILOG (Thissen, 1991), which allow item parameters to be either constrained or estimated freely in both groups. This approach is described more fully later, in the context of the likelihood ratio test for DIF.

When item parameters are estimated separately in the reference and focal groups, the linking of metrics is required to obtain this transformation. Two general methods for linking are the mean and sigma methods (e.g., Bejar & Wingersky, 1981; Cook, Eignor, & Hutton, 1979; Linn, Levine, Hastings, & Wardrop, 1981; Loyd & Hoover, 1980) and the characteristic curve methods (e.g., Divgi, 1980; Haebara, 1980; Stocking & Lord, 1983). Mean and sigma methods use the mean and standard deviation of the distribution of item parameter estimates to calculate the linear equation for transforming item parameter estimates from one metric to another. Characteristic curve methods obtain the linear equation by minimizing some measure of the difference between the test characteristic curves estimated in each sample.

DIF and Metric Transformation

An important problem in DIF detection arises in the metric transformation context. The concern about DIF in this context arises because the presence of DIF items themselves affects the quality of the metric transformation (Shepard, Camilli, & Williams, 1984). In addition, the detection of some or all DIF items may be affected by the particular metric transformation (Kim & Cohen, 1992). This circularity arises for DIF detection under IRT.

Lord (1980) recommended that the set of items be purified of DIF items before the metric transformation is calculated. The approach described by Lord involves identifying all DIF items, removing them, and then reestimating item parameters in the reference and focal groups. The purification procedure is somewhat laborious. A simpler approach is described by Kim and Cohen (1992) based on a method due to Candell and Drasgow (1988) called iterative linking. In this latter method, item parameters are estimated once followed by calculation of a metric transformation and then detection of DIF items. The procedure iterates between the removal of DIF items and the recalculation of the metric transformation until either no DIF items are detected or the pattern of DIF items recurs.

Lord's Chi-Square

Lord (1980) described a chi-square statistic for DIF detection that compares the item parameter estimates obtained for an item in the reference and the focal groups. The number of parameters for this chi-square statistic is determined by the particular IRT model. In the case of the three-parameter model, estimates for the a_j and b_j parameters are compared. The guessing parameter, c_j, in the three-parameter model is not compared in this particular DIF statistic. The degrees of freedom of the statistic are determined as the number of parameters that are compared in the model. In the case of the two-parameter and three-parameter models, the degrees of freedom for a comparison between the reference and focal groups for a single item are two. Research with **Lord's chi-square** has demonstrated that type I errors are controlled for the one- and two-parameter models but not for the three-parameter model (Kim, Cohen, & Kim, 1994; McLaughlin & Drasgow, 1988).

Likelihood Ratio Test

The problem of type I error control for the dichotomous models, including the three-parameter model, is resolved with the **likelihood ratio test** (Thissen, Steinberg, & Gerrard, 1986; Thissen, Steinberg, & Wainer, 1988, 1993). In this test, item parameters are estimated concurrently in both the reference and focal groups. The test compares the likelihood estimated in the reference group with that estimated in the focal group. When these likelihoods are estimated for one item at a time, the likelihood ratio test provides a test for DIF that maintains type I error control and is similar to Lord's chi-square. Systematic studies of the power of this statistic, however, have not been reported.

The likelihood ratio test for DIF described by Thissen and colleagues (1986, 1988, 1993) compares two different models—a compact model and an augmented model. In the compact model, the item parameters are assumed to be the same for both the reference and the focal groups. The computer program MULTILOG (Thissen, 1991) has an option that permits equality constraints between the focal and reference groups to be placed on items for estimation of the compact model. In the augmented model, item parameters for all items except the studied item are constrained to be equal in both the reference and the focal groups. These constrained items are referred to as the common, or anchor, set. In a single-item DIF comparison, only the item parameters for the studied item are estimated separately in the reference and focal groups.

The likelihood ratio test statistic (G^2) is calculated as the difference between the values of −2 times the natural log (ln) of the likelihood for the compact model (i.e., $−2 \times \ln L_C$) and −2 times the natural log of the likelihood for the augmented model (i.e., $−2 \times \ln L_A$). G^2 is distributed as a chi-square under the null hypothesis with degrees of freedom equal to the difference in the number of parameters estimated in the compact and augmented models (Rao, 1973). When all parameters in a three-parameter model are tested, G^2 is distributed as a chi-square with three degrees of freedom. The values of the quantity $−2 \times \ln L$ can be calculated if the item parameter estimates and the distributions of ability are known for the focal and reference groups. It can also be obtained from the output of the calibration runs from computer programs such as MULTILOG (Thissen, 1991).

Raju's Area Measures

Another measure of DIF is obtained by calculating the difference between IRFs estimated in the reference and focal groups. Raju (1988, 1990) developed two measures and a statistical test that can be used for this purpose. These two tests are based on the signed area and the unsigned area between two IRFs. Type I error rates have not been reported for these statistics; however, Cohen and Kim (1993) suggested that these measures, like Lord's chi-square, are subject to problems with control of type I errors.

Lord's chi-square and **Raju's area measures** make excellent sense from a theoretical perspective, and both control type I errors well for the one-parameter and two-parameter models. Unfortunately, both methods provide inflated estimates of DIF for the three-parameter model.

Selecting a DIF Detection Technique

Each DIF detection technique has advantages and disadvantages. The nonparametric techniques, such as the Mantel-Haenszel and the SIBTEST, should be considered when the assumptions of a parametric model are not satisfied. Most DIF is reflected in the item or task difficulty. Consequently, a nonparametric technique such as the Mantel-Haenszel is simple and should be adequate. It can be computed using software such as SPSS. The SIBTEST procedure is more powerful than the Mantel-Haenszel, but requires the use of special software (e.g., Stout & Roussos, 1996). When model assumptions are satisfied, then one of the parametric techniques, such as the likelihood ratio test, Lord's chi-square, or Raju's area measures, may be considered. Specialized software is also required for these methods. The likelihood ratio test can be computed using the computer program MULTILOG (Thissen, 1991). Lord's chi-square and Raju's area measures can be obtained using a program such as IRTDIF (Kim & Cohen, 1992).

MEASURES OF THE AMOUNT OF DIF

In addition to detecting DIF using a statistical test, most measurement experts recommend using some measure of the amount of DIF. This is because large samples often yield statistically significant DIF results for very small differences. A simple measure of the amount of DIF can be calculated using MH D-DIF (given in equation [7.1]). Ziecky (1993) provided the following guidelines used by the Educational Testing Service (ETS) for categorizing the DIF as category A, category B, or category C, based on the size of MH D-DIF:

Category A: Defined as "negligible DIF" because either the DIF test is not significant or the absolute value of MH D-DIF is less than 1.0.

Category B: MH-D DIF is significantly different from 0.0, and the absolute value is at least 1.0 and either less than 1.5 or not significantly greater than 1.0.

Category C: MH D-DIF is significantly greater than 1.0, and the absolute value is 1.5 or greater.

Roussos and Stout (1996) provided the following guidelines for interpreting the magnitude of a significant SIBTEST result:

1. Negligible DIF: when the absolute value of $\hat{\beta}_i$ is less than .059.
2. Moderate DIF: when the absolute value of $\hat{\beta}_i$ is equal to or greater than .059 and less than .088.
3. Large: when $\hat{\beta}_i$ is equal to or greater than .088.

AREAS OF FUTURE DIF RESEARCH

DIF is assumed to be due to undesired multidimensionality in the set of item responses (Ackerman, 1992; Roussos & Stout, 1996). The causes of this DIF, however, have been difficult to determine. Bolt (2000) described one interesting methodology. Using a modification of the SIBTEST procedure, Bolt experimentally manipulated features of items (e.g., open-ended vs. multiple-choice format and abstract vs. concrete problem type) and then studied the items to determine which manipulations would cause DIF. The Bolt method is important because it illustrates an approach to the identification and then the testing of factors that may be causing particular types of DIF.

Related to the Bolt method is an effort to identify the characteristics of items that were identified as related to DIF. This approach was illustrated by Ibarra and Cohen (1999) in a study of reading comprehension items. Gender DIF in favor of females was found to occur, for example, when the reading passage dealt with relationships between people. Likewise, when the passage dealt with abstract or impersonal topics, gender DIF in favor of males was found to occur. Using this approach on a mathematics test, Li (2001) and Li, Cohen, and Ibarra (2004) found that items with practical application, female-oriented topics, strictly symbolic manipulation (such as in figure 7.1b), or verbal content were more likely to function differentially in favor of females. Gender DIF in favor of males was more likely to occur when the item consisted of male-oriented content, had less of a practical application, or consisted of figural content (e.g., as in figure 7.5).

One problem with DIF research is that it is often based on a comparison of groups defined on the basis of manifest characteristics such as gender or ethnicity. Such distinctions are only loosely related to DIF. Recent research has suggested that this kind of distinction provides little information about who is really affected by DIF in the items. Using IRT models designed to isolate groups that are latent rather than manifest in the data is one way to identify those examinees for whom an item is truly functioning differentially. In this way, it is possible to determine the causes of the detected DIF. This approach, described by Bolt, Cohen, and Wollack (2001), locates latent groups of examinees that are influenced by nuisance dimensions that cause an item to function differentially for one group of examinees. These types of models are useful for detecting latent groups that use specific response strategies (Bolt, Cohen, & Wollack, 2002; Cohen & Bolt, 2005; Kang & Cohen, 2003) and then for studying characteristics of individuals in each group. Identifying groups that use different response strategies is central to determining what it is about an item that may be causing some examinees to respond differentially.

The study of DIF can often provide insight into how a particular measurement procedure functions in particular contexts. One of the more interesting contexts is that provided among cultures. The detection of DIF among different cultural

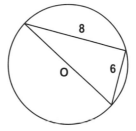

In the figure, O is the center of the circle. What is the radius?

a. 5
b. 7
c. 10
d. 14
e. $\sqrt{39}$

Figure 7.5 Example of an item with figural content.

groups serves to highlight patterns of thinking or behavior that differ among cultures. Studies by Ibarra and Cohen (1999); Li (2001); and Li, Cohen, and Ibarra (2004) suggest that cultural differences can lead to DIF. Zhu (2000) explored the detection of DIF in cross-cultural research on physical activity. Zhu described the way differences in wording can lead to DIF in questionnaires on physical activity. This is an important and growing area of research as scholars try to understand how differences in physical activity can be measured and compared among different cultures.

SUMMARY

Bias in measurements can lead to erroneous conclusions about the behavior being measured and also to unfairness in the decisions that are made based on results from the test or measure being used. The study of DIF is important because it can shed light on the quality with which the activity or skill is measured. The methods presented in this chapter are important because they provide tools that can help researchers determine whether construct-irrelevant variation may be intruding on their measures. It is important to recognize, when using statistical methods and significance tests such as those illustrated in this chapter, that they can be very useful but they need to be viewed as aids in making a decision about retaining or removing a DIF item. That is, tools, and statistical tools in particular, can help detect effects leading to a better understanding of behavior, but they need to be combined with judgment. It would be inappropriate, for example, to reject the use of an item or task simply because a significant DIF statistic was observed. Other evidence such as effect sizes and a judgmental analysis of the item (neither of which was discussed here) should also be considered.

CHAPTER 8

Computerized Adaptive Testing

Richard C. Gershon and Betty A. Bergstrom

Reprinted from *Introduction to Rasch measurement.* by Evertt V. Smith, Jr. and Richard M. Smith (Eds.) 2004, pp. 601-629. With permission from Richard M. Smith.

Sitting in the sports medicine center's waiting room is a woman awaiting treatment. She is handed a handheld computer and asked to answer a few short questions related to her current symptomology. She is able to answer the 25 to 30 questions in the short period of time prior to being called into the examining room. Her clinician enters the examining room holding a report detailing the current status of six health-related quality of life variables related to the patient's treatment, including recommendations for treatment—all based on data collected just a few minutes earlier in the waiting room. This capability is empowered by computerized adaptive testing, or CAT. Rather than requiring patients to answer several hundred questions regarding their current functioning level (a task that a patient might voluntarily complete once in the course of treatment, but certainly not prior to every visit), CAT uses a computer-driven branching scheme that zeroes in on a person's symptom level in as few as five questions with accuracy more typically associated with instruments consisting of 36 to 100 questions (items) per scale. **Computer-based testing** (CBT) powered by CAT represents the future of assessment with the potential for increasing the efficiency of clinical trials and providing new diagnostic tools for clinicians.

INTRODUCTION TO COMPUTER-BASED TESTING

It is more than a simple coincidence that the growth of CBT as we know it coincides with the expansion in use of the personal computer. To be historically accurate, CBT was used on mainframe systems dating back to the early 1960s. But its use was primarily limited to the military and a few large corporate and private training companies who could afford to purchase their own hardware. Fast forward to the 21st century: Computers are readily available in almost any setting, high-speed access is (relatively) easy to acquire, and Internet access is a necessary component of many facets of our lives. Today, computers are ubiquitous, and CBT is a fait accompli.

The advantages of CBT begin with the capability of offering tests with increasing frequency. The opportunities for test development and administration are limitless.

High-stakes tests can now be offered anytime, anywhere. For example, the certification test to practice as a nurse anesthetist used to be given on paper only twice a year. If a test taker couldn't get time off from work that day, or was ill, he or she had to wait another six months to try again. Today this exam is available at proctored test sites open six days a week, in approximately 200 locations across the United States. This real ease of access is only possible because of the advent of CBT powered by CAT.

CBT allows for real-time (immediate) scoring, and consequently real-time reporting, whether directly on the screen when the examinee completes the test, directly available to an instructor, or immediately delivered to a hiring manager. Computer-based tests are administered over computer networks or over the Internet, allowing for easy updating of examination content and fast centralization of data. For many testing programs, CBT is made available on a continuous basis, and despite the dramatic increase in test dates available, CBT is typically more secure than paper-and-pencil testing formats. There are no paper documents to be copied, and computerization allows for greater randomization in the test items presented to any single examinee.

Given the numerous advantages of CBT, it is further comforting to note that in general the "digital divide" does not apply to test performance. Regardless of commonly considered demographic variables (ranging from socioeconomic status to gender and cultural affiliation), research studies have repeatedly shown no performance differences between examinees using CBT and those using the original paper-and-pencil versions of a test. Indeed, in some cases minority groups have been found to demonstrate better performance on the computer-based versions of test compared to the paper-and-pencil versions (Zara, 1992).

The similarities between paper-and-pencil tests and CBT at the macro level are also evident at the micro level. Many item types presented on paper perform similarly on the computer screen, whereas others are enhanced by CBT delivery. For example, graphical items may be more vibrant and more easily reproduced in color on the screen than in print format. One primary exception to comparability of item performance appears when exploring the use of items that include long reading passages. The computer administration of a reading passage forces text to be scrolled and therefore the text is not wholly visible at any one time. Older examinees may find this disconcerting relative to traditional paper tests in which the entire reading passage is displayed side-by-side with the related questions.

Given the overall comparability of CBT with paper-based testing, it should not be surprising that the inherent risks are similar as well. Issues of test development, administration, reporting, and general program evaluation are almost identical and are better covered in a general text dedicated to test development (although some unique concerns for test validity and reliability for CAT will be discussed later in this chapter). As with all testing, the best way to keep out of trouble is to establish a strict protocol of quality standards and associated quality control procedures for use throughout the test development and administration life span. Early concerns that computer-based tests would be riddled with problems such as power failures and other computer errors have never materialized because software and hardware quality has rapidly improved.

A number of features inherent in CBT enhance test validity. Graphical images and multimedia presentations enable the measurement of concepts not possible with text-only formats (Bergstrom & Lunz, 1999). CBT, in most cases, offers the advantage of allowing the examinee to view and confirm the highlighted answer on screen. This is typically compared to the examinee who has the opportunity to "mis-bubble" the paper-based answer sheet, even though he or she knows what the correct answer is. Given that some test takers are known to make this type of error on paper-based bubble-sheet test answer forms (Daniel, 1983), it is little wonder that CBT test performance is typically consistent with, or even better than, performance on the same test content administered in a traditional paper-and-pencil format.

HISTORICAL BACKGROUND OF ADAPTIVE TESTING

CAT is a form of computer-based test administration in which each test taker takes a "customized" test. Test-taker competence is assessed after each item is administered, and the difficulty of the next item administered is targeted to the current estimate of ability (Gershon & Bergstrom, 1995). Thus, CAT mimics, electronically, what a wise examiner would do (Wainer, 2000). The Stanford-Binet intelligence test (Binet, 1908) is an example of an early oral adaptive test. Other examples of initial forays into targeted testing include two-stage testing (Angoff & Huddleston, 1958) and pyramidal testing (Krathwohl & Huyser, 1956). These modes of administration placed items in a particular arrangement based on p-values and developed fixed paths through the items to match the test to the test taker (Reckase, 1989).

In the 1960s, the development of the Rasch model (Rasch, 1960; Wright & Stone, 1979) and other item response theory (IRT) models (Lord, 1952) provided a theoretical structure for building large-scale calibrated item banks (Choppin, 1985; Lord, 1980; also see chapter 4 by Weimo Zhu). The ability to order all of the items on the same scale is essential for CAT. Because all items are on the same scale in a calibrated item bank, the particular items that are administered to a given test taker become immaterial. Each individualized adaptive test created from the calibrated bank is automatically equated to every other test that has been or might be drawn from the bank (Masters & Evans, 1986; Wright & Bell, 1984). When items are calibrated to the same scale, a pass–fail point (criterion-referenced standard) can be established for the entire item bank; thus, test takers are measured against the same criterion-referenced standard regardless of the group of test takers with whom they are examined, the particular set of items they are administered, or when they take the test (Gershon & Bergstrom, 1995).

One of the first adaptive tests to be developed was the Armed Services Vocational Aptitude Battery (ASVAB) (American Psychological Association, 1997). The armed services produced a CAT for personnel selection and classification to reduce testing time and to increase test security.

Other work from the late 1970s to the early 1990s revealed the promise of adaptive testing. A meta-analysis of 20 studies published from 1977 to 1992 compared results from paper-and-pencil administrations to CAT administrations, and consistently found that both modes of test administration yielded similar results (Bergstrom & Lunz, 1992). Each of these studies (despite differences in test content, the age of test takers, the latent trait model used, and study design) demonstrated the comparability of measures obtained using CAT and pencil-and-paper test versions. Indeed, what is most remarkable in reviewing the literature comparing these two test modalities is the marked absence of any significant studies demonstrating the *inability* of CAT to capture the same trait levels originally assessed using paper tests. Even the minor decrement in performance realized with long reading passages in CAT may in practice prove that the CAT format better captures reading comprehension. The paper format may benefit the test taker who is quick to rescan the material, whereas the CAT version may benefit the examinee who is better able to commit the material to memory.

ADVANTAGES OF CAT

Basic metrics for testing include reliability (see chapter 3 by Ted Baumgartner), validity (see chapter 2 by David Rowe and Matthew Mahar), fairness (see chapter 5 by Patricia Patterson and chapter 7 by Allan Cohen), and feasibility. CAT takes advantage of technology and modern measurement theory to deliver tests that are often more reliable than their paper-and-pencil equivalents. Because only items of appropriate difficulty are administered to test takers, lower measurement error and higher reliability can be achieved using fewer items. When items are targeted to the ability level of the examinee, the standard error of measure (SEM) is minimized and test length can be minimized without loss of precision. Thus CAT can substantially reduce test length compared to paper-and-pencil tests (Olsen, Maynes, Slawson, & Ho, 1986; Weiss & Kingsbury, 1984).

Common definitions of validity include the fact that (a) the test measures what it purports to measure; (b) the inferences made from the test scores are meaningful and useful; and (c) the content of the test reflects critical aspects of the crucial skills or knowledge. Shorter tests with acceptable precision, possible with CAT, can enhance validity by decreasing examinee fatigue or test anxiety that may introduce construct-irrelevant variation (Gershon & Bergstrom, 1991; Huff & Sireci, 2001). Sophisticated item-selection algorithms built into CAT can ensure that content is balanced for each test taker.

Computerized adaptive tests also have characteristics that enhance fairness. Because tests are administered via the computer from a large bank of items, there is no human intervention in the selection of test forms. Assuming a well-constructed item bank, each test taker has the same opportunity to demonstrate ability or achievement as any other test taker. Recent improvements in electronic test publishing ensure that banks can be easily swapped in and out such that compromised items can be removed from circulation in real time.

From a cost perspective, CAT is feasible for many organizations. The cost of administering tests is spread out over several areas that roughly conform to the test development and administration cost structure of any exam at a comparable level of security: test content development, test administration, scoring, and reporting. Test content development for CAT differs in terms of the number of items required to create an item bank large enough to cover the range of abilities, and also large enough to ensure overall bank security. For criterion-referenced mastery tests, the test may only need to have a large number of items near a pass point; but for a norm-referenced test, a large number of items may be required across the ability or trait continuum. For high-stakes tests administered to thousands of examinees, it may be necessary to have a large number of items to merely ensure test security. At the other extreme are low-stakes and self-assessment tests in which a very small bank of fewer than 50 items may be sufficient.

The cost consideration for item development is primarily of concern for high-stakes norm-referenced testing programs. Once items have been written, the next cost relates to calibrating IRT parameters for every item. In the case of an established testing program using previously administered items, the calculation of bank parameters may simply require reanalyzing old data sets. At the other extreme, all newly written items may have to be piloted on hundreds of examinees. Although many organizations will experience increased up-front costs to create their CAT programs, they may similarly encounter decreased costs in the future, as the necessity to write completely new tests each year is replaced by lesser bank-maintenance tasks such as ensuring the currency of existing items (getting rid of items that are now outdated) and writing a greatly reduced number of new items each year to ensure content coverage and to further increase security by keeping the bank fresh.

The cost of test administration is also related to the security level of the test. High-stakes tests must be administered in proctored settings. Third-party test delivery vendors, with test administration centers located in thousands of cities throughout the United States and around the world, act as subcontractors to provide a secure high-stakes test environment. Alternatively, a test administration organization can set up its own private centers on a full-time or part-time basis. Lower-stakes CAT exams can now be administered over the Internet. Although testing time for a CAT is typically shorter than that for its fixed-length test equivalent, test administration time at a testing vendor is often paid for based on the maximum time allotted for testing. The cost of scoring a computerized adaptive test is basically nonexistent because the scoring burden is borne by the test administration process itself. There are no bubble sheets to collect and scan, and indeed, for many organizations, the final score report is produced on screen or on paper at the time of testing, removing the cost of generating and distributing reports altogether.

PROCESS OF CAT

CAT is appropriately used for both knowledge tests and attitude surveys. This is because the adaptive process is essentially the same when estimating a test taker's ability level, or when attempting to determine a person's level on a particular trait. For the purpose of the following discussion, the reader can exchange the term *ability* or *trait level* for the term *person measure*, and the term *item level* or *item severity* for the term *item difficulty*.

The basic process of administering a CAT exam is very similar to that of conducting a simple binary search. For example, if you are asked to think of a number between 1 and 100, a typical linear testing process would have to ask up to 100 questions to determine the correct answer: Is the correct answer 1?, Is it 2?, Is it 3? and so on. A binary search provides the same result in as few as five questions. If the correct answer is 82, the questioning would go something like this (see figure 8.1): Is the number greater than 50? Yes. Is it greater than 75? Yes. Is it greater than 81? Yes. Is it greater than 83? No. Is it 82? Yes. By using a binary search, we never need to ask about each of the first 50 numbers because we immediately know that the unknown value was greater than 50. Similarly, we didn't need to ask about each of the numbers above 87 after the second question.

In the CAT process, each time an examinee responds to a question, we can converge on an estimate of a person's measure (zeroing in on his or her ability level). On a pass–fail test, we would typically administer the first item at the pass point. If that item is answered correctly, a more difficult item is administered. If that item is answered incorrectly, an easier item is given. This process is iterated until specific stopping conditions are met (such as testing until a specific measurement precision is obtained). Many testing options, such as various stopping conditions, will be discussed in greater detail later in this chapter.

The true process of CAT is typically a bit more complicated than a simple binary search. Because estimates of ability contain error components, we cannot simply administer a single item and "know" that a person's measure is higher than the

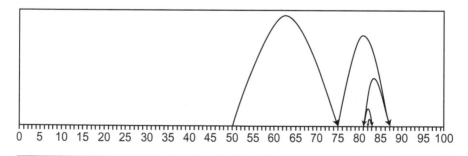

0   5   10   15   20   25   30   35   40   45   50   55   60   65   70   75   80   85   90   95   100

Figure 8.1 CAT process.

Reprinted from *Introduction to Rasch measurement.* by Evertt V. Smith, Jr. and Richard M. Smith (Eds.) 2004, pp. 601-629. With permission from Richard M. Smith.

difficulty of that one item. Also, examinee performance is not perfectly predictable. Typically, the item administered would be one very close in difficulty to the true person-measure. Because we don't know what the true person-measure is, the CAT process is repeatedly reestimating the person's measure after each item is administered. The response to each successive item offers additional information, subsequently allowing us to converge on an estimate of the person's measure.

It is interesting to note the differences between rating-scale items and typical ability-based items. For an ability item, the test taker is forced to choose a response that is either correct or incorrect. From a CAT perspective, we assume that the examinee is likely to be higher in ability than is represented by the current item if he or she answered that item correctly and lower in ability if he or she answered the item incorrectly. For a rating scale item, the threshold between each option has its own measure. For example: I am tired (a) All of the time (–3.0 logits); (b) Usually (–2.0 logits); (c) Sometimes (–1.0 logits); (d) Never. The number between each option is the IRT *difficulty* parameter associated with the threshold between those two responses. The administration of a rating scale item has even more impact in CAT than in dichotomously scored ability items, which are only scored right or wrong. Effectively, the impact on standard error of each four-choice rating scale item may be able to reduce the standard error typically associated with the administration of *three* dichotomous items.

CAT and IRT

Modern adaptive tests use IRT as their mathematical foundation. Numerous models, including the one-parameter model (also known as the Rasch dichotomous model [Rasch, 1960]), the two-, and three-parameter models (Hambleton, Swaminathan, & Rogers, 1991), and the rating scale and partial-credit models (Andrich, 1978; Bock, 1972; Wright & Masters, 1982), have been used for adaptive testing.

In practice the Rasch model works well for adaptive tests and has been used extensively. In Rasch measurement models, the underlying construct or latent trait described by the items on a test is a continuous variable extending to negative and positive infinity on an abstract continuum. All possible test item difficulties and all possible test-taker ability levels lie on this continuum. The measure estimated for a test taker on a set of items on a test is the result of the interaction between the ability of the test taker and the difficulty of the items administered. In practice, the measurement of the construct is bounded by the range of scores obtainable given the range of calibrated items administered on the test.

The idea of a single continuous scale for test takers and items implies that there is a point at which the ability of the test taker equals the difficulty of the item, a point at which the difference between the estimate of ability and item difficulty is as close to zero as possible. Although this point can only be approximated in practice, the idea is crucial to CAT. CAT algorithms "target" the difficulty of the test to the current ability estimate of the test taker by attempting to present an item at the point at which the difference between test-taker ability and item difficulty is zero. By targeting the difficulty of the items to the ability of the test taker, CAT maximizes the information from each item. Items that are too easy or too hard are not administered. When test information is maximized, the SEM is minimized. Thus, administering items adaptively can reduce test length and improve measurement precision. If the item exceeds the test taker's ability, the Rasch model predicts that the test taker will have less than a 50% probability of correctly answering the item. If the test taker's proficiency exceeds the difficulty of the item, the Rasch model predicts that the test taker will have a greater than 50% probability of correctly answering the item (Bergstrom & Lunz, 1999; Wright & Stone, 1979). Two- and three-parameter IRT models incorporate additional parameters that model discrimination and guessing, respectively (Wainer, 2000). Thorough overviews of the use of these models in adaptive testing are provided by Parshall (2001) (also see chapter 4 by Weimo Zhu).

Estimating Ability

CAT has several methods for estimating test-taker ability. Given an IRT-calibrated item bank, each method looks at the characteristics of the items administered so far, combined with the answers given by the respondent, and determines a current estimate of the test taker's ability. Once this estimate is obtained, the remaining items in the item pool can be evaluated in terms of which item will lend the greatest efficiency to the overall testing process.

Two common methods for estimating test-taker ability are maximum likelihood estimation (MLE) and **Bayes estimation.** In the case of MLE, the test taker's ability is updated by using the difficulty of the items already administered, and the response to the most recent item. The next item selected is based on maximum information, and thus the difficulty of the item selected closely matches the current estimate of test-taker ability. MLE with two- and three-parameter models is unstable for short tests and thus problematic for CAT based on these models.

Bayesian algorithms are based on the normal distribution or on knowledge of the typical population distribution of test takers taking a particular examination. Given that we do not initially have any knowledge of a particular person's ability, the algorithm administers items as if the true ability estimate is close to the mean or mode of the population, updating each "prior" distribution with new information based on the test-taker response. Bayes estimation is often used with two- and three-parameter CAT (see Parshall, 2001, for a more detailed explanation).

Selecting Items

Although estimating test-taker ability and matching that ability to the next best item from the pool represents the basic item selection process, CAT typically considers numerous additional factors when selecting items. The selection of the very first item on a test is dictated by the goal of that test. For a criterion-referenced mastery test (in which the goal is to determine whether a test taker has obtained a threshold of knowledge), the first item administered is typically selected at or close to the pass point of the test. If on the other hand, the goal of the test is to ascertain a specific ability (or trait) level, it is usually more efficient to administer the first item at the mean of the population being tested. This can be accomplished by simply selecting an item from that point in the pool or, as described earlier, by using a Bayesian estimation algorithm.

The choice of the second item is dictated by the examinee's answer to the first item, but in the case of an ability test, all we now know is that it is likely that the test-taker measure is above the difficulty of the item (because the person answered the item correctly) or less than the difficulty of the item (because the person answered the item incorrectly). Indeed, if the person answers the first few items correctly, the estimate of his or her ability is theoretically infinitely high. Although we know this is not true, at that point in the test, we do have to admit that the test taker has answered "all" of the questions correctly. This situation is typically remedied after a few items, when a test taker (finally) answers an item incorrectly and we can start bouncing back and forth to find his or her probable ability level. The reverse is also the case. If the test taker answers the first item incorrectly, theoretically his or her ability level is infinitely low. Because it is unlikely that the true ability is plus or minus infinity, we need to constrain the ability estimate at this point in a manner that is efficient for the type of test being administered.

On a trait-based test, in which the goal is to estimate the test taker's true ability or trait level, we might consider constraining the first couple of items to the population mean and then letting the test take a natural CAT approach to zeroing in on the test taker's true ability. The problem of selecting an extreme response on a rating scale item is much like the situation on an ability test after only one item has been answered. The examinee who responds to the first question on a survey: "I am in pain—Never, Some of the time, Most of the time, All of the time," by responding "All of the time" is, mathematically speaking, in *infinite* pain. It will take the administration of additional rating scale items to converge on the true trait level.

Item Enemies

Item selection in CAT may take into account item enemies. An **item enemy** refers to one or more items that should not appear on the same test as an item already administered. On a paper-and-pencil test, it is a relatively simple task for the test author to review the entire test to ensure that one item does not cue the answer to another item on the same test. In CAT the entire item pool must be reviewed for item enemies in advance of the testing session, because final test construction only takes place while the examinee is sitting at the computer.

Grouping Items

The test author must consider what to do with items that are typically administered together (e.g., a series of items that follows a long reading passage, or a series of items that refer to same the same illustration). Some test authors will opt to administer these items in preconstructed groupings frequently referred to as **testlets** (Wainer, Bradlow, & Du, 2000). Within the CAT process, a testlet is typically selected by the average difficulty most closely matching that of the current test-taker measure estimate. Testlets are often associated with two-parameter CAT exams to easily enable the administration of chunks of content-balanced items (Wainer & Kiely, 1987). Testlets are also useful when a test developer wants to review a short series of items.

Test Length

The next major question to address in CAT administration is, How long does the test have to be? In many instances, the true question is, When can we stop? CAT can be fixed length or variable length. If the test is terminated after a specified number of items, it is a fixed-length test. Variable-length CAT is terminated by one of two stopping rules: when a specified level of precision (SEM) is reached, or when a specified level of confidence in a pass–fail decision is achieved (Bergstrom & Lunz, 1999). In many cases, a computerized adaptive test is actually designed to be a combination of variable length and fixed length. For instance, a test could be designed to be at least 70 items in length given the satisfaction of a confidence-interval-based stopping rule, or the test may be primarily variable length with an absolute maximum length of 100 items regardless of whether the confidence rule is satisfied.

A standard-error-type stopping rule is also used in CAT with rating scale and partial-credit models. The Center on Outcomes, Research and Education (CORE) at Evanston Northwestern Healthcare in Evanston, IL is researching the way physicians are able to assess various outcomes of medical treatments. Ultimately, health care professionals would like to be able to assess patient progress for numerous treatment outcome variables such as fatigue or pain. In a traditional test environment, gaining an accurate assessment of how a test taker is feeling might require the administration of hundreds of items across numerous dimensions. With CAT, the same symptom measures can be obtained in the short period of time that one sits in the waiting room prior to seeing the doctor. Given the strength of a good set of rating scale or partial-credit items, this is likely to take place with as few as 6 items, instead of the 16 to 75 currently used by typical paper-based surveys.

Reporting

Organizations that report scaled scores will probably not have to change their reporting procedures at all when using CAT. For Rasch measurement models, the logit measure range is typically -3 to $+3$. A simple scale score transformation would be to multiply the obtained measure by 100 and add 500. This would result in a scale score range of 200 to 800. Alternatively, a scale could be constructed with a mean of 50 (multiply the measure by 10 and add 50), or the range of the scaled score could be stretched by multiplying the test-taker measure by a larger number. Scale scores can also be reported for content areas within an adaptive test, providing increased levels of feedback for test takers, test sponsors, or program directors. It should be noted that it does not make sense to report percentage scores with CAT. In most CAT applications using one-parameter IRT, everyone answers approximately 50% of the items correctly regardless of ability. In multiple-parameter IRT-based CAT everyone typically answers 60% of the items correctly. It should also be noted that these percentages will hold only when the item pool is sufficient in size to administer appropriately targeted items. The percentage-correct target is likely to fail when an examinee is deliberately attempting to answer items incorrectly (as might be the case when a test is being used to place examinees in a course and the examinee purposefully wants to be placed in a lower level section).

CRITICAL ISSUES IN CONSTRUCTING AND MAINTAINING CAT

Several critical issues are associated with constructing and developing CAT tests. Eight of these are discussed in this section.

Content Balancing

A frequently considered constraint for item selection is that of **content balancing.** Although a typical requirement for CAT is the measurement of a unidimensional construct (i.e., the CAT process can test only *one* thing at a time), that one dimension may be made up of subareas that the test sponsor believes are important. For example, the certification examination for podiatric surgeons includes items that demonstrate knowledge of basic science as well as knowledge of surgical principles. CAT can accommodate these concerns by selecting the next item to be one that maximizes information, while also adhering to the desired content scheme (Gershon, 1995).

Exposure Control

One cited advantage of CAT is that it can be used to administer tests more frequently than is typically practical with paper-and-pencil test forms. For example, the National Council Licensure Examination for years was administered only twice per year because of the expense of creating an original test for administration to over 100,000 people. Today it is administered on a daily basis year-round. This is accomplished without breaching the security of the exam because each examinee is administered an individualized exam, custom tailored to his or her ability level. Given a large enough item bank, it is extremely unlikely that any two people will see the same set of items. To further ensure this per-examinee customization, most CAT formats also provide a randomization factor. Rather than always picking the "absolute best" item given all of the other item selection constraints previously discussed, the algorithm might choose the "best 10" choices from the pool and then randomly select the next item from among those 10 (Gershon, 1995). In theory this will decrease the efficiency of the test, but in practice most item pools for CAT are populated with sufficient items so that no noticeable loss in efficiency takes place (Stocking & Lewis, 2000). **Exposure control** is rarely an issue for low-stakes tests such as mood inventories or quality of life surveys.

Test Difficulty

To maximize efficiency in the CAT process, items are typically targeted such that the examinee has only a 50% likelihood of answering the item correctly for one-parameter models, or close to a 60% likelihood for two- or three-parameter models. Although there is no question that this is the most efficient way to administer the items, it is clearly surprising to many examinees. When piloting the concept of CAT for certification in the early 1990s, the first author was startled to see an examinee walk out of the testing session in tears. The candidate was the smartest test taker in the class and had never performed poorly on a major test throughout her graduate school training. One can easily imagine the shock that this student must have had when she was confronted with a test specifically designed so that she was able to answer only half of the items correctly. The reality is that the student passed the test with flying colors. Yet how much easier it would have been for her had she understood the adaptive process prior to taking the test.

In the future, psychometricians may explore the possibility of manipulating test difficulty to maximize performance. For example, low test-anxious examinees typically underestimate the difficulty of tests and appear to lose attention at the beginning of a test (Gershon, 1996). For them, administering items of increased difficulty at the beginning of the test may help them to focus quickly on the task at hand. Conversely, the typical high test-anxious examinee is overwhelmed by the relative difficulty of most tests. By manipulating the test difficulty early in the test, high test-anxious examinees could be given an early success experience.

Pretest Items

Test length will also be affected by the addition of experimental or field-test items. These are typically new items added to a computerized adaptive test, but not scored, in order to collect enough data to eventually be able to estimate item calibrations. This is the most common way to incorporate new items into the test, because most organizations that administer CAT no longer have paper-and-pencil tests with which to field-test new items. Pretest items may be seeded into the CAT experience or administered at the end of the test. In any case, their score is not included in the active adaptive algorithm, and the items are not included in the final test-taker ability estimate. An exception to this rule could be made when CAT is administered over a brief period of time and immediate results are not needed. In this case, experimental items are still seeded into the test but not used as part of the targeting process. At the end of the test, results are not given to the examinee. When all exams are complete, the experimental items are calibrated. The complete tests (adaptively administered items and the now calibrated field-test items) are rescored in their entirety. Following additional reliability checks the answers to the field-test items can then be included in the candidate's ability estimate.

Review

A somewhat controversial option in CAT is to enable examinee review. In paper-and-pencil testing a common testing strategy is to review one's answer at the end of the test, time permitting. This process can be replicated in CAT. Following the completion of the test, examinees can be given time to review and modify any of their answers. Research conducted with high-stakes tests has found that allowing the review option increases examinee satisfaction and typically enables nominal test improvement (Gershon & Bergstrom, 1995; Lunz, Bergstrom, & Wright, 1992; Stocking, 1997; Vispoel, Hendrickson, & Bleiler, 2000; Vispoel, Rocklin, Wang, & Bleiler, 1999).

Skipping Items

Another CAT option is that of "skip." In theory, a CAT algorithm can be constructed to enable examinees to skip items they do not wish to answer. The reason this works is that, given a large enough item bank, examinees should be able to be given another item of similar difficulty (and similar content) to that of the skipped item. Practical research conducted with high-stakes tests has found that enabling skip does not change final test-taker measure estimates—although it can result in significantly greater test lengths and time required for test completion if the number of items allowed to be skipped is not limited (Bergstrom, Lunz, & Gershon, 1994). At present, we are not aware of any organization that allows for a skip condition in any type of criterion- or norm-referenced CAT ability test. This option cannot be so quickly discarded, however, when CAT is used for rating scale tests such as a test of sexual functioning. Although it would be preferable to insist that everyone who takes a survey fill out all items, it is quite typical that people opt to skip questions they may deem as being too personal or not appropriate for a variety of other reasons.

Quality Control

It is often thought that CAT requires extremely large item banks consisting of thousands of questions. This is somewhat mythical. The size of the item bank depends on a number of factors including the size of the population being tested, the time frame for test administration (e.g., continuous, in windows, and so on), the number of content areas and other constraints, the stakes of the test, exposure rules, the minimum and maximum length of the test, and the use of testlets. Most high-stakes item banks contain only a few hundred questions; others that test tens or even hundreds of thousands of people each year maintain item banks with 2,000 to 5,000 questions. In general, most high-stakes testing organizations aim to have item banks with close to 1,000 items, although this is often a goal only reached several years after an organization has commenced with CAT. For low-stakes rating scale tests, a suitable bank may consist of as few as 30 questions.

New Technologies

Zenisky and Sireci (2002) outlined over 15 innovations in task presentation, including drag and drop, graphical modeling, concept mapping, sorting, passage editing, multiple correct response, and problem-solving vignettes. These interactive item types change the way test takers answer items. Many item types make increasing use of a computer mouse, and some of them allow responses to be collected with touch screens, light pens, joy sticks, and speech recognition software. Many items are now being developed that incorporate graphics, video, and audio (Bergstrom & Cline, 2003). Another change enabled by computers and the Internet is the extent to which test takers are allowed to reference auxiliary material (Zenisky & Sireci, 2002). Test takers are allowed to use on-screen calculators, reference compact discs, and Web pages. These changes in presentation require analogous changes in test scoring, and consideration of consequences for CAT item selection and ability estimation algorithms.

Given the growth of CAT, one would think that a plethora of software options would be available for test developers. The opposite is the case. Almost without exception, CAT developers write their own test-delivery software or license test-delivery software from a provider who maintains a network of secure test sites for high-stakes testing. Well-known vendors in this regard include Promissor, Pearson Vue, and Prometric. Promissor also provides CAT via the Internet. To the best of our knowledge, the only "off the shelf" CAT software system is FastTESTPro published by Assessment Systems Corporation. All of the aforementioned groups support only dichotomous item types. The following section describes current CAT programs across several industry sectors. Many of these organizations would be more than happy to share their expertise, and potentially their software or online technology, with interested organizations.

CAT TODAY

CAT is seeing increased popularity in such diverse areas as information technology (IT) certification, state licensure, college entrance, and more recently in health-related quality of life. Novel® and Microsoft® initiated programs in the mid-1990s to administer adaptive tests to a peak of over 1 million examinees per year. These examinations were designed to assess minimal competence in the installation and maintenance of myriad software products. IT certification tests take particular advantage of the strength of CAT to administer unique tests to many people at locations worldwide. If fixed-length tests were to be used, there would be no practical way to release the examinations for simultaneous administrations worldwide without serious threats to security. CAT is also seeing increased use in state licensure, once again because of its ability to provide frequent testing with preserved security. In addition to pioneering work by the National Council of State Boards of Nursing mentioned earlier, programs exist for adaptive administration by the National Association of Securities Dealers, the National Association of Boards of Pharmacy, and the National Board of Medical Examiners. Allied-health board certifications using CAT originated by the American Society of Clinical Pathologists are now used by the American Board of Podiatric Surgeons and the American Association of Nurse Anesthetists.

In college entrance, the College Board now administers the Graduate Record Examination almost exclusively using CAT. They also administer the Test of English as a Foreign Language adaptively in locations worldwide where computer access is available. In education, CAT is used from elementary school practice tests to college entrance and placement. For example, Renaissance Learning provides thousands of schools with adaptive reading and math tests, allowing for fast placement and assessment of current and continuing skill levels. Northwest Evaluation Association has developed large item banks for use in high school proficiency tests, often custom tailored to specific school objectives (Vispoel, 1999). In 1986 the Psychological Corporation published an adaptive version of the Differential Aptitude Test, a series of eight tests used since 1947 for placement and vocational guidance in grades 7 through 12 (McBride, Corpe, & Wing, 1987). Indiana University Purdue University Indianapolis uses adaptive tests for testing high school students for potential acceptance, and then again for placement in certain classes.

The University Cegeps de Jonquier in Quebec researched an interesting CAT variant that adaptively assesses entering students for placement in English as a Second Language (in the predominantly French-speaking program) (Stahl, Bergstrom, & Gershon, 2000). Their exam stands out in two ways: (a) for its use of hundreds of audio prompts (instead of the typical on-screen text) and (b) for its use of CAT and IRT to help identify students who are purposefully cheating on the poor side (i.e., attempting to incorrectly answer items in order to be placed in an easier class).

The most recent cutting-edge applications in CAT are taking place in rating scale surveys. The first author is working with David Cella at the Center for Outcomes Research and Education (CORE) to calibrate numerous IRT-calibrated item banks for use in CAT on stand-alone desktop or laptop computers, over the Internet, on PDAs, and even using integrated voice response over the telephone. Recent work has focused on using the results of IRT-based assessment systems (including CAT) in regular clinical practice. Using their SyMon (Symptom Monitoring) program, patients are regularly assessed for quality of life symptoms using CAT. Significant changes in symptomology (up or down) are flagged for immediate clinical review.

Additionally, clinicians are provided with treatment recommendations relative to current functioning. CORE's early CAT research focused on the health-related outcome of fatigue. Subsequent item banks have been written and calibrated to assess physical function, pain, emotional distress, cognitive complaints, illness impact, social support, and social role participation. Additional IRT and CAT projects in progress include the creation of additional item banks for spiritual well-being and disease or treatment-specific symptoms such as nausea, appetite, anorexia, pulmonary symptoms, and endocrine symptoms (see www.facit.org) (Gershon et al., 2003).

Qualimetric also specializes in CAT-based assessments at www.amihealthy.com. This group has been focused primarily on the development of the CAT-delivered DYNHA Headache Impact assessment. Additional computerized adaptive tests have been created for osteoarthritis, rheumatoid arthritis, asthma, and the SF-36. New projects include creating computerized adaptive tests to assess the impact of rhinitis, pediatric asthma, congestive heart failure, diabetes, depression, and pain (Ware, 2003).

The American Academy of Orthopaedic Surgeons (AAOS) has developed a series of instruments designed to collect patient-based data within clinical practices to assess the effectiveness of treatment regimens and in musculoskeletal research settings to study the clinical outcomes of treatment. The AAOS is now assessing the use of CAT to reduce assessment length while maintaining overall validity and reliability. The new assessments will be available to proprietary software companies to create paperless work environments. The organization has also considered creating a central, national data repository to provide real-time analysis of patient data across numerous treatments for similar conditions (AAOS, 2004).

Numerous grants are in progress to create additional CAT opportunities. For example, Stephen Haley at Boston University is working under a National Institute of Child Health and Human Development grant to build a CAT system and to examine the usefulness of CAT to assess execution of daily routines and involvement in life situations. Similarly, Dijkers (2003) has simulated CAT for the FIM instrument motor component, and Alan Jette is developing CAT tools for patients with musculoskeletal and neurological disabilities to assess mobility, activities of daily living, and community functioning.

Two other significant CAT grants have recently been awarded. The National Institute of Neurological Disorders and Stroke (NINDS) is building a single set of item banks and subsequent computerized adaptive tests for assessing quality of life outcomes in neurological disorders. The primary goal is to have a single set of instruments for use in all NINDS-supported clinical trials. Similarly, the Patient-Reported Outcomes Measurement Information System (PROMIS) is now under development, the cornerstone of the $25 million National Institutes of Health (NIH) "Roadmap" initiative on Dynamic Assessment of Patient-Reported Disease Outcomes. PROMIS will create item banks and enable the necessary technology for a central CAT system to assess health-related quality of life data across NIH agencies (see www.nihpromis.org).

SUMMARY

With each new advance in technology, the potential for CAT grows exponentially. Each step forward increases the reliability of the tools we can use to measure competence in numerous areas and for multiple ability levels. Yet, in contrast to other health-related professions, the application of CAT in the field of kinesiology has been minimal. Since the introduction of CAT to kinesiology by Spray (1987), few applications have been forthcoming. Notable exceptions are the Zhu, Safrit, and Cohen (1999) IRT-calibrated physical knowledge test and a proposed interactive-video-technology-based CAT testing model in which real-life scenes (e.g., accidents for a first-aid test or basketball competition for testing referee competency) are selected and presented based on test takers' previous responses and decisions (Zhu, 1992). No real progress, however, has been made in the development of CAT applications in kinesiology; this is likely due to the lack of understanding of CAT advantages, the lack of technical support, and high start-up costs. Perhaps the best way to start such an application is to develop it from an existing assessment practice (e.g., computerized neuropsychological testing of sport-related concussion). CAT applications can be further promoted after demonstrating the advantages of the application (e.g., shorter, more accurate assessment with real-time administration, scoring, and reporting).

PART III

Advanced Statistical Techniques

CHAPTER 9

Structural Equation Modeling and Its Applications in Exercise Science Research

Fuzhong Li and Peter Harmer

The past decade has seen a significant increase in interest in the application of structural equation modeling (SEM) techniques to the field of exercise science. Much of the interest can be attributed to the growing popularity of SEM in social sciences, but, more important, to the comprehensive, flexible approach of SEM to modeling relations among variables that may be observed or latent. In exercise science, typical examples of SEM applications include developing and validating scales and inventories, describing the change or developmental trends of an outcome, and testing theoretical models of structural relations among independent and dependent variables. Examples of these applications can be found in periodicals such as the *Journal of Sport and Exercise Psychology, Research Quarterly for Exercise and Sport, and Measurement in Physical Education and Exercise Science,* among others.

This chapter outlines a practical approach to the use of SEM in exercise science. It offers a "low-tech" introduction to basic concepts, common issues, and applications of SEM specifically tailored toward graduate students and applied researchers in the field. Readers who are interested in gaining a deeper understanding of the statistical underpinnings of SEM should consult more comprehensive texts, including those by Bollen (1989); Hayduk (1987); Kaplan (2000); Kline (1998); Loehlin (1998); Maruyama (1998); and Schumacker, Randall, and Lomax (1996).

We begin with an overview of the major components of SEM (i.e., **measurement models** and **structural models**) and proceed to a brief discussion of why and when to use SEM, logical and sequential modeling steps, statistical assumptions, model-fit evaluation, and missing data treatment. For the sake of completeness, information on currently available SEM programs is also provided. We then illustrate how SEM can be used to investigate problems such as validating surveys, scales, and inventories and hypothesis-testing of relationships among theoretical constructs measured by multiple indicators. Sample problems are provided detailing the measurement model in the context of **confirmatory factor analysis** (CFA) and examining structural relations among variables using the structural model.

Given the evolving developments in SEM, we conclude the chapter with information on, and applications of, extended SEM models, including **growth curve models, growth mixture models, multilevel models,** and latent variable models with interaction effects.

STRUCTURAL EQUATION MODELS

In a nutshell, a general structural equation model consists of two parts: a measurement part and a structural part, each with a specific set of equations. The measurement equations describe the relationship of each of the **observed variables** to an underlying **latent variable,** whereas the structural equations describe the relationships among independent and dependent variables, either observed or latent. These relationships are described later following a brief description of model **path diagrams** commonly used in SEM.

Figure 9.1 displays path diagrams illustrating elements of measurement and structural equation models. Following SEM convention, boxes represent observed variables (or indicators), and ovals represent unobserved (or latent) variables. Single-headed arrows represent the impact of one variable on another, and double-headed arrows represent covariances or correlations between pairs of variables. In the left panel, a measurement model is presented with one latent variable (X) and three observed variables (x_1 through x_3; observed variables in a measurement model are indicated by lowercase letters). The arrows emanating from the latent variable (oval) to the observed variables (boxes) represent factor loadings (i.e., the relationship between the observed indictors and the latent variable), which are interpreted similarly to regression coefficients in multiple regression. The unattached arrows going to each of the observed variables (boxes) represent measurement error (which contains both measurement error variance and the variance that is not explained by the latent variable).

The model shown in the middle panel of figure 9.1 is a path model. The two observed variables $(X_1, X_2)^1$ on the left side of the model are labeled **exogenous** (or independent) **variables,** and the two observed variables (Y_1, Y_2) on the right side are labeled **endogenous** (or dependent) **variables** (for path models, the observed variables are indicated by uppercase letters). The unattached arrows going to the two endogenous variables (Y_1, Y_2) represent **residual error variances** or disturbances (i.e., variance unaccounted for by the independent variables).

The right panel in figure 9.1 represents a general structural equation model—that is, a structural model with latent variables (two latent exogenous variables [Xs] and two latent endogenous variables [Ys]), each defined by three observed indicators. As in the path model, the unattached arrows going to the two endogenous variables (Y_1, Y_2) represent residual variances or disturbances. This model essentially integrates the measurement model with the path model described in the left and middle panels of figure 9.1.

Assuming a priori knowledge about a factor structure, a measurement model (see the left panel of figure 9.1) can be specified to test a hypothesis about the relations

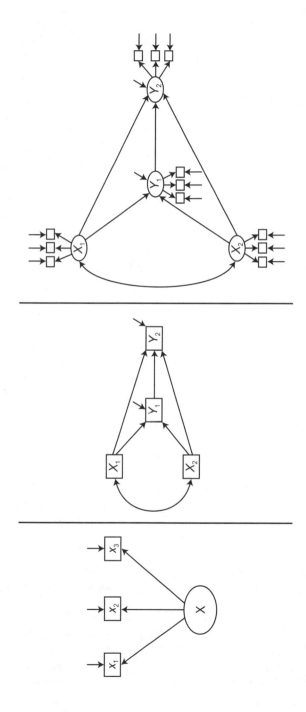

Figure 9.1 Path diagrams illustrating elements of a structural equation model. Measurement model (left panel); path model (middle panel); structural model (right panel).

among a set of observed variables, such as ratings or questionnaire or survey items, and the unobserved (latent) variables or theoretical constructs they were designed to measure. For example, a researcher may be interested in developing and validating a measure of exercise self-efficacy. The construct of self-efficacy in this case is a latent variable that cannot be measured directly. Therefore, a set of measurement indicators, or items—in this case, perhaps three items on a questionnaire—is typically developed to define the latent variable. Each of the indicators, however, is assumed to be an imperfect measure, and therefore contains measurement error. The goal here is to model whether the variation and covariation among the three indicator measures can be explained by a single underlying latent variable—self-efficacy. The common variance shared by all three indicators of self-efficacy provides a good representation of the self-efficacy latent variable. From this hypothetical example, we can see that the researcher needs to specify a measurement model based on an a priori hypothesis. For this reason, the measurement model is often labeled a "confirmatory," rather than an "exploratory," technique (Long, 1983).

Exploratory factor analysis (EFA) plays an important role in identifying the underlying dimensions of a measurement instrument when there are no a priori theoretical expectations with regard to links between the observed and latent variables. In contrast, CFA is more restrictive in the sense that it imposes theory-guided restrictions on measurement parameters (i.e., factor loadings, residual variances) and structural parameters (i.e., factor variances and covariances), resulting in a more parsimonious and substantive model. CFA is most useful in the later stages of instrument development for statistically evaluating whether theoretical relations among the observed measures and the underlying factors exist. A measurement model typically involves multiple latent constructs (or common factors in the sense of factor analysis), and allows for correlational (nondirectional) relationships among latent variables. It does not contain the directional influences (described later) that are included in general SEM.

The structural portion of structural equation models comes in two basic forms: (a) path models (middle panel of figure 9.1), which contain structural parameters representing hypothesized relationships among a set of observed variables; and (b) structural models (right panel of figure 9.1), in which a measurement model is linked with a path model. The hypothesized relationships among latent variables in either can be directional or nondirectional (MacCallum & Austin, 2000). Directional relationships (i.e., single-headed arrows) imply some sort of directional influence of one variable on another (e.g., $X_1 \rightarrow Y_1$), whereas nondirectional relationships (i.e., double-headed arrows) are correlational (e.g., $X_1 \leftrightarrow X_2$) and imply no directed influence.

In the exercise psychology literature, the work by McAuley (1991) exemplifies a typical path model that focuses on structural parameters representing hypothesized relationships among a set of observed variables (i.e., middle panel of figure 9.1). McAuley examined, through a path model framework, the relationships proposed by Bandura (1986) and Weiner (1985) with respect to the influence of self-efficacy and causal attributions on affective responses. Greenockle, Lee, and Lomax (1990) provided an example of a typical structural equation model (e.g., right panel of figure 9.1) in which, on the basis of Fishbein's Behavioral-Intention Model (Fishbein & Ajzen, 1975), the authors examined relationships among several theoretical constructs, each measured by multiple observed indicators.

Structural equation models are often referred to as "causal" models because the researchers are interested in making causal inferences of this kind. Some scholars have greeted attempts to make causal inferences with SEM analysis with cross-sectional data with a good deal of skepticism, and applied researchers are warned that results using correlational data do not necessarily indicate causation. We refer the reader to the articles by Breckler (1990), Cliff (1983), Mulaik (1987), and Pearl (2000) for a discussion on making causal inferences in SEM.

Why SEM?

A question naturally to be asked is, Why should one consider using SEM? Why not use conventional multivariate procedures such as EFA for measurement problems and multiple regression analyses for testing structural relationships among variables of interest? Summarizing from the general SEM literature, we consider the following five advantages of SEM:

1. SEM has the capability of correcting estimates by separating measurement error from the equations, thereby providing a means of modeling the latent variables, directly accounting for measurement error.

2. SEM incorporates elements of factor analysis and regression analysis that allow a simultaneous test of within- and between-constructs hypotheses.

3. SEM allows the use of multisample modeling to address invariance issues such as testing invariance of critical measurement parameters (i.e., factor loadings, **error variances,** correlations) across groups or testing differences in structural parameters (relationships) across groups.

4. SEM allows simultaneous tests of means and covariance structures.

5. Structural equation models can be readily extended to explore more complex research hypotheses (e.g., developmental change, multilevel influences, population heterogeneity) in social and exercise science research.

When SEM Should Be Considered

After gaining an understanding of some of the main advantages of SEM, the next question is, When should one consider using it? It is evident that SEM can be used for developing measurement scales that are either empirically driven or hypotheses driven, including measures of unitary and multifaceted psychosocial, physiological, and behavioral **constructs.** Its use should be considered when examining a set of relationships simultaneously between one or more independent variables, either continuous or discrete, and one or more dependent variables, either continuous or discrete.[2] Both independent variables and dependent variables can be either latent or observed.

The Process Involved in SEM Model Testing

All SEM analyses follow a sequential logical process (McDonald & Ho, 2002): (a) specification, (b) identification, (c) estimation, (d) evaluation, and, if necessary, (e) respecification. Model specification is most often driven by theory or a conceptual framework, which guides the development of a full model structure specifying the relationships among the observed variables, latent factors, and measurement errors. Once the hypothesized model is laid out, the components in the model are then operationalized in the form of a full measurement or structural model. The model may consist of a series of measurement or structural equations defining the model's parameters. Parameters may be specified, on the basis of theory, in the form of either "free" (to be estimated) or "fixed" (constrain to zero or some fixed value). The specified model is then path diagrammed (as we have shown in figure 9.1), outlining how the various elements of the model relate to one another and providing an overall view of the model's structure.

After the model is specified, it is necessary to ensure that the model is statistically identified. By **model identification** we mean that the information provided by the observed data (i.e., observed sample variances and covariances matrix) is sufficient for model estimation; in other words, the researcher must be able to obtain a single unique value for every specified (unknown) parameter from the observed data under investigation. A model is considered "identified" if there is one "optimal" value for each unknown parameter in the model. In SEM, the known parameters are usually the variances and covariances of the measured variables; the unknown parameters depend on the type of model fitted to the data (the so-called model parameters).

Once this necessary condition is fulfilled, estimation to derive parameter estimates for the model through the use of a SEM computer program can proceed. Common estimation procedures include **maximum likelihood estimation** and generalized least squares. The general objective in estimating a model is to generate an implied (i.e., model-based) covariance matrix that is as close as possible to the observed (i.e., actual) covariance matrix. This is analogous to the situation in multiple regression in which regression beta weights are sought that will reproduce the original outcome values as closely as possible.

In SEM, the distance or discrepancy between implied and observed matrices defines the **goodness-of-fit** of the model under investigation and is evaluated using a series of model-fit indexes including the chi-square statistic, fit measures based on population error of approximation, information measures of fit, or any number of the numerous comparison-based fit indexes. Based on the evaluation of model-fit statistics or indexes, the quality and soundness of the model are evaluated in line with the model's operationalization and theory-based hypotheses.

It is often the case that the a priori specified model does not describe or fit the data satisfactorily. In the presence of an unfit model, researchers are then tempted to engage in post-hoc **model respecification,** or modification. By adding or deleting certain parameters, researchers can make a significant improvement to the model. This practice is fairly common in applied research; however, it often leads to deviation from the original model specification. This step is not generally recommended unless one can demonstrate that the alterations made to the model are justifiable (e.g., based on theory).

The final step in model testing is cross-validation to demonstrate model generalizability. This involves fitting the model to a new sample from the same population or a sample from a different population. This step is often used in a confirmatory mode and is particularly important when substantial modifications have been made to the original model, as described previously.

Statistical Assumptions

There are two important issues related to statistical assumptions in SEM applications: sample size and distribution characteristics. First, SEM is a parametric statistical methodology grounded in large-sample theory. The goal is to draw inferences to a large, but usually finite, population based on estimates from a sample obtained from that population. Because SEM is a large-sample-based technique, deciding an appropriate sample size for a study is always a practical concern that depends on a number of factors, including data distributions and model complexity. However, there is no clear rule on sample sizes governing the use of SEM. Small sample sizes (less than 200) are likely to yield unreliable results (e.g., negative variances) but avoid problems related to sensitivity of the chi-square statistic. Large sample sizes (greater than 1,000), in contrast, are favorable to the underlying statistical assumptions but certainly make the model easy to reject. As a rough rule, it is generally recommended that minimum sample sizes of at least 200 be considered (Boomsma, 1985; Boomsma & Hoogland, 2001). Alternatively, sample sizes can be assessed through conducting power analyses (MacCallum, Browne, & Sugawara, 1996; Muthén & Muthén, 2002). Another frequently used rule in SEM applications involves the ratio between sample size and the number of parameters (to be estimated) in the model (Bentler & Chou, 1987; Bollen, 1989). The general rule of thumb on the use of this ratio is that it not be less than 10:1 (Bentler, 2000; Kline, 1998). Results from a recent study appear to indicate that higher values of observations per parameter ratio have a positive effect for some measures of model fit (Jackson, 2003). It is recommended, however, that the sample size issue be considered with other important issues such as reliability of indicators, number of indicators per latent variable, and multivariate normal distributions of variables.

The second statistical assumption associated with the use of SEM requires normality of observed variables. This means that the observed, or measured, variables be continuous and follow univariate normal distributions or a multivariate normal distribution. In practice, many SEM applications involving real data in behavioral research often violate these assumptions (Micceri, 1989), and some researchers fail to examine whether these assumptions have been violated (Breckler, 1990). Thus, there has been growing interest in determining the robustness of SEM techniques to violations of the normality assumption and in developing alternative remedial strategies when this assumption is seriously violated (West, Finch, & Curran, 1995). For example, there is some evidence that parameter estimates remain valid even when the data are nonnormal, whereas standard errors do not (Satorra & Bentler, 1994). A number of simulation studies suggest that maximum likelihood and generalized least squares estimation methods can produce biased standard errors and incorrect test statistics in the presence of excessive skewness and/or kurtosis in the data (West et al., 1995). Some SEM programs have provided remedies for addressing nonnormality problems, such as estimation procedures based on asymptotically distribution-free methods (Browne, 1984), but they come with a cost—a sample size of at least 1,000 is needed to obtain reliable weight matrices (West et al., 1995). A promising alternative is the use of the Satorra-Bentler chi-square statistic (S-B χ^2, Chou & Bentler, 1995; Satorra & Bentler, 1994). This statistic serves as a correction for the chi-square statistic when distributional assumptions are violated. Another alternative to handling the presence of multivariate nonnormal data is a procedure known as "the bootstrap" (Arbuckle & Wothke, 1999; West et al., 1995; Young & Bentler, 1996). With bootstrapping, empirical standard errors for parameters of interest in SEM can be generated without having to satisfy the assumption of multivariate normality (see Byrne, 2001, for an application).

Model-Fit Evaluation

Conventional practice in SEM relies on using chi-square statistics to judge model fit, incorporating probability as characterized by type I and type II errors. Because the goal is to test a model that fits the observed data, a nonsignificant chi-square is desired. It is well known, however, that chi-square values depend directly on sample size, making the model easy to reject under large sample size conditions (see Hu & Bentler, 1995). For this reason, Jöreskog (1969) recommended that the chi-square statistic be used more as a descriptive index of fit rather than as a statistical test. Accordingly, many researchers in SEM have recommended the use of alternative model-fit indexes (Bollen & Long, 1993). Commonly recommended indexes include (a) the nonnormed fit index (NFI; also known as the Tucker & Lewis index [TLI; Tucker & Lewis, 1973]), (b) the comparative fit index (CFI; Bentler, 1990), (c) the root mean square error of approximation (RMSEA; Steiger, 1990), and (d) the standardized root mean square residual (SRMSR). Hu and Bentler (1999) suggested the following fit-index cutoff value guide for good fitting models: TLI > .95, CFI > .95, RMSEA < .06, and SRMSR < .08. However, the model may be rejectable on other grounds such as producing nonsensical, out-of-bounds parameter estimates (e.g., negative variances or correlations over 1.00), or poorly explained variance.

Missing Data Problems

Possibly the most difficult estimation problem the researcher faces is that of missing data. This is particularly salient in longitudinal studies. The longer the study, the greater the probability of cases being incomplete or missing as a result of differential attrition and death. Therefore, researchers usually must deal with missing data problems, even when they take appropriate measures to avoid them. Incomplete data are often dealt with by listwise or pairwise deletion methods, which omit entire records, or pairs of variables containing missing values. The conventional listwise deletion approach for treating incomplete data can dramatically reduce the sample size available for analysis. Similarly, pairwise deletion can lead to input matrices that behave poorly in statistical terms. Thus, analysis based on listwise and pairwise solutions can be both inefficient and potentially biasing (Brown, 1994; Little & Rubin, 1987).

Little and Rubin (2002) provided a thorough discussion of the issues related to assumptions, estimations, and imputations of missing data. Within the SEM framework, the multisample maximum likelihood estimation procedure is a well-known method (Allison, 1987; McArdle, 1994; Muthén, Kaplan, & Hollis, 1987; Wothke, 2000) that determines whether patterns of covariation are similar for subgroups with different missing data structures. When the number of types of missing data are small and few distinctive patterns of "missingness" can be identified, the multisample approach can be useful. If, however, the number of subsamples is large and the sample size for each group is too small to estimate a positive definite observed covariance matrix (which happens if the number of variables exceeds the number of subjects), this approach may break down. For this reason, several alternative methods, including direct maximum likelihood methods (e.g., Arbuckle, 1996) and multiple imputation methods (Little & Rubin, 1987), are proposed. It is generally recommended that likelihood-based procedures (Schafer & Graham, 2002) be used in the presence of missing data.

At present, a number of SEM computer programs (discussed next) can perform direct (or full information) maximum likelihood estimation using both complete and incomplete cases or data points. This method does not place a limit on the number of missing data patterns, and does not require the user to take elaborate steps to accommodate missing data. Users, however, should be aware that all of these programs assume that data are missing at random (commonly known as MAR) in the sense defined by Little and Rubin (1987, 2000).

SEM Software

Currently, several commercial SEM software programs are available, with some of the most well-known being (a) Amos (Arbuckle & Wothke, 1999; www.smallwaters.com/index.html), (b) EQS (Bentler, 2000; www.mvsoft.com), (c) LISREL (Jöreskog & Sörbom, 1996; www.ssicentral.com), and (d) M*plus* (Muthén & Muthén, 1998-2004; www.statmodel.com). Each of these programs is available in a free student version that can be downloaded from the developer's Web site and used to test simple models that involve a limited number of observed variables. Other commercial packages include CALIS (Hartmann, 1992), RAMONA (Browne, Mels, & Cowan, 1994), and SEPATH (Steiger, 1995). Available freeware includes

COSAN (Fraser & McDonald, 1988) and Mx (Neale, Boker, Xie, & Maes, 1999; www.vcu.edu/mx).

SEM programs share many basic features, but each is unique in terms of its programming details and its functional capacity for handling more complicated models and data structures. For example, EQS (version 6), unlike other programs, provides a variety of robust test statistics (Gold, Bentler, & Kim, 2003). At the time of this writing, M*plus* is the only one to offer mixture modeling (a combination of modeling continuous and categorical variables) and random slopes models of either continuous or categorical outcome. However, in general, any of the software packages can handle most research problems in exercise science.

SAMPLE APPLICATIONS

The following sections provide examples of two SEM applications (a measurement model and a structural equation model) that are considered most applicable to many research problems in exercise science. The sample data, along with selected computer program output, are available on the Internet for downloading at http://healthyaging.ori.org/SEM/intro.htm. The reader is referred to articles by Boomsma (2000); Breckler (1990); Hoyle and Panter (1995); MacCallum and Austin (2000); McDonald and Ho (2002); Raykov, Tomer, and Nesselroade (1991); and Steiger (1988) on how to report details of SEM applications.

Measurement Model Example

In this example, we consider a measurement model with four measurement indicator variables and a single latent variable.[3] These indicators come from a scale designed to measure levels of confidence of older persons in performing activities of daily living while avoiding falls. The measurement question is whether the four observed (item) indicators adequately measure the underlying latent construct—fall confidence. The four indicators of the hypothesized construct form the measurement model.

The analysis is conducted following the steps outlined earlier. Figure 9.2 shows the specification of this model, which represents the hypothesized measurement model of fall confidence. The text at the bottom of figure 9.2 shows the content of the scale items used. This model has four observed variables (y_1 through y_4) and one latent variable named fall confidence. In addition, there are four measurement error variables (e_1 through e_4) in the model, with sources of error assumed to be either measurement-related error or item-specific variance. These errors are not allowed to covary (i.e., they are independently distributed per "classical" assumptions).

The next step involves checking model identification. Recall that a model is considered identified if there is a unique numerical solution for each of the parameters in the model. Model identification can be assessed through the use of covariance algebra to calculate equations (Bollen, 1989; Davis, 1993; Long, 1983; Rigdon,

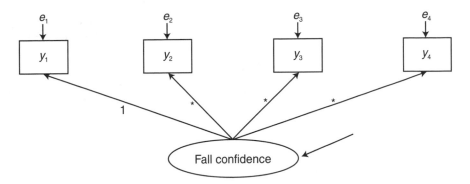

Item content of the fall confidence scale:
y_1: Walk up or down a ramp
y_2: Walk in a crowded mall where people rapidly walk past you
y_3: Are bumped into by people as you walk through the mall
y_4: Walk outside on icy sidewalks

Figure 9.2 Measurement model for a single latent construct—fall confidence.

1995). For the sake of simplicity, we use the easiest test—the t-rule (Bollen, 1989, p. 93)—which holds that the number of nonredundant elements (data points) in the covariance matrix of the observed variables must be equal to or greater than the number of unknown parameters in the model to be estimated.[4] The number of data points refers to the number of sample variances and covariances and is calculated as

$$p(p + 1) / 2, \tag{9.1}$$

where p equals the number of observed variables. Applying this simple rule to the model shown in figure 9.2, we have a total of four measured variables, p. Thus, we obtain 10 data points in the covariance matrix: $4(4 + 1) / 2 = 10$ (i.e., four variances, six covariances). There are eight parameters to be estimated in this hypothetical model (three factor loadings [indicated by asterisks], four unique variances $[e_1 - e_4]$, and one factor variance [unattached arrow]). Note that one of the loadings is fixed (fixed parameter; denoted by 1 in the figure; refer to the explanation in the next paragraph), leaving three to be estimated. Therefore, the model has two fewer parameters than data points, so the model is considered identified using the t-rule.

As an additional step for model identification, the scale of the latent factor needs to be established because latent variables are unobservable and have no inherent scale (or unit of measurement). In this case, the metric of the latent factor (fall confidence) is set by fixing its loading on the first item (the reference variable) to be 1.0. This reference (item scale) is measured on a 10-point scale, and thus the factor is in the metric of a 10-point scale. Alternatively, we can standardize the latent variable so that it has unit variance in the population (see Jöreskog & Sörbom,

1996, p. 339, for a full explanation). The remaining three factor loadings (indicated by the three asterisks) are considered "free" parameters and are estimated by the program. It was hypothesized that variation among the four items is explained by a latent variable construct—fall confidence.

After identification has been completed, model estimation can proceed. As we alluded to earlier, there are several common estimation procedures (or fitting functions) in SEM, including maximum likelihood (ML), generalized least squares (GLS), and asymptotic distribution free (ADF) methods (Bollen, 1989). In this chapter we will focus on the most common one, the ML method, which is the default estimation procedure in most SEM software.

Following common practice in SEM, the covariance matrix of the scale items is used as data input for estimation. The model is estimated using M*plus* with ML estimation, which results in a chi-square value of 1.848 with 2 degrees of freedom (*df*) ($N = 214, p = 0.393$), indicating that the model fits the data very well. Other goodness-of-fit indexes also indicate good support for the model: TLI = 1.00, CFI = 1.02, RMSEA = 0.00, SRMSR = 0.02. The *df* for the chi-square distribution is given by

$$df = q(q + 1) / 2 - t, \qquad (9.2)$$

where q is the number of measured variables and t is the number of estimated (free parameter) values. Thus, in the preceding example, the chi-square is evaluated with 2 *df* (i.e., 4[4 + 1] / 2 − 8 = 2), $\chi^2(2, N = 214) = 1.848, p = 0.393$.

As a further step in model evaluation, the relationships between observed indicators and the latent factor (fall confidence) are examined. This is accomplished by examining the statistical significance of each parameter in the model, which is determined by the *t*-values (defined by the ratio of the parameter estimate and its standard error). Values ≥ 1.96 are generally considered to be statistically significant. Inspection of the factor loadings indicated statistical significance for all estimates ($p < .001$). Both the unstandardized and standardized regression coefficients (loadings) for the model are presented in table 9.1. Because the first loading is fixed to 1.0 for latent variable scaling purposes, a standard error is not calculated for that estimate.

The quality of the four measurement indicators can be judged by looking at the squared multiple correlation, R^2 (squaring the standardized factor loadings), with its values ranging from 0.0 to 1.00. The R^2 provides an indication of the reliability of each observed measure. In this example, R^2 values for the second and third observed variables (0.593 and 0.745, respectively) are reasonable. However, less than 50% of variance in the first item ($R^2 = .31$) and the fourth item ($R^2 = .25$) is accounted for by the latent variable, indicating a high amount of variance that is not explained by the latent variable. This is an example of a model that fits almost perfectly to the empirical data but with certain low-quality parameter estimates.

To look into this further, we test whether the four items are equally loaded on the latent variable. This is a test of tau-equivalent measures, where all items have equal true-score variance but different error variances (Jöreskog & Sörbom, 1996).

Table 9.1   Selected Parameter Estimates Derived From the Measurement Model

Observed indicator	Unstandardized factor loading	Standard error	Unique variance	Standardized factor loading	R^2
y_1	1.000[a]	0.000	2.214	0.559	0.312
y_2	1.350*	0.178	1.255	0.770	0.593
y_3	1.681*	0.222	0.971	0.863	0.745
y_4	1.425*	0.247	6.220	0.497	0.247

[a]This factor loading is purposely fixed to 1.00 for model identification purposes.
*Factor loadings are statistically significant at $p < .001$ (i.e., t-value ≥ 1.96).

The test is accomplished by constraining all four of the factor loadings to be equal. This model is considered nested in the unconstrained model; that is, the unknown parameters in the constrained model are a subset of the unknown parameters in the unconstrained model. The constraint is evaluated using a chi-square difference test, which provides a statistical test of whether the constraints that produce the **nested models** are justified. The chi-square difference test is calculated by subtracting the chi-square and *df* associated with the less restricted model (in this case, the original model) from the chi-square and *df* associated with the more restricted model (the constrained factor-loading model). A significant chi-square relative to the difference in *df* means that the constraints that produce the more restricted model are not justified.

The resulting model with constraints produces a chi-square value of 6.72, $p = 0.15$ with 4 *df*, TLI = 0.98, CFI = 0.99, RMSEA = 0.056, and SRMSR = 0.048. The chi-square difference that results from comparing this model to the model with no constraints is not statistically significant, $\chi^2(df = 2, N = 241) = 4.872$, $p = 0.088$, leading to a nonrejection of the tau-equivalent hypothesis at the 0.05 α-level. The conclusion is that the four measurements are tau-equivalent (i.e., equal true score variances, but possibly different error variances).

Judging from the overall model-testing results and their parameter estimates, we conclude that the original model fits the empirical data satisfactorily. Therefore, no attempt is made to modify the model. It should be noted that various measurement issues can be addressed within this framework. For example, a measurement model may involve multiple latent constructs in which within-construct and between-constructs measurement properties can be evaluated (e.g., Li, 1999), and in which the question of how well a higher-order (e.g., second-order) factor structure accounts for the lower-order factors (e.g., Li & Harmer, 1996). A measurement model may also be used to assess the comparability of factor structures (factor loadings, factor variance or covariance, error variance) across groups (e.g., Marsh, 1993).

Six key points to take from this section are as follows:

1. A minimum of 200 observations is recommended for SEM analysis.
2. Always use the covariance matrix as input for analysis.
3. Use the t-rule to check for model identification. In some situations, models are known to be identified—for example, one factor models with three observed indicators, or models with two correlated factors each with two indicators.
4. Use multiple model-fit indexes for model evaluation.
5. R^2 values of ≥ 0.50 on factor loadings are desirable.
6. Keep observed variables on a similar scale to facilitate model estimation.

Structural Equation Model Example

We now turn to a structural equation model consisting of a measurement model involving a set of observable variables as multiple indicators of a smaller set of latent variables and a structural path model describing relationships among the latent variables. Data come from an exercise intervention trial examining falls and functional ability in the elderly (Li et al., 2005). We use the data to describe relationships among four latent variables, each measured by two observed indicators. Thus, in this example, we have two exogenous latent variables (functional balance, fall confidence) and two endogenous latent variables (instrumental activities of daily living, quality of life).[5] This is shown schematically in figure 9.3.

The research question in this hypothetical example is whether functional balance (balance) and fall-related confidence (fall confidence) are related to the abil-

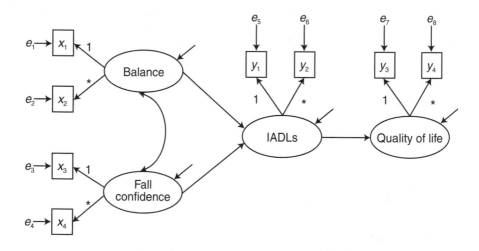

Figure 9.3 Structural equation model with latent variables on quality of life.

ity to perform instrumental activities of daily living (IADLs) and whether IADLs are associated with health-related quality of life (quality of life). The model has eight observed variables (x_1 through x_4 [x-variables are used to define the latent exogenous variables] and y_1 through y_4 [y-variables are used to define the latent endogenous variables]). The variables of x_1 and x_2 are observed variables defining the independent (exogenous) latent variable (balance), and x_3 and x_4 are observed variables defining the second exogenous latent variable (fall confidence). Similarly, the variables of y_1 and y_2 are observed variables defining the dependent (endogenous) latent variable (IADLs), and y_3 and y_4 are observed variables defining the second endogenous latent variable (quality of life). As was the case previously, the metric of the factors is set by fixing the first factor loading in each construct to 1.0. In addition to the structural paths indicated in the model, the covariance (correlation) between the two exogenous latent variables and the unique variances (e_1 through e_8) in the model are also estimated. The model is known to be identified because the number of parameters to be estimated is less than the number of known data points.

Following recommendations by Anderson and Gerbing (1988), testing is conducted in two steps, the measurement model and the structural model, using M*plus* with ML estimation. The measurement model fits the data well, resulting in a chi-square value of 18.399 with 14 df ($N = 214$), $p = 0.189$, TLI = 0.988, CFI = 0.994, RMSEA = 0.038, and SRMSR = 0.026. All factor loadings are statistically significant and in the hypothesized direction. All loadings, with the exception of y_4 ($R^2 = 0.18$), have an R^2 value greater than 0.50. The subsequent structural model testing yields a chi-square value of 22.623 with 16 df ($N = 214$), $p = 0.124$. The 16 df derives from the fact that there are 20 parameters estimated in the model: eight unique variances, four factor loadings, two variances for the exogenous latent variables, three path regression coefficients, one covariance between the two exogenous latent variables, and two residual variances (one for each of the endogenous latent variables). The model has 16 fewer parameters than data points, which total 36 (8[8 + 1] / 2). Other indexes also support the fit of the model to the data: TLI = 0.991, CFI = 0.984, RMSEA = 0.044, SRMSR = 0.035. Parameter estimates for the relationship among the latent variables are presented in the bottom section of table 9.2. Results indicate that both balance and fall confidence are related significantly to IADLs, which in turn are related to quality of life. The R^2 values indicate that balance and fall confidence explain 65.4% of the variance in the construct IADLs, which, in turn, account for 25.5% of the variance in the construct quality of life.

It is important to note that plausible competing models need to be considered (Hoyle & Panter, 1995; MacCallum, 1995). For instance, in the preceding example, it is possible that both balance and fall confidence are directly related to quality of life. To examine this possibility, a model including the two omitted paths (the path from balance to quality of life and the path from fall confidence to quality of life) should be tested against the original model using the chi-square difference test.

Table 9.2   Parameter Estimates for the Structural Model on Quality of Life

Measurement model	Unstandardized factor loading	Standard error of estimate	Standardized factor loading	T value	P value
x_1 Balance	1.000[a]	0.00	0.711	NA	
x_2 Balance	1.421	0.234	0.912	6.066	< 0.001
x_3 Fall confidence	1.000[a]	0.00	0.847	NA	
x_4 Fall confidence	1.234	0.095	0.875	13.153	< 0.001
y_1 IADLs	1.000[a]	0.00	0.893	NA	
y_2 IADLs	0.755	0.055	0.830	13.658	< 0.001
y_3 Quality of life	1.000[a]	0.00	0.904	NA	
y_4 Quality of life	0.742	0.223	0.433	3.253	0.001

Structural model[b]	Unstandardized path coefficient		Standardized path coefficient		
Balance → IADLs	0.478	0.245	0.128	1.960	0.05
Fall confidence → IADLs	0.666	0.068	0.747	9.818	< 0.001
IADLs → quality of life	0.655	0.098	0.505	6.683	< 0.001
R^2 IADLs	0.654				
R^2 quality of life	0.255				

[a]This factor loading is purposely fixed to 1.00 for model identification purposes.
[b]Refer to the model shown in figure 9.3.

Many additional research questions can also be pursued through SEM. For example, hypotheses about group differences in a population with respect to the magnitude and direction of relationships among variables across various known groups can be tested using a multiple group comparison approach (Jöreskog, 1971). This may involve estimating the same model structure for two or more groups and testing the equality of estimates of particular structural parameters in the different groups. In the previous quality of life example, one may wish to test whether the hypothesized structural paths in the model are tenable across, for example, gender. Examples of applications can readily be found in the social science literature and increasingly in the exercise science literature.

Key points to remember from this section are as follows:

1. The same key points as for measurement models apply.

2. Establish an appropriate measurement model before examining structural relationships in the structural model.

EXTENDED TOPICS

The remainder of this chapter touches briefly on some of the recent advances in SEM, including growth curve modeling, growth mixture modeling, multilevel modeling, and modeling interactions among latent continuous variables.[6] Because these topics are considered more advanced, we offer only a light walk-through by providing a general idea of each topic, its analytic relevance to research questions, and samples of applications.

Growth Curve Modeling

Traditionally, longitudinal data are analyzed using the repeated measures analysis of variance (ANOVA) model or some variation of it. However, this approach has various limitations arising from the group differences viewpoint of all ANOVA models, in which a group mean depicts an overall tendency in the sample data and variation around the group mean is considered error. This orientation is not consistent with the goal of researchers interested in change who seek to model the very individual differences in change or growth that the repeated measures ANOVA relegates to the error mean square. There are other restrictions in this methodology as well, such as the requirement for equal temporal spacing of observations and time-structured data for all individuals.

Growth curve models extend ANOVA models by combining elements of repeated measures ANOVA, CFA, and SEM to analyze changes in a construct or outcome over time. In particular, this modeling approach takes into account both means and variances, thus describing both the group-level change and individual variation in change over time (McArdle & Epstein, 1987; Meredith & Tisak, 1990; Muthén, 1991). The method is therefore useful for researchers who are interested in assessing within-person and between-persons change, and the question of the shape of the developmental trajectory or the rate of change in the person over time. Willett and Sayer (1994) outlined details of how issues of change can be addressed using the analysis of means and covariance structures within the general SEM framework. The key feature of this approach to investigating change is that it fuses within-person and between-person models of individual growth within the same structural framework. The reader is referred to Duncan, Duncan, Strycker, Li, and Apert (1999) for a general, "low-tech" introduction to latent growth curve analysis (see also chapter 10 by Ilhyeok Park and Robert Schutz); to Byrne and Crombie (2003); to Li, Duncan, McAuley, Harmer, and Smolkowski (2000) for a full exploration of a growth model involving time-varying and time-invariant covariates; and to Li, Duncan, Duncan, McAuley, and colleagues (2001) for an example of the application of a growth curve model in the context of experimental studies.

Growth Mixture Modeling

In growth curve analysis, questions about the rate of change and its variability are explored. With respect to variability, the growth curve analysis models heterogeneity corresponding to different growth trajectories across individuals and captures that by variation in the continuous growth factors (e.g., intercept, slope). However, the method cannot capture heterogeneity that corresponds to qualitatively different development (e.g., individuals classified as having no change, little change, or significant change in levels of physical activity) (Muthén, 2001). When group membership is known (e.g., gender), multisample SEM can be used to examine qualitatively different development across individuals. However, when group membership is unknown but needs to be inferred from the data, the growth mixture modeling methodology can be applied. Fundamentally, this technique combines several conventional models, such as growth curve modeling previously discussed, latent class analysis (Clogg, 1995), latent transition analysis (e.g., Collins, Graham, Rousculp, & Hansen, 1997), and latent class growth analysis (Nagin, 1999), into a single statistical framework labeled general growth mixture modeling (Muthén, 2001).

The growth mixture modeling approach, introduced by Muthén and Shedden (1999), offers researchers a new perspective on the issue of change, particularly with respect to population heterogeneity in change over time. In essence, growth mixture modeling relaxes the single population assumption (as in conventional growth curve modeling) to allow identification of qualitatively different classes of individuals across unobserved subpopulations, (what are referred to as latent trajectory classes; i.e., categorical latent variables). Thus, instead of considering individual variation around a single mean growth curve, growth mixture modeling allows different classes of individuals in the study population to vary around different mean growth curves. The combined use of continuous and categorical latent variables in a single analytic model provides a very flexible analysis framework that is capable of discerning and testing hypotheses about clusters of developmental trajectories (Muthén, 2004).

Currently, M*plus* is the only SEM program that has the capacity for growth mixture modeling. Interested readers should consult Muthén (2001, 2003, 2004), Muthén and Shedden (1999), and Bauer and Curran (2003). Li, Duncan, Duncan, and Acock (2001) provided a full exploration of growth mixture modeling, whereas Li, Fisher, Harmer, and McAuley (2002) provided an example of its use in the context of an experimental study.

Multilevel Modeling

In exercise science, it is common practice to analyze data at the individual level. However, with increasing attention to social and physical environment influences on health behavior in the physical activity literature (e.g., Bauman, Sallis, & Owen, 2002; Li et al., 2005), it is important that data be observed from more than a single source (e.g., people nested in families in neighborhoods). This leads to multilevel research in which a model may contain information from a set of variables characterizing individuals and a set of variables characterizing groups or clusters. However, if data are hierarchically arranged or nested (e.g., units at one level of observation are nested within units at another, such as students nested within classes within schools) and variables are measured at each level, analysis using conventional statistical modeling techniques (e.g., ordinary least squares regression) is not appropriate because these techniques require that all observations be independent and identically distributed. Failure to consider this basic assumption will result in biases in parameter estimates, standard errors, and associated tests of parameter significance. Therefore, if the data are hierarchically structured, multilevel analysis techniques should be considered to examine both within- and between-level relationships. In brief, multilevel analysis simultaneously examines groups (or samples of groups) and individuals within them (or samples of individuals within them). Variability at both the group level and the individual level can be examined, and the role of group-level and individual-level constructs in explaining variation among individuals and among groups can be investigated within a single framework (Diez-Roux, 2003). Interested readers should refer to chapter 10 by Ilhyeok Park and Robert Schutz and the work by Muthén (1994, 1997) for more details on multilevel analysis within the SEM framework. Examples of SEM applications in exercise and physical activity research can be found in papers by Li and his colleagues (Fisher, Li, Michael, & Cleveland, 2004; Li & Fisher, 2004; Li, Fisher, & Brownson, 2005).

Modeling Interaction Among Latent Variables

Many researchers in exercise science fields deal with complex models involving interaction effects that are typically evaluated through multiple regression methodologies (Aiken & West, 1991). However, the traditional approach has been criticized for introducing measurement errors in regression coefficients and lowering the power of statistical tests for interactions effects (Jaccard & Wan, 1995, 1996). By using multiple indicators and SEM, much of the concern related to measurement errors and power issues can be eased. Unfortunately, many procedures developed in the 1980s (e.g., Kenny & Judd, 1984) were difficult to apply because they required complicated parameterization. More recent advances in SEM have made the estimation of nonlinear models more accessible to applied researchers. There are currently two classes of estimation methods: (a) full-information maximum likelihood (e.g., Jaccard & Wan, 1996; Jöreskog & Yang, 1996) and (b) limited-information two-stage least squares (Bollen, 1995). These techniques are applicable in situations in which both interacting variables are continuous (see Algina & Moulder, 2001; Jaccard & Wan, 1996; Schumacker & Marcoulides, 1998). However, when one or both of the interacting variables is discrete, a "multisample" approach can be applied in a straightforward fashion (Rigdon, Schumacker, & Wothke, 1998). With this approach, the different samples are defined by the different levels of one or both of the interacting variables. If interaction effects are present, then certain parameters should have different values in different samples. Applications of SEM with interaction among latent variables are rare in exercise science. A didactic example of using maximum-likelihood-based methods with exercise data can be found in Li and colleagues (1998), and a similar example of using the two-stage least squares method can be found in Li and Harmer (1998).

SUMMARY

The goal of this chapter was to introduce, in a nontechnical way, the basic concepts, applied issues, and applications of various structural equation models. Although we acknowledge that numerous SEM books and articles are available for the social sciences, few have been considered accessible to applied researchers or students. This chapter is intended to bridge this gap by presenting basic and extended latent variable models, including conventional structural equation models and more complex ones, such as latent variable interaction models. Like information technology, SEM is an ever-evolving methodology with advances occurring on a yearly basis. In this regard, we strongly encourage readers to check SEM software Web sites for updates and upgrades to keep pace with new developments in this area.

Notes

1. In contrast to the measurement model, in this chapter, uppercase letters are used to identify observed variables in the path model and latent variables in the structural model. In applied research, descriptive names are used to represent each variable or construct.

2. SEM can be used to analyze noncontinuous data such as dichotomous and ordered categorical variables (indicators) in addition to continuous variables (see Muthén, 1984). Both LISREL and M*plus* SEM software provide this feature.

3. Both the raw data and computer output for this example are available on our Web site (http://healthyaging.ori.org/SEM/measurement/cfa.htm).

4. In general, models that have more unknown parameters than pieces of information are called underidentified models. Suppose that four variables are measured and that a model is to be fitted to the observed covariance matrix. Because this matrix contains $4(4 + 1) / 2 = 10$ nonduplicated ele-

ments, the model will not be identified if the number of parameters to be estimated is more than 10. Models with just as many unknowns as pieces of information are referred to as just-identified models. Models with more information than unknowns are called overidentified models (or identified models). If, for example, a one-factor model with a diagonal error matrix is fitted to the data, the number of unknown parameters is eight (the four factor loadings and the four error variances), and it can be shown that this model is identified. The interested reader is encouraged to consult Bollen (1989) on this topic.

5. Details of the computer output generated by M*plus*, LISREL, Amos, and EQS, as well as the raw data used for this example, can be found at http://healthyaging.ori.org/SEM/structural/sem.htm.

6. We have provided useful references on these topics at http://healthyaging. ori.org/SEM/reference/references.htm, as well as examples of these techniques at http://healthyaing.ori.org/SEM/intro.htm.

CHAPTER 10

Repeated Measures
and Longitudinal Data Analysis

Ilhyeok Park and Robert W. Schutz

Repeated measures and **longitudinal data analyses** encompass an extremely large body of knowledge and constitute some aspect of the majority of research studies published in the exercise science literature. With respect to research design, repeated measures includes everything from the simple pretest–posttest experiment, to longitudinal growth studies consisting of 5 to 20 measurement points, to an EMG recording comprised of hundreds or thousands of sequential measures. The statistical analysis of longitudinal data is even broader in scope; it includes the correlated t-test, repeated measures (RM) analysis of variance (ANOVA) or multivariate analysis of variance (MANOVA), time series analysis, stochastic models, curve fitting, latent curve models, and hierarchical linear models, to name a few. In this chapter we make no effort to summarize this vast field, choosing instead to focus on a few topics. First, we have selected the most commonly used method for analyzing RM in experimental studies, ANOVA, and we introduce two recent developments, namely post hoc tests for RM factors and power for RM designs. We also review the long-standing issue of MANOVA versus ANOVA for RM designs. Second, we introduce and compare two similar models for analyzing longitudinal data that are quite different from ANOVA: **hierarchical linear models** (HLM) and **latent curve models** (LCM).

Statistical methods for the analysis of repeated measures and longitudinal data have advanced rapidly over the last decade. The most notable aspect of this advance is the increasing number of applications of structural equation models (see chapter 9 by Fuzhong Li and Peter Harmer) and multilevel models (e.g., Collins & Sayer, 2001; Gottman, 1995; Moskowitz & Hershberger, 2002). Although the traditional methods such as ANOVA and time-series models focus on analyzing one of either interindividual variation or intraindividual variation, recent methods use all the available information from both interindividual and intraindividual variations. As a result of this advance, researchers now are able to examine the shape of the change at the individual subject level as well as at the group level, the interindividual variation in change, the relationship between the change and other relevant variables (e.g., predictors of change), the change of a latent construct, and more. Scholars in exercise science have already started applying these advances in their research (e.g., Duncan & Duncan, 1991; Schutz, 1998; Zhu & Erbaugh, 1997).

ANALYSIS OF VARIANCE FOR REPEATED MEASURES DESIGNS

The research questions posed by exercise and sport scholars frequently involve the component of change over time (e.g., learning, aging, growth, acceleration). The most common method of statistical analysis for data arising from such studies is the repeated measures analysis of variance (RM ANOVA). This is especially evident in experimental studies. Indeed, a review of papers published in the *Research Quarterly for Exercise and Sport* and the *Journal of Exercise and Sport Psychology* for the 3-year period 1998 to 2000 revealed that over 50% of experimental studies (studies that employ any kind of treatment) involved RM ANOVA as an analysis procedure. Researchers appear to be quite competent in conducting and interpreting the basic components of such analyses; however, three questions are often raised with respect to specific issues:

- Should one use the MANOVA or ANOVA to analyze RMs, and with multiple dependent measures should one use a MANOVA or multiple ANOVAs?
- How can one estimate sample size for a given power, a priori, for repeated measures designs?
- What are the most appropriate post hoc procedures for following up an RM ANOVA?

MANOVA or Multiple ANOVAs?

Most of us are aware of the inflation in the **familywise (FW) error rate** (referred to as the experimentwise error rate in older texts) brought about by conducting a *t*-test or *F*-test (ANOVA) on each one of a large number of dependent variables. A possible, and commonly used, solution to this problem is to conduct **multivariate tests,** such as Hotelling's T^2 and MANOVA instead of multiple *t*-tests and multiple ANOVAs. These multivariate procedures purport to provide an initial overall test of the equality of the mean vectors at some probability level. If the probability value associated with our multivariate test statistic is sufficiently small (e.g., < .05), we reject the null hypothesis of equal cell (trials, conditions) means for the set of dependent variables and proceed with some type of univariate procedure for each dependent variable. Which procedure we use as a follow-up to the multivariate test is a point of some debate, but univariate ANOVAs or *t*-tests, step-down *F*, simultaneous confidence intervals, and discriminant analysis are some of the choices. Most multivariate statistics texts cover this issue (Tabachnick & Fidell, 2000, is probably the most readable).

Not that long ago the choice between multiple **univariate tests** and one multivariate test was a nonissue for most researchers and applied statisticians. When possible (theoretically justifiable and with adequate sample size), the multivariate procedure was recommended for the purposes of data interpretation and to reduce the problem of "probability pyramiding" inherent in multiple t-tests or ANOVAs (Schutz & Gessaroli, 1987). However, that philosophy was challenged 17 years ago and is gradually giving way to a less automatic use of MANOVA. The essence of the case in support of multiple univariate analyses over a multivariate analysis is well presented by Huberty and Morris (1989). They reviewed the 1986 issues of six psychological journals and identified 222 articles that used multiple ANOVAs, MANOVA, or both, as the primary analysis. The majority, 59%, used multiple ANOVAs without a preliminary MANOVA, and 40% of the studies applied a preliminary MANOVA before using ANOVAs as a follow-up. Huberty and Morris asserted that the "MANOVA-ANOVAs approach is seldom, if ever, appropriate" (p. 302), and that the type I error rate is not fully controlled by first conducting an overall MANOVA. When control of the FW error rate is desirable (and we think it always is), Huberty and Morris suggested using either a simultaneous test procedure or some type of error-splitting protocol (e.g., the Bonferroni procedure in which the per-comparison error rate used for each ANOVA is the FW error rate divided by the number of separate ANOVAs). In both cases the researcher is faced with a considerable loss of power for each dependent variable test, but that is the cost of minimizing the FW error rate. Huberty and Morris reminded us that the multivariate and univariate analyses address different questions, and the choice of which to use must be based on the nature of the research question. A common research question takes the form, Is there change over time, and do the groups change differentially, on one or more of these dependent variables? In this case the multiple univariate ANOVA approach is preferred. If, however, the primary research question is of the form, What combination of these dependent variables distinguishes the groups with respect to the amount or pattern of change, and which variables contribute most to the between-group variance? then a preliminary MANOVA is appropriate. The follow-up analyses in this case should be a **discriminant analysis,** with the F-to-remove or the standardized discriminant function weights providing an indication of the relative importance of the dependent variables.

A second situation in which MANOVA is preferable to ANOVA is when the data analyst has a choice of treating the RM as univariate or multivariate. The usual method is to treat RM as univariate measures and perform an RM ANOVA when there is only one dependent variable, and an RM MANOVA (called a multivariate mixed model [MMM] MANOVA) when two or more dependent variables are analyzed simultaneously. Another approach is to treat the RM multivariately and thus avoid the restrictive assumptions of **sphericity** inherent in a repeated measures ANOVA. In this case the single dependent variable repeated measures design is analyzed using the "MANOVA method," and the multiple dependent variable repeated measures design is analyzed with a procedure referred to as the "doubly multivariate (DM)" MANOVA. Hertzog and Rovine (1985) and Schutz and Gessaroli (1987) provided extensive discussions and examples of these procedures, and Hand and Taylor (1987) provided a very thorough treatment of all aspects of MANOVA for repeated measures. Most computer software programs automatically provide both MANOVA and ANOVA test statistics in their repeated measures analyses.

Statistical Power With Repeated Measures ANOVA

Statistical power is the probability of obtaining a statistically significant test statistic, given alpha (α), sample size (n), and variance (σ^2), assuming some magnitude of true differences in the population means. Setting power and estimating the n needed to achieve this power a priori is an important part in the planning and formulation of experimental research. It allows researchers the opportunity to evaluate whether their experimental design has a sufficient probability of showing a statistically significant effect, thus warranting the time, money, and effort necessary to conduct the research. In light of the benefits of statistical power in experimental planning, a number of computer programs have been produced to aid researchers in estimating the n required to achieve a given power before conducting a study (e.g., Bradley, 1988; Erdfelder, Faul, & Buchner, 1996; Gorman, Primavera, & Allison, 1995). There are statistical tests, however, for which power analysis procedures have been less extensively described. This is particularly the case for complex RM ANOVA designs. Many of the programs noted earlier do not include options for computing n for the F-test in an RM ANOVA, and those that do, do so only for designs with a single RM factor (or for two or more RM factors under very restrictive conditions of equal correlational patterns for all factors). Although a considerable amount of research has been conducted on power analysis for RM ANOVA, much of this work has focused on comparing power values between univariate and multivariate RM tests (for a single-factor RM design) under varying conditions of nonsphericity (Muller, LaVange, Ramey, & Ramey, 1992).

In an RM ANOVA the power of the F-test of the repeated measures factor (and all its interactions) is a function of the magnitude of the correlations between the levels of the RM factor(s). These correlation coefficients can range from near-zero to near-one, depending on the nature of the study and the reliability of the dependent variable. Consequently, the power of the test can vary considerably from study to study, even with constant ns, magnitude of effect, and within-cell variances. Within a single study, if there are two RM factors, the power to detect a difference of, say, d units could be high for one factor and very low for the other.

To determine power for a univariate RM ANOVA, a **noncentrality parameter** must be computed based on the experimental conditions of the design. The noncentrality parameter, lambda (λ), a function of effect size and n, is a measure representing the factor by which the F ratio departs from the central F distribution when a difference between treatment means actually exists (Winer, Brown, & Michels, 1991). Although three other statistics are also used as noncentrality or effect size measures; ε^2 (Davidson, 1972), ϕ (Winer et al., 1991), and f (Cohen, 1988), all are closely related to λ, and like λ, essentially represent the magnitude of difference between population means in relation to the population error variance of the dependent variable involved. For within-subjects RM ANOVA designs with q levels on the only RM factor, the noncentrality parameter for the repeated measure factor can be computed as follows:

$$\lambda = \frac{n\sum(\mu_j - \mu)^2}{\sigma_e^2}, j = 1, \ldots, q, \tag{10.1}$$

where μ_j represents the marginal means for the levels of the RM factor, μ is the grand mean, n is the sample size, and σ_e^2 represents the error variance for the specific effect (Bradley, 1988; Winer et al., 1991). Potvin and Schutz (2000) provided formulas for computing λ for more complex designs.

As shown in equation (10.1), the numerator of λ for each test can be computed simply by inputting the desired n and an estimate of the means. Approximating the denominator of λ, the error variance for these tests, is often much more complex. The error variance (MS_{error}) is a function of σ^2, the common within-cell variance (i.e., the estimated average variance among scores of the dependent variable within a group), and $\bar{\rho}$, the estimated average of the $q(q-1)/2$ correlation coefficients among repeated trials (Winer et al., 1991). For one-way RM tests, and for both the RM main effect and the interaction test of the two-way mixed model, the expected error variance is

$$E(MS_{error}) = \sigma_e^2 = \sigma^2(1 - \bar{\rho}). \tag{10.2}$$

Thus, the larger the average correlations among the repeated measures, the smaller the MS_{error} and the greater the power for the F-test. For the group main effect test of a two-way mixed model, the relationship involves a third variable, q, the number of trials of the RM factor. In this case, the error term is expressed as

$$E(MS_{error}) = \sigma_e^2 = \sigma^2\left[1 + (q-1)\bar{\rho}\right]. \tag{10.3}$$

Note here that the larger the average correlation between the levels of the RM factor, the *larger* the error term for the grouping factor, and the lower the power for the grouping factor F-test. Figure 10.1 provides a graphic example of the effects, on power, of the correlations among the RMs (trials). When $\bar{\rho} = 0$, the power is the same for the group (G) and trial (T) effects (assuming equal magnitudes of differences among marginal means for the two factors). However, as $\bar{\rho}$ increases, the power increases dramatically for the T effect but decreases considerably for the G effect. For the G × T interaction the power is lowest at $\bar{\rho} = 0$, but because the F-test for the G × T effect uses the same error term as for the T effect, the power increases with an increase in $\bar{\rho}$.

Once λ (or any of its equivalents) has been determined, power can then be estimated for a specific RM ANOVA test by inputting λ values directly into software programs that include algorithms for computing the cumulative distribution function of the noncentral F distribution (e.g., DATASIM [Bradley, 1988]). For designs with

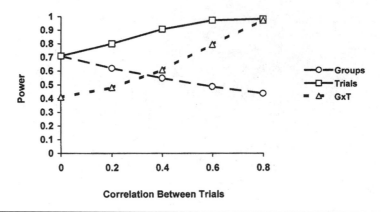

Figure 10.1   Effect of correlation on power of groups, trials, and G × T effects (all effect sizes are equal, $f = .33$ at $r = 0.0$).

one RM factor, an alternative method offered by some computer power programs (e.g., PASS [Number Cruncher Statistical Software, 1991]) is to input the conditions of the experimental design (p, q, n, means, σ^2, $\bar{\rho}$) directly and have the computer calculate values of λ and power automatically. A suitable estimate of power for the one-way RM ANOVA can be computed rather easily using a standard regression equation developed by Park and Schutz (1999).

When two or more RM factors are present in a design, a problem exists in determining λ, and more specifically, in computing its denominator term or error variance. In the case of the two-way RM ANOVA, three average correlation values are possible, one for each independent matrix of the design (within-factor A, within-factor B, and the AB submatrices). Until recently it was not clear how these values affect the error variance and thus the power of a particular test. Thus there are no computer programs that will compute power for designs with two RM factors. Potvin and Schutz (2000) developed equations for estimating error variance for the two-factor RM design, and established their accuracy through a series of simulations. Using their equations, researchers can now calculate λ and then use a program such as DATASIM (Bradley, 1988) to estimate power. Potvin and Schutz showed that, similar to what happens to the power of the group factor in the two-factor mixed model design, the greater the magnitude of the correlations among the levels of factor A, the larger the error variance for factor B (and thus the lower the power). Table 10.1 illustrates this effect.

The body of table 10.1 contains power values computed by using the Potvin and Schutz equations and DATASIM. The effect size used here (.50) is expressed in an easy-to-use form so readers can follow this table to estimate power for their own studies. Effect size is equal to the difference between the largest and the smallest marginal means divided by the average within-cell variance. Using $n = 20$ as an example, note that with correlation pattern (1) in table 10.1 the power is higher for A (.97) than for B (.74), but only because of the fewer number of levels for A

Table 10.1   Effect of the Magnitude of the Between-Trials Correlations on the Power of the A, B, and AB Effects for a 3(A) × 6(B) ANOVA With Repeated Measures on Both Factors (medium [.50] effect size and α = .01)

		Correlation pattern			
		(1)	(2)	(3)	(4)
	$\bar{\rho}$ for A:	.4	.4	.8	.8
	$\bar{\rho}$ for B:	.4	.8	.4	.4
	$\bar{\rho}$ for AB:	.4	.4	.4	.8
n	**Effect**				
10	A	.59	.09	.99	.99
	B	.29	.91	.09	.91
	AB	.02	.07	.07	.07
20	A	.97	.27	1.0	1.0
	B	.74	1.0	.26	1.0
	AB	.05	.22	.22	.22
30	A	1.0	.47	1.0	1.0
	B	.95	1.0	.47	1.0
	AB	.08	.42	.42	.42

(3 vs. 6 for factor B). In pattern (2), in which the correlations among the levels of B are high (.8), the power for B is very high (approaching 1.0 but the power for A is greatly reduced (down to .27). The reverse happens in pattern (3), in which the high average correlation among the levels of A raises its power to near 1.0 but has a negative effect on the power of B (down to .26). For more details on this phenomenon, see Potvin and Schutz, but the basic message is that researchers should examine the correlations of their RM factors and be aware of their differential effects on the power of these factors.

Post Hoc Comparisons for Repeated Measures ANOVA

The concepts of per comparison (PC) and FW type I error rates are well-known. Common practice is to minimize the FW error rate as much as possible while retaining a reasonable level of power—that is, do not set the PC α level too low. Thus we often use ANOVA and follow up a significant omnibus F-statistic with a post hoc multiple comparison procedure (for between-levels tests); this protects us against too high an FW type I error rate. We usually do this with one of the following tests: Tukey HSD, Newman-Keuls (now often referred to as the Student-Newman-Keuls procedure), or Scheffé (but, hopefully, never the very liberal Duncan Multiple Range Test). Alternatively, we may decide a priori that we wish to test only a small subset of all the possible pairwise contrasts and use a Bonferonni procedure by using a PC error rate that is the ratio of the desired FW error rate and the number of contrasts, k (i.e., PC = FW / k).

Researchers may be confronted with a problem when attempting to perform post hoc **multiple comparison tests** on a repeated measures factor. First, some commonly used software packages do not have an option for testing differences between the levels (e.g., trials) of repeated measures factors (but STATISTICA [Statsoft, 2001] does provide some tests). Preplanned contrasts, such as polynomials (trend analysis), Helmert, or deviation contrasts, are available and may be appropriate if indicated by the research question. If all pairwise contrasts, or combination contrasts of some form (e.g., the average of conditions A and B vs. the average of conditions C, D, and E) are desired, then the researcher must use hand computation. When this is done, common practice is to use a Tukey or Scheffé test and use the same procedures and formula as for a non-repeated measures factor. The error term used for the omnibus F-test for the repeated measures effect is often used as the error term for the post hoc test. However, psychometricians, most notably Keselman and his colleagues (Keselman, 1982; Keselman, Algina, & Kowalchuk, 2001; Keselman, Keselman, & Shaffer, 1991) have pointed out that using a pooled error term (the error term used to test the repeated measures main effect) is not appropriate and generally leads to an inflated FW error rate. In a typical growth study or learning experiment, use of a pooled error term could result in an inflated FW error rate for some contrasts, but loss of power for others. In these studies a common situation is to have high between-trials correlations for adjacent trials and gradually decreasing correlations as the trials become further apart. Thus a pooled error term, which uses the average of all between-trials correlations, will underestimate the correlations and overestimate the error term for adjacent trial contrasts (with a consequent loss of power), and overestimate the correlations and underestimate the error term for contrast of trials far apart (with a consequent inflation in the type I error rate). A rather simple, but quite adequate, solution to this is to use a separate error term for each contrast, derived by running an ANOVA on only those two trials to be compared in the post hoc test. One does not use the resultant F-test from this analysis but rather just uses the error term in the Tukey or Scheffé post hoc tests, or tests this difference with a simple paired t-test and applies a Bonferroni α adjustment.

Keselman and colleagues have made a number of recommendations over the years with respect to multiple comparisons for repeated measures designs in the presence of nonsphericity. Based on a series of simulation studies, Keselman advocates the use of a Bonferroni step-up test (called the Hochberg step-up Bon-feronni procedure) in conjunction with an adjusted degrees of freedom (df). This df adjustment, the Welch-James adjustment, is similar but not identical to the more familiar Satterthwaite df adjustment. Keselman, Keselman, and Shaffer (1991) and Keselman (1998) provided details of these methods. These procedures provide an optimum solution with respect to minimizing the FW type I error rate in the presence of nonsphericity and maximizing power when these assumptions are met. However, if there is no violation of the sphericity assumption, then standard post hoc test procedures can be applied using a pooled error term. Furthermore, if the sphericity assumption is violated, using a unique error term for each contrast and standard post hoc procedures leads to results very close to those obtained by the more complex procedures advocated by Keselman (1998).

HIERARCHICAL LINEAR MODELING WITH LONGITUDINAL DATA

The ANOVA procedure has serious limitations in analyzing the change of a variable as it relates to time (e.g., physical growth). First, ANOVA requires balanced data. Generally, each subject is required to have the same number of repeated measurements, and the spacing between any two adjacent time points should be the same if polynomial contrasts are used (i.e., trend analysis). This requirement is difficult to satisfy in practice because of testing schedules, dropouts, and other research conditions. Second, ANOVA relies heavily on group statistics. The information concerning the individual level of change and the interindividual variation in change is not analyzed or interpreted, although it is partly available in the form of the within-subjects error term. Third, ANOVA requires one to satisfy the sphericity assumption. This is also difficult to achieve, especially in a longitudinal study in which the covariance (correlation) between any two time points generally becomes smaller as the interval between the time points increases.

The development in advanced statistical models over the last 20 years, along with corresponding development in computer programs, allows one to overcome the limitations of ANOVA procedures in analyzing change. One of the most well-recognized statistical procedures for the analysis of change is the application of a hierarchical linear model (HLM) to repeated measures data. The general HLM can be viewed as an extension of a linear regression model. The most important aspect that makes HLM quite distinctive from traditional statistical models is that the statistical models are formed at several levels. In a two-level HLM, for example, linear regression models are formed at two levels, and the parameter estimates at the first level of analysis (i.e., intercepts and slopes) are used as data at the second level of analysis. Because of this characteristic, HLM is often called a "multilevel model" (Goldstein, 1995).

The multilevel modeling feature of HLM provides a great deal of flexibility in examining various research questions, especially when the structure of data is multilevel in nature. The HLM has been applied in many educational studies (e.g., Lee & Bryk, 1989) in which students are nested within classes, the classes are nested within schools, and the schools are nested within districts. One of the most useful flexibilities of HLM is the application to repeated measures data for examining change. Laid and Ware (1982) and Strenio, Weisberg, and Bryk (1983) were among the first to apply the concept of multilevel modeling to the analysis of change. Bryk and Raudenbush (1987, 1992) later formalized the application of multilevel modeling to the analysis of change within the framework of the general HLM. In exercise science, Zhu (1997) introduced the general model, and Zhu and Erbaugh (1997) showed the application of this model for repeated measures data.

The basic idea of multilevel modeling of repeated measures data is that the change trajectory for each subject is modeled at the first level (level 1), and the interindividual variations in change are modeled at the second level (level 2). The application of HLM in analyzing change has several strengths over traditional models. First, both individual and group levels of change parameters are estimated. This further enables one to examine the interindividual variation in change. Second, individual change can be described by either a linear or curvilinear function. Third, the number of measurements, the measurement occasions, and the interval between measurement occasions can vary across individuals. Fourth, the relationship between the initial status (i.e., the score at the initial time point) and the change rate is estimated. Fifth, predictors or covariates of change can be easily included in the model (Bryk & Raudenbush, 1987). In the following sections, these merits are presented in detail with examples. We first explain the basic concept of two-level modeling with a simple linear change model followed by a brief discussion of extensions to the basic model. An example and interpretation of results follow the conceptual explanation.

Data for Examples

The data used for examples in the following sections were part of a large data set from the Motor Performance Study at Michigan State University, provided by J. Haubenstricker, V. Seefelt, and C. Branta. We selected the variable jump-and-reach (JAR) measured from 210 male children once a year from the age of 8 to 12 (five time points). The children were measured at approximately the same time in terms of their age, with approximately the same time interval (1 year). We intentionally used balanced data with no missing values. This is useful in clearly explaining both the HLM and the latent curve model (introduced later), and comparing the two models in the later sections. Using unbalanced data in the analysis is discussed in the later sections. Another variable, the number of measurements before age 8, is included in the example. Children were recruited at different ages and measured every 6 months. Some children had been measured as often as 11 times before age 8 (the initial time point in this example), whereas some had not been measured at all before age 8. We employed this variable to represent a *"test practice"* effect, and later included this variable in the model to examine the effect of test practice on the initial status and change rate.

Level-1 Model: Within-Subjects Model

The first step of HLM analysis is modeling individual change and estimating the change parameters in the level-1 model. Figure 10.2 shows the trajectories in JAR over time for four selected children. The dotted line is the mean trajectory of all 210 children. A notable characteristic of individual change from figure 10.2 is the considerable variation among children in both initial status (the JAR score at age 8) and change rate. Child 1 shows a relatively low level of JAR performance at age 8 but rapidly improved over the years, whereas child 210 started at a high level but did not improve much. Child 3 showed a decline in JAR performance over time. On average, however, the children showed a steady improvement in JAR performance over 5 years, as indicated by the mean trajectory. The raw data for these four children are presented in the first four columns of table 10.2. Each child has an identification number and five JAR records from age 8 (time 0) to age 12 (time 4). The age variable is rescaled so that the time variable represents the relative time change from the initial time point (age 8) with 1 year as a unit. Using this rescaled time variable instead of children's actual age is useful in interpreting estimated change parameters. Assuming a linear change, we can represent an observed JAR score as a function of time by using a linear equation:

$$Y_{it} = B_{0i} + B_{1i}\,(time) + R_{it} \tag{10.4}$$

where Y_{it} is an observed score of individual i at time t, B_{0i} is the predicted Y_{it} score at time 0, B_{1i} is the coefficient associated with time variable, and R_{it} is the residual at time t for individual i. It is assumed that Y_{it}, B_{0i}, and B_{1i} are normally distributed, and the level-1 predictor "time" is not correlated with R_{it}. In addition, it is assumed that R_{it} is normally distributed with a mean of zero and a variance of σ^2, which is assumed to be constant across time points and across subjects, and the correlation between the residuals of any two subjects is zero for any time point (see Goldstein, 1986, for relaxations of these assumptions). This model is the same as a simple linear regression model with the Y as a criterion (dependent) variable and the time as a predictor (independent) variable. Thus, with the rescaled time variable as the predictor, B_{0i} represents the estimated true JAR score at age 8 (intercept) for child i, and B_{1i} represents the linear change rate per year (slope) for child i. Note that each child has his or her own intercept (B_0) and slope (B_1), and these vary across children. This is the level-1 model; it is often called a "within-subjects model" because the change parameters are estimated within each subject (Bryk & Raudenbush, 1987).

For the data, we conducted ordinary least square (OLS) linear regression analyses for each child with JAR as a criterion variable and time as a predictor variable for the presentation purpose. In the actual HLM analysis, different methods (described later) than OLS are usually used in the estimation of these parameters. The OLS

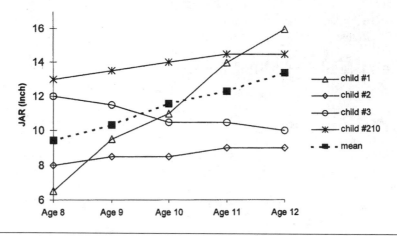

Figure 10.2   Trajectories for jump-and-reach over time for four selected children.

estimates for intercepts (B_{0i}) and slopes (B_{1i}) for the four selected children are presented in the fifth and sixth columns of table 10.2 along with another individual characteristic variable, test practice (the number of measurements taken before age 8) in the last column. For child 1, the B_{0i} and B_{1i} are 6.7 and 2.35, respectively. This means that child 1 recorded 6.7 inches at age 8 (time 0), and improved 2.35 inches per year. Thus, the predicted JAR score at time 3 (age 11) for this child can be calculated as

$$\hat{Y}_{13} = 6.7 + 2.35(3) = 13.75 . \tag{10.5}$$

The difference between the observed and predicted values ($14.00 - 13.75 = .25$) is a residual R_{13}, the residual of predicted JAR for child 1 at time 3. In general, we are not interested in the significance of estimated intercepts and slopes at this level-1 model in longitudinal analyses (Kenny, Bolger, & Kashy, 2002).

Level-2 Model: Between-Subjects Model

As mentioned earlier, each subject has his or her own intercept (B_{0i}) and slope (B_{1i}), and there are between-subjects variations in these individual change parameters. This means that we can treat these parameters as variables representing certain characteristics of each subject, and attempt to explain the between-subjects variations in intercept and slope by some predictor variable (X). These variables are used as criterion variables in a different set of regression models. The level-2 models are formulated as

$$B_{0i} = \gamma_{00} + \gamma_{01}X_i + U_{0i} \tag{10.6}$$

$$B_{1i} = \gamma_{10} + \gamma_{11}X_i + U_{1i} \tag{10.7}$$

Table 10.2 Jump-and-Reach (JAR) and Estimated Individual Change Parameters for Four Children

Child ID	JAR (Y_{it})	Age	Time	Intercept (B_{0i})	Slope (B_{1i})	Test practice (X_L)
1	6.5	8	0	6.7	2.35	1
1	9.5	9	1			
1	11.0	10	2			
1	14.0	11	3			
1	16.0	12	4			
2	8.0	8	0	8.1	.25	9
2	8.5	9	1			
2	8.5	10	2			
2	9.0	11	3			
2	9.0	12	4			
3	12.0	8	0	11.9	−.50	7
3	11.5	9	1			
3	10.5	10	2			
3	10.5	11	3			
3	10.0	12	4			
:	:	:	:	:	:	:
210	13.0	8	0	13.1	.40	4
210	13.5	9	1			
210	14.0	10	2			
210	14.5	11	3			
210	14.5	12	4			

Note: JAR is measured in inches.

where B_{0i} and B_{1i} are the intercept and slope from the level-1 model, respectively (see equation [10.4]), γ_{00} and γ_{10} are the intercepts of the level-2 models, γ_{01} and γ_{11} are regression coefficients associated with the predictor variable X, and U_{0i} and U_{1i} are residuals of these two regression models. It is assumed that predictor X is not correlated with U_{0i} and U_{1i}, and that U_{0i} and U_{1i} are normally distributed with a mean of zero. The U_{0i} and U_{1i} have variances, τ_{00} and τ_{11}, respectively, and covariance, τ_{01}. Thus, two regression models are formed for the two coefficients that were estimated in the level-1 model. Because the regression parameters are now estimated based on between-subjects variations, this level-2 model is called a "between-subjects model." With our sample data, X can be the test practice variable, the number of measurements taken before age 8. That is,

$$B_{0i} = \gamma_{00} + \gamma_{01} \, (\textit{test practice})_i + U_{0i} \qquad (10.8)$$

$$B_{1i} = \gamma_{10} + \gamma_{11} \, (\textit{test practice})_i + U_{1i} . \qquad (10.9)$$

Thus, in this level-2 model, the between-subjects variations in B_{0i} and B_{1i} are explained by a time-invariant variable, test practice. The parameters γ_{00} and γ_{10} are the values of B_{0i} and B_{1i} when the value of test practice is 0. The γ_{01} and γ_{11} represent the effect of test practice on B_{0i} and B_{1i}. An example of estimation and interpretation for these level-2 parameters is presented in the example section. As in a general regression model, multiple predictors can be included in the model regardless of their measurement scale. However, a categorical variable has to be treated properly (e.g., with dummy coding).

Extensions

There are a few notable extensions of the basic model. First, subjects may have various scores for the time variable. That is, the scores for the time variable for a subject do not have to be 0, 1, 2, 3, 4; they can be 1, 1.5, 2, 4.5, or any other numbers in one data set. This means that subjects are not required to be measured within the same time period, and the number of repeated measurements can be different across subjects as long as they are measured at two or more time points. However, one has to be cautious about the scale of the time variable across different subjects, so that all subjects have the same time scale. The anchor point and the unit of the time length have to be the same across all subjects. Second, the level-1 model can be easily extended to a curvilinear model by adding additional term(s). For example, a quadratic change can be represented at level 1 as

$$Y_{it} = B_{0i} + B_{1i} \, (\textit{time}) + B_{2i} \, (\textit{time})^2 + R_{it} \qquad (10.10)$$

where B_{2i} is a coefficient associated with the quadratic change. Because there are now three change parameters in this model (B_{0i}, B_{1i}, and B_{2i}), there should be three regression models at level 2. Third, any number of time-varying predictors can be included in the level-1 model. For example, a child's body weight that also changes over time can be used as a predictor for the JAR score in the same way as the quadratic term is used in the model. Fourth, in the level-2 model, different predictor variables can be included for each model. For example, two different variables can be used as predictors, one for the intercept, B_{0i}, and another for the slope, B_{1i}. Fifth, a level-3, or higher, model is possible. For example, individuals may be nested within different schools, and the characteristics of schools can be used as predictor variables in the level-3 model, explaining the parameters of level-2 models (i.e., γs).

Estimation and Model Evaluation

The estimation issue in HLM analysis is complex. Selecting one of many available estimation methods depends on various conditions such as sample size, the characteristics of data, and the availability of computer programs. We briefly explain the most generally used estimation methods in this section. An in-depth discussion of estimation theory in HLM can be found in Bryk and Raudenbush (1992, pp. 32-59) or Kenny and colleagues (2002).

In a two-level HLM analysis, the intercept, the slope, and the variance of the residuals of the level-1 model (Bs and σ^2) are estimated, although we are not directly interested in these estimates and their statistical significance in most cases. The intercepts, regression coefficients, variances of the residuals, and covariance between the two residuals of the level-2 model (γs and τs) are also estimated, along with their corresponding standard errors. One may use OLS estimation methods for both level-1 and level-2 models. This requires one to estimate the parameters of the level-1 model first and then use these estimates as data in the next stage of analysis for the level-2 model. When the data are not balanced, however, OLS cannot provide reliable estimates, especially for variance and covariance components of the parameters (Kenny et al., 2002).

Typically, two estimation methods, empirical Bayes (EB) and maximum likelihood (ML), are used in a single two-level model analysis. All level-2 model parameters (and higher-level parameters for a higher-level model) and the variance of level-1 model residuals (i.e., σ^2) are estimated with ML, and level-1 coefficients, B_{0i} and B_{1i}, are estimated with EB (Raudenbush, 2001). These two methods are interconnected and produce better estimates than OLS with sufficient sample size. Similarly, with a weighted least squares or generalized least squares method, EB weights data based on all the information that is available from the data in the estimation of B_{0i} and B_{1i} (Morris, 1983). This is especially important when the data are unbalanced and the reliability of the data varies across subjects in the context of modeling change. The idea of ML, in principle, is maximizing the likelihood of observing a particular sample of data. The ML estimation is computationally heavy, requiring an iteration procedure, and the whole estimation procedure is usually accomplished using a commercially available computer program (e.g., HLM 5). Generally, these estimation methods and OLS produce similar results when the data are balanced as in our sample data (Kenny et al., 2002).

One needs to examine the fitted model before the parameter estimates are interpreted. If the model is misspecified at any level or one of the statistical assumptions is seriously violated, interpreting the estimated parameters has little value. The idea of evaluating the fitted model in an HLM analysis is similar to that of general regression analysis. Model assumptions are examined with both raw data and residuals, and particularly, the misspecification of models is examined using residuals. Usually, computer packages provide residuals along with parameter estimates. These residuals are the residuals of level-1 coefficients (i.e., B_{0i} and B_{1i} in a linear model), representing differences between the estimated values and the predicted or fitted values based on the level-2 model (Raudenbush, Bryk, Cheong, & Congdon, 2001). Residuals should be scattered around zero without any specific pattern. Residuals with a skewed distribution, with a certain relationship with a predictor variable, or both, suggest that the specified models may not be correct at one or both of two levels. For example, a quadratic model, instead of a linear model, may be a more appropriate level-1 model, or one or more important predictors could be missing from the model. Thus, one has to carefully examine the residuals along with data and other components of the models. A more detailed discussion about model evaluation is given in Bryk and Raudenbush (1992, pp. 197-229).

Analysis Example

We analyzed the data in table 10.2 using the program HLM 5 (Raudenbush et al., 2001) and present the results in this section. Before specifying any model, we first examined the raw data. The descriptive statistics and correlations between the time points are presented in table 10.3. There were no notable extreme values, and the univariate **skewness** and **kurtosis** statistics were less than 1.0 for all time points. Thus, the distributions of JAR at all five time points are approximately normal. According to mean scores, children showed a steady improvement in JAR performance over a 5-year period. The correlations between the time points are relatively high considering the 1-year time interval. Generally, the correlation between the time points becomes smaller as the interval between the time points increases. This indicates that there was a considerable variation in the change rate of JAR performance over time across children.

Based on the examination of the scatter plot between JAR and time as well as the examination of mean JAR scores over time, we hypothesized a linear change. Thus, our level-1 model has exactly the same form as equation (10.4):

$$JAR_{it} = B_{0i} + B_{1i}(time) + R_{it}. \qquad (10.11)$$

For level 2, we first examined the simplest models that have no predictors. These are

$$B_{0i} = \gamma_{00} + U_{0i} \qquad (10.12)$$

$$B_{1i} = \gamma_{10} + U_{1i}. \qquad (10.13)$$

Table 10.3   Descriptive Statistics of Jump-and-Reach (JAR) at Five Time Points

	Age 8	Age 9	Age 10	Age 11	Age 12	Test practice
Age 9	.59					
Age 10	.52	.66				
Age 11	.52	.62	.69			
Age 12	.49	.60	.64	.70		
Test practice	.27	.26	.27	.23	.25	
Mean (in.)	9.42	10.34	11.58	12.33	13.40	4.85
SD	1.78	1.81	1.89	1.85	2.08	2.00
Skewness	−.23	−.004	.11	−.12	.31	−.71
Kurtosis	.70	.81	.22	.09	.54	.30

This model, encompassing both regression equations, is called an unconditional model. In most cases, an HLM analysis begins with fitting this unconditional model (Bryk & Raudenbush, 1992, p. 135). The unconditional model provides useful information about the individual level change in variable Y (JAR). Without any predictor in the model, the estimated γ_{00} and γ_{10} represent the mean initial status, B_{0i}, and the mean change rate, B_{1i}, respectively. The estimated variances of the residuals, τ_{00} and τ_{11}, represent the variance of initial status (B_{0i}) and change rate (B_{1i}), respectively. In addition, the results from this unconditional model can be compared with those of the model(s) that includes a predictor variable. This provides the degree of the predictor variable's contribution to the explanation of the between-subjects variation in initial status and change.

   The results of this two-level model without a predictor variable at level 2 are presented in table 10.4. The examination of residuals did not show any indication of model misspecification. The estimated mean JAR score at age 8 was 9.43, and on average, JAR performance improved .99 inches per year. These two estimates were significantly different from zero (p < .001). The significant variances of both the initial status (1.93) and change rate (.08) imply there was a considerable between-subjects variation in the performance level at age 8 and in the rate of improvement in JAR performance, respectively. Thus, children started at different levels at age 8, and some showed rapid improvement whereas some showed slow improvement as shown in figure 10.2. The correlation between the initial status and change rate was −.06, and was not statistically significant at an α level of .05.

Table 10.4   HLM Analyses Results for the Model Without a Predictor Variable at Level 2

	Estimated value	Standard error	t	p-value
γ_{00}: Mean initial status	9.43	.113	83.70	<.001
γ_{10}: Mean linear change rate	.99	.031	31.94	<.001
	Estimated value	df	χ^2	p-value
τ_{00}: Variance of initial status	1.93	209	767.5	<.001
τ_{11}: Variance of linear change rate	.08	209	351.6	<.001
σ^2: Variance of level-1 model residual, R_{it}	1.21			

Thus, unlike the general belief that a negative relationship exists between the initial performance level and the change rate (Schutz, 1989), no relationship existed between the JAR performance at age 8 and the improvement rate. We also fitted a quadratic change model to see whether children showed a quadratic change. Both the mean and variance of the quadratic change rate were not significant at an α level of .05. Thus, the linear model was tenable in explaining the individual change in JAR performance.

In the next analysis, the predictor variable, test practice, is included in the level-2 models. Thus, the level-2 regression models are the same as equations (10.8) and (10.9). The HLM analysis results for this model are presented in table 10.5. As in the first model, the examination of residuals did not show any indication of model misspecification. Test practice significantly explained the between-subjects variation in the JAR performance level at age 8 (.24, $p < .001$). This means that the children who experienced the measurement more often showed a better performance level at age 8. On average, each additional measurement experience resulted in a .24 in. increase in the performance level at age 8. Note that the variance of residual U_{0i} (τ_{00}) was reduced from 1.93 (the model with no predictor) to 1.71. This significant effect of test practice on the initial status results in the reduced variation in residuals, although the difference between the two is small. Test practice accounted for only 11% ([{1.93 − 1.71} / 1.93] × 100) of between-subjects variation in initial status. The effect of *test practice* on the change rate was not statistically significant (.002, $p = .905$). This implies that although the number of measurements had a small positive effect on the JAR performance level, this did not affect the rate of improvement in the JAR performance. The variances of the residual U_{1i} (τ_{11}) were about the same (.08) between the models with and without the predictor variable.

Table 10.5   HLM Analyses Results for the Model With Test Practice As a Predictor Variable at Level 2

	Estimated value	Standard error	*t*	p-value
Model for initial status				
γ_{00}: Intercept	8.28	.303	27.35	<.001
γ_{11}: Regression coefficient for test practice	.24	.056	4.22	<.001
Model for change rate				
γ_{10}: Intercept	.98	.083	11.92	<.001
γ_{11}: Regression coefficient for test practice	.002	.016	.12	.905

	Estimated value	df	χ^2	p-value
τ_{00}: Variance of residual U_{0i}	1.71	208	703.43	<.001
τ_{11}: Variance of residual U_{1i}	.08	208	351.57	<.001
σ^2: Variance of level-1 model residual R_{it}	1.21			

STRUCTURAL EQUATION MODELING WITH LONGITUDINAL DATA

Since the early 1970s several models have been suggested for the analysis of change within the framework of structural equation modeling (SEM) (e.g., Jöreskog, 1970; Meredith & Tisak, 1984, 1990; see chapter 9 by Fuzhong Li and Peter Harmer), and the application of these models has been rapidly growing for the last 15 years. These models differ from traditional approaches (which model observed variables) in that they are largely based on modeling with latent variables. These models can be viewed as regression models with both observed and latent variables. In this section, the two most widely used models, autoregressive models and latent curve models (LCMs), are introduced. After a brief overview of autoregressive models, we introduced LCMs in detail with examples because LCMs have now become the most versatile statistical method for studying change.

Brief Overview of Autoregressive Models

In the **autoregressive model** a variable, measured at different occasions, regresses on itself. That is, a variable is repeatedly measured, and the variable at a later time point is explained or predicted by a preceding variable. Figure 10.3 shows an example of an autoregressive model with a variable Y that is measured repeatedly at five time points. Following the general rules of SEM, boxes represent observed variables and ovals represent latent variables. An observed variable is a directly measured variable such as JAR, 12 min run, or weight; and a latent variable or a factor is a theoretical (conceptual) variable that is not directly measurable but is represented by one or more observed variables. For example, power of the lower leg could be regarded as a latent variable, and may be represented by the observed variable JAR. In figure 10.3 the latent variables $\eta_{i1} \sim \eta_{i5}$ are represented by the observed variables $Y_{i1} \sim Y_{i5}$ (however, conceptually, the underlying latent variable causes the observed variable). The arrows represent explanatory (or causal) relationships; the variable at the tail of an arrow explains (causes) the variable at the head of the arrow. The magnitude of the relationship is represented by a path coefficient, which is equivalent to a B coefficient (regression coefficient) in linear regression. In the context of SEM, the path coefficient between a latent and an observed variable is called a factor loading. In figure 10.3, all factor loadings are fixed at 1 (because each latent variable is assessed with only one observed variable). The $e_{i1} \sim e_{i5}$ are the errors or uniqueness, the part of the observed variable that is not explained by the latent variable. This model is called a quasi-simplex model and is a special case of an autoregressive model (Jöreskog, 1970). In this model, the latent variable at a time point is explained by the latent variable at the immediately preceding time point. The βs are path coefficients showing the degree to which a preceding latent variable explains the latent variable at the next time point. Thus, the higher the magnitude of β, the stronger the relationship of variables between

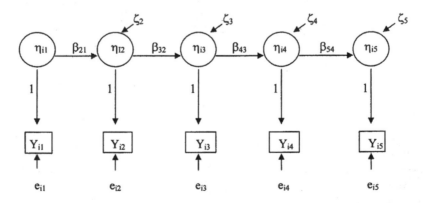

Figure 10.3   An example of an autoregressive model: a quasi-simplex model with five time points.

the two time points, meaning that children are stable between time points in their underlying latent attribute η. Because a β coefficient represents the degree of stability of a variable (or an attribute) over time, it is often called a stability coefficient. A notable feature of this model is that the observed variable Y is composed of two elements, the latent variable η and the error e (i.e., $Y = \eta + e$). In terms of classical test theory, this relationship implies that an observed variable (Y) equals the sum of a true score (η) and an error (e), and the true score at any time point is, in part, a function of preceding true scores.

This autoregressive model can be extended to a multivariate model in which multiple observed variables form a latent variable at each time point, and the stability of this multivariable latent construct is estimated (see Schutz, 1998, for a detailed explanation of this model). The simplex model has been used by many in the study of stability (e.g., Marsh & Grayson, 1994; Wheaton, Muthén, Alwin, & Summers, 1977) and in the reliability estimation for repeated measures data (e.g., Heise, 1969; Werts, Linn, & Jöreskog, 1978). This model, however, focuses mainly on the examination of the stability, rather than on the change, of a variable. Although Kenny and Campbell (1989) and Bast and Reitsma (1997) emphasized the usefulness of this model in the study of stability, Rogosa and Willet (1985) and Park (2001) showed that the stability coefficient (β) is often overestimated, especially when a change occurs, and when the assumptions that are imposed in the model are not tenable (e.g., independence among errors). Moreover, this model provides limited information about the change of a variable compared to other models such as HLM or LCM. These are serious shortcomings in analyzing change.

Latent Curve Models

Although the LCM was first introduced more than 20 years ago (Meredith & Tisak, 1984, 1990), it has received little attention from research practitioners until recently. Conceptually, this model is based on the idea that change is an unobservable latent trait; thus, the change of a variable is represented by latent variables in the model. Although mathematically it is an extension of SEM, the LCM is different from general SEM in that it takes into account both the means and covariances of the measured variables, whereas SEM is based on only covariances among variables. The LCM has several merits over traditional analysis methods in analyzing change, and many of these are similar to those of HLM. First, LCM describes change at both group and individual levels. Second, any type of curvilinear change, as well as linear change, can be examined at the individual level. Third, measurement occasions need not be equally spaced. Fourth, because measurement errors are accounted for by the statistical model, the model represents true individual change. Fifth, multiple predictors or covariates of change can be easily examined in the model. Sixth, the statistical model is very flexible, allowing one to extend the basic model in several ways to examine various research hypotheses (Meredith & Tisak, 1990; Stoolmiller, 1995). In the following sections, we explain the simplest linear LCM with examples that were used in the previous sections and discuss the extensions of the model.

Linear LCM

A diagram showing a linear LCM is presented in figure 10.4. The same notations that are used in the HLM section are used and presented in parentheses to facilitate understanding. For the moment, disregard the predictor variable X that is presented with a dotted line. As in the previous sections, the same variable Y is measured at five equally spaced time points. There are two latent variables or factors, ("intercept" and "slope") that explain the measured variable Y at five time points. $R_{i1} \sim R_{i5}$ are errors or residuals that are not explained by the two latent variables. Note that the path coefficients between the latent variables and observed variables (factor loadings) are all fixed at 1 for the intercept factor, and at 0, 1, 2, 3, 4 for the slope factor. With these fixed factor loadings, the two latent variables have specific meanings. That is, the intercept factor represents the initial status, and the slope factor represents a linear change rate over time. An individual's (i) score Y at each time point (t) can be represented as follows:

$$\text{Time 1: } Y_{i1} = Intercept_i + (0) \times (Slope_i) + R_{i1}$$
$$\text{Time 2: } Y_{i2} = Intercept_i + (1) \times (Slope_i) + R_{i2}$$
$$\text{Time 3: } Y_{i3} = Intercept_i + (2) \times (Slope_i) + R_{i3} \qquad (10.14)$$
$$\text{Time 4: } Y_{i4} = Intercept_i + (3) \times (Slope_i) + R_{i4}$$
$$\text{Time 5: } Y_{i5} = Intercept_i + (4) \times (Slope_i) + R_{i5}.$$

Apart from error, which is unique at each time point, these equations clearly imply a linear individual change over time. Thus, an individual's score at time t, Y_{it}, is explained as a function of the individual's own intercept (initial status) and slope (change rate). Similarly with HLM, the intercept and slope factors have means (γ_{00} and γ_{10}) and variances (τ_{00} and τ_{11}), representing the average initial status and change, and the between-subjects variations in the initial status and change, respectively. The double-headed arrow between the intercept and slope factors is the covariance (τ_{01}) between the two, showing a nondirectional relationship between the initial status and the change rate.

In the model it is easy to include one or more predictor variables that are hypothesized to explain the between-subjects variation in the intercept and slope. In figure 10.4, a predictor variable X is hypothesized to explain the two latent variables as indicated by the two arrows from this predictor variable to the intercept and slope factors (dotted line). The two path coefficients (γ_{01} and γ_{11}) associated with this predictor variable represent the strength of this predictor variable in explaining the intercept and slope.

Extensions

The basic model can be extended several ways based on the characteristics of the data and the research questions posed. First, the interval between the time points

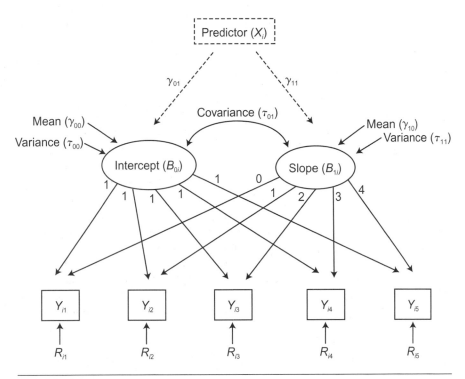

Figure 10.4 A linear latent curve model.

may not be equal over time. To specify a linear model with varying lengths of time intervals, one changes the factor loadings of the slope factor so that each measurement occasion corresponds with the elapsed time from the initial time point. Second, by adding one or more change factors, one may examine curvilinear individual levels of change. For example, if one adds another latent change factor to the linear model (see figure 10.4) with factor loadings of 0, 1, 4, 9, and 16 that are the squares of the linear factor loadings, this model now represents a quadratic change. Another interesting curvilinear model is an "unspecified curve model," which is specified by freely estimating the last three factor loadings of the slope factor in figure 10.4. This model describes the change rate that is different for each time interval. In this model, the first two factor loadings still have to be fixed (e.g., 0 and 1) to provide a scale to the change factor (change rate). Third, various types of predictors or covariates of change may be included in the model. The predictor, or a covariate, may be another latent variable that is measured by several observed variables. In general, modeling the relationship between the change factors and other variables is very flexible in LCM. Fourth, multivariate LCM is possible. This model is especially useful when one is interested in the change of an attribute that is measured by several observed variables. That is, several observed variables represent a latent variable (e.g., perception or intelligence) at each time point, and the change of this latent variable is examined by a multivariate LCM. See McArdle (1988) and Meredith and Tisak (1990) for more detailed discussions on the various LCMs.

Estimation and Model Evaluation

The ML method is the most widely used procedure for the parameter estimation in LCM analysis. The means and variances of observed variables, and covariances among the observed variables, are used as data in the analysis. The means and the variances of the intercept and slope, the covariance between the two, and error variances for each time point (σ_t^2) are estimated with corresponding standard errors. The estimations and related statistics are usually computed by a commercially available computer program (e.g., LISREL, EQS, or M*plus*).

One of the strengths of SEM (and thus of LCM) is that the estimation procedure provides goodness-of-fit indexes, allowing one to examine whether the specified model (e.g., linear model) is tenable given a sample data set. The most widely used indexes are the chi-square statistic (χ^2) with associated *df*, the comparative fit index (i.e., CFI), the non-normed fit index (i.e., NNFI; also known as the Tucker-Lewis index), the root mean square error of approximation (RMSEA), the standardized root mean square residual (SRMR), and the expected cross-validation index (ECVI). If the ratio χ^2 / df is smaller than 2, the CFI or NNFI is close to or larger than .95, the RMSEA is smaller than .06, and the SRMR is smaller than .08, the specified model is tenable (Hu & Bentler, 1999). The ECVI does not have any meaningful magnitude, but it is used when one compares two or more competing models. For more detailed information about various fit indexes, see any recent SEM text (e.g., Bollen, 1989). A statistical test comparing two models is possible when one model is nested within another (when one specifies a model by fixing one or more parameters of another model). This is usually done by examining the difference between the two χ^2 statistics with the difference in corresponding *df*s. This is a useful tool when one wants to compare and select the best fitting but most parsimonious model among several competing change models (e.g., linear vs. quadratic).

Analysis Example

In this example, the LCM was applied to the same data set that was used in the previous sections. The LISREL 8 program (Jöreskog & Sörbom, 1996) was used for estimation of parameters and related statistics. Interested readers may use the correlation coefficients, means, and standard deviations presented in table 10.3 as data for a replication of the analyses. Two models, the linear LCM without the predictor variable and the linear LCM with the predictor variable, were examined as for the HLM example. As an unconditional model in HLM, the model without the predictor variable provides information about the individual and group level of change, and the model with the predictor variable provides information regarding the effect of the predictor variable on the initial status and change rate. The error variances across time points are set to be equal to present equivalent models with those of HLM (it is assumed that the residual variance σ^2 of the level-1 model is constant across time points in HLM). In addition, a quadratic model was examined to compare this model with the linear model.

The goodness-of-fit of the linear model was very good. The χ^2 statistic was not significant (χ^2 [14] = 17.31, p = .240), CFI (.99) and NNFI (.99) were larger than the suggested criteria (i.e., .95), and RMSEA (.030) and SRMR (.038) were smaller than the suggested criteria (i.e., .06 and .08, respectively). This implies that the linear model described the data very well. The quadratic model also fit the data very well (χ^2 [10 = 11.60, p = .313; RMSEA = .021). Because the linear model is nested within the quadratic model (i.e., one may obtain a linear model by eliminating the quadratic term from the model), a significance test comparing these two models was employed using χ^2 statistics and associated dfs. The difference in χ^2 statistics between the two models was 5.71 (17.31 – 11.60) with 4 df (10 – 6). This was not significant at α = .05, meaning that the linear and the quadratic model were not significantly different. That is, adding an additional change factor, quadratic, did not help in explaining the individual change in JAR. Thus, we may conclude that the children showed a linear individual change in JAR scores over the 5-year period. The model with the predictor variable "test practice" also showed a very good model fit. The χ^2 statistic was 18.03 with df of 17 (p = .387), both CFI and NNFI were .99, RMSEA was .007, and SRMR was .033.

The estimated parameters are presented in figure 10.5. All parameter estimates wcre from the linear model without the predictor variable except the two path coefficients showing the effect of the predictor variable on the intercept and slope factors. These two path coefficients were from the model with the predictor variable. The results were almost identical with those of HLM analyses. The true mean score at age 8 (initial status) was 9.43 inches (p < .001), and on the average, the children improved .99 inches (p < .001) per ycar. The variances of the intercept (1.94, p < .001) and slope (.08, p < .001) factors were significantly different from zero, indicating that there was considerable interindividual variation in the rate of improvement as well as in the initial status. This implies that children improved in jumping ability at different rates. The covariance between the intercept and slope factors was not significant. The covariance in a standardized unit (i.e., correlation) was –.06. The initial status and the rate of change did not show a significant relationship.

COMPARISON AMONG STATISTICAL MODELS: ANOVA VS. HLM VS. LCM

The ANOVA model, HLM, and LCM have similarities and differences, and each model has its own strengths and weaknesses compared to other models. In terms of analyzing change, at least four issues need to be discussed in comparing these three models: (a) data requirement, (b) statistical assumptions, (c) modeling flexibility, and (d) feasibility of application. These are interconnected rather than independent issues.

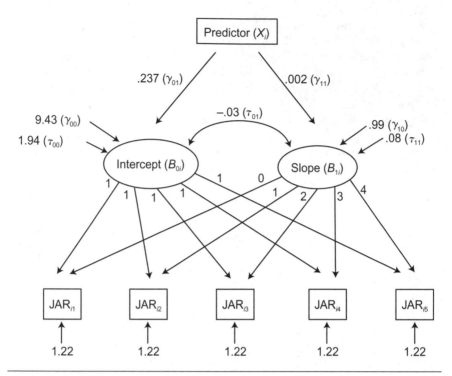

Figure 10.5 Estimated parameters of the linear model for jump-and-reach.

Most flexibility in the data structure is afforded by HLM. In HLM, the number of measurements and measurement occasions may be different across subjects, and the occasions do not have to be equally spaced, while LCM requires that subjects be measured at approximately the same time. Missing time points can be modeled (i.e., missing by design) using a multigroup modeling feature in LCM. However, this is a limited method because one may not include too many groups in the model. Including too many groups in the model is not practically useful and may cause problems in the estimation procedure (Duncan, Duncan, Li, & Strycker 2002; see also McArdle & Hamagami, 1991). The ANOVA model, in general, is most restricted in terms of data structure requirements in that it requires balanced data. Measurement occasions have to be the same across subjects, and the spacing between time points is supposed to be equal when polynomial contrasts are used for the analysis of change (i.e., trend analysis). Obtaining balanced data is especially difficult in a longitudinal study in which change and between-subjects variation in change are usually expected. However, the ANOVA model is still preferred to HLM and LCM in an experimental study in which one may not obtain a large sample size. Both HLM and LCM require a relatively large sample size (e.g., 200), especially when the estimation is conducted through ML procedures. The optimal sample size is not well-known because it depends on too many factors, such as the number of repeated measurements, the number of free parameters, and even the covariance structure. This is a major shortcoming of HLM and LCM. More research is needed on this issue.

The LCM provides the most flexible and informative capability in modeling, evaluating a model, and comparing two or more models. The LCM can estimate a change rate that is different at each time interval (i.e., unspecified curve model), whereas in HLM, one has to have an a priori hypothesis about the shape of change (e.g., linear or quadratic). The LCM's capability of estimating a change rate that is different at each time interval is useful when one is exploring the change pattern of a variable that is not well-known. The capability of including latent variable(s) in the model is another strength of LCM. One may include a multivariable latent variable as a predictor or a covariate of change in the model, or one may examine the change of a multivariable latent variable with LCM (i.e., curve-of-factors model; McArdle, 1988). This is not available in HLM, nor is it in ANOVA (the term *latent variable* in HLM does not mean a multivariable latent variable; see Raudenbush et al., 2001). Modeling the errors of measurement (residuals) is also available only in LCM, allowing one to examine various hypotheses (e.g., correlated errors, equal error variance over time). Once one specifies the individual change with LCM, the full capacity of modeling with general SEM can be used (Raudenbush, 2001). In addition, various goodness-of-fit indexes are provided in LCM, whereas these are not available in HLM and ANOVA. These goodness-of-fit indexes are useful in evaluating the overall fit of a specified model, and in comparing and selecting the best model among several competing models. When one is interested in analyzing change, the ANOVA model certainly is limited in modeling. This is because the ANOVA model is based mainly on the mean differences, whereas HLM and LCM are based on the individual level of change and its variation. Although the between-subjects variation in change is represented as the interaction between the time and subjects (and used as the within-subjects error term) in ANOVA, this error term is often not interpreted, nor is it regarded as a part of the model. In summary, one specifies the individual-level change in HLM and LCM, whereas one hypothesizes mean differences across time points in ANOVA (Bryk & Raudenbush, 1992, p. 133).

The statistical assumptions are also most restrictive in the ANOVA model. The ANOVA model requires one to satisfy the sphericity assumption. The variances of the variable should be approximately equal across time points, and the covariances among the time points should be approximately equal. This can be viewed as a special case of a change model in which the individual change rates are constant across subjects (Bryk & Raudenbush, 1987). As shown in the sample data, in general, the magnitude of covariance decreases as the interval between the time points becomes larger with longitudinal data. Thus, the sphericity assumption is not practical or reasonable in the analysis of change in which a changing structure of covariances among time points is an important aspect. Other assumptions, such as distributional assumptions and measurement scale, are similar in all three statistical models.

Given the availability of all these methods to analyze change, a logical question is, What method should I use? A statistical model for the analysis of change should be selected based on several aspects. The most important aspect to consider is the research questions asked. Although the logical order is to first build the research questions and then choose an appropriate statistical model, building a research question is limited without the knowledge of various statistical models. Certainly, the flexibility in building research questions has been improved with the development of more sophisticated statistical models. A more advanced statistical model, however, is not always the answer for examining the research questions one has. For example, if one is merely interested in the difference in a physical performance level before and after an exercise program, one may simply use an ANOVA model. If one is interested in the pattern of change in a physical performance over time, in the covariates of change, or both, a multilevel model such as HLM or LCM will be more appropriate.

Most other aspects to consider in choosing a statistical model are related to the characteristics of the data. First, one has to consider the measurement scale of the dependent (outcome) variable. Most of the parametric statistical models, including the models introduced in this chapter, require that the dependent variable be measured at an interval or ratio scale. Statistical models for categorical variables are available (e.g., von Eye, 1990), but they are seldom used in our field. Second, the number and the timing of measures are important factors in analyzing change. When there are only two time points, one certainly cannot choose a model that examines the change pattern of the variable. If subjects are measured at different occasions with different spacing between the time points, one has to consider using a more advanced model such as HLM or LCM. Third, sample size is a limiting factor in choosing a statistical model because many advanced multivariate models require a relatively large sample size (e.g., 200). However, often in an experimental study, it is impossible to collect data from a large number of subjects. Small sample size limits the range of research questions one can ask. With a very small sample size one has to use a simple univariate statistical model.

The feasibility of understanding a new statistical method and how to use associated computer programs are practical but important issues for applied researchers. Having comprehensive knowledge about any statistical model is a necessary prerequisite to making full use of the model and to accurately interpreting the results. In general, most applied researchers are familiar with the ANOVA model, and most of the standard statistical computer packages have algorithms for the ANOVA analysis (e.g., SAS, SPSS). Understanding HLM is relatively easy if one is knowledgeable about the general regression model. HLM analysis can be conducted using an available general statistical package (e.g., SAS) or a specialized program (e.g., HLM 5). Making full use of the options available with LCM requires a comprehensive knowledge of SEM in general. Once one understands the model, using an available SEM computer program (e.g., LISREL, M*plus*) for an LCM analysis should be relatively easy. Applied researchers are encouraged to continue searching and learning the new methods that will help to build and assist in answering scientifically meaningful and useful research questions.

ANALYSIS OF CHANGE IN THE NEAR FUTURE

The area of statistical methods is rapidly changing in a way that allows researchers to have more flexibility in building and examining research questions. Because of the flexibility discussed in the previous sections, the application of multilevel modeling including the SEM approach for the analysis of change will be dominant over the next decade. Moreover, these rather complex statistical methods are becoming more flexible and more accessible to applied researchers. Studies related to robustness of the estimated statistics (e.g., Curran, West, & Finch, 1996), sample size (e.g., Jackson, 2001), various residual structures (e.g., Goldstein, 1986), and missing data (e.g., Graham, Taylor, & Cumsille, 2001) will result in the relaxation of the associated statistical assumptions. The development in modeling with categorical and binary variables (e.g., Muthén, 1996, 2001; Rodriguez & Goldman, 1995) and the development of more efficient and robust estimation methods are also notable (e.g., Muthén, 1994; Muthén & Muthén, 1998). Combining and integrating two or more different models into one by adopting the strengths of each model (e.g., Curran & Bollen, 2001; Muthén, 2002) will provide more flexibility in modeling. The development of computer programs is continuing, and implementing these complex procedures is just a mouse click away. Thus, theoretical aspects of these models are becoming more complicated, but also more flexible, and the applications of these models in practice will become easier.

SUMMARY

In this chapter we presented an overview of some recent developments in analyzing repeated measures and longitudinal data that have potential usefulness in exercise science research. RM ANOVA, despite its long history and its place as one of the most frequently used methods of analysis, still presents researchers with a number of difficult issues and decisions. We summarized these issues and made practical recommendations with respect to (a) choosing between a univariate or a multivariate approach, (b) selecting the most appropriate post hoc multiple comparison procedure, and (c) determining a method to compute statistical power or sample size. Although still useful in many experimental studies, RM ANOVA has limitations in analyzing change. Many new statistical models have been developed to overcome these limitations. The basic concepts of two of these recently developed statistical models, HLM and LCM, were introduced with examples. These methods focus on the change of a variable at two levels, individual and group, and provide considerable flexibility in modeling change (e.g., the capability of incorporating predictor(s) of change in the model). The theory and utility of these statistical models will continue to evolve, and exercise science researchers are encouraged to take advantage of the additional insight they provide in an attempts to understand the processes of change.

CHAPTER 11

Analyzing Very Large
or Very Small Data Sets

Weimo Zhu and Anre Venter

We are living in the Age of Information due mainly to our easy access to computers and the Internet. Information about our daily life and activities is collected and stored automatically by electronic point-of-sale sites (e.g., supermarkets, gas stations, and retail stores), banks, credit-card companies, and travel reservation systems. As a result, the volume of collected data grows rapidly, and databases with many millions of records and hundreds, or even thousands, of fields have become common. Many national studies (e.g., National Health and Nutrition Examination Survey [NHNES] and Behavioral Risk Factor Surveillance System [BRFSS]) have hundreds of variables and thousands of cases. In addition, huge data sets can be downloaded from a Web site with a single mouse-click in a matter of minutes. Finally, new technologies have created many new data formats and patterns with immense amounts of digital information. For example, a single satellite image can easily create a data set having $10^8 - 10^{12}$ bytes. Only a very small proportion of such mass data (about 5-10% according to Welge & Bushell, 2000), however, is usually analyzed. With continued improvements in computer power and technology, data volume is expected to increase rapidly and traditional query languages may soon no longer be able to detect patterns or knowledge within the masses of available data.

In contrast, traditional exercise science laboratory research is constrained by the number of participants because of the difficulty in recruiting participants or the high cost of experimental materials. Small sample sizes, such as 8 or 10 participants are still commonly used in laboratories, and a lack of sufficient statistical power is often a concern. Analyzing very large or very small data sets has often been a challenge to researchers in exercise science, and analytical methods that are appropriate to very large or very small data sets have not been systematically described in typical statistics books or classes at the graduate level.

The purpose of this chapter is to provide an overview of the appropriate analytical techniques for very large and very small data sets. Data mining, a state-of-the-art method for analyzing large data sets, is introduced first, followed by a detailed description of how to effectively design small-n studies and appropriately analyze the data. Finally, some useful references and resources on both topics are provided.

ANALYZING VERY LARGE DATA SETS
WITH DATA MINING

Data mining is a process that uses a variety of data analytical tools to discover patterns and relationships in data to enable more accurate and valid descriptions and predictions (Two Crows Corporation, 1999). Earlier "data mining" was considered only as a very important component of "knowledge discovery in databases" (KDD), which is defined as a non-trivial process of identifying valid, novel, potentially useful, and ultimately understandable patterns in data (Fayyad, Piatetsky-Shapiro, Smyth, & Uthurusamy, 1996). Some statisticians even referred to data mining simply as "secondary data" analysis (Hand, 1998). Because the data (e.g., information on Web sites) in data mining are no longer limited to data stored in a database, the broader definition of data mining is more appropriate. To distinguish, the term "database mining" is used in this chapter to refer to the more narrowly defined term, while "data mining" is reserved for the broader definition.

Many of the ideas and concepts used in data mining, such as machine learning and neural networks, have a long and rich history. Data mining as a discipline, however, has a rather short history. Piatetsky-Shapiro and Frawley published the first book about data mining in 1991. In 1996, Fayyad and others published the proceedings of the first international conference on KDD and data mining; a benchmark publication of data mining and KDD. In 1997, the first data-mining journal *Data Mining and Knowledge Discovery* was published. Shortly thereafter, the interests, research efforts, and applications in the field expanded rapidly. Successful applications of data mining have been reported in almost all related disciplines (Delmater & Hancock, 2001), and it is expected that an even greater expansion will occur in the near future.

Types of Data Mining

Based on the characteristics of the data, data mining can be conveniently classified into three categories: database, text, and Web mining.

Database Mining

The most commonly used type, **database mining,** comprises three components: data sources, from which the data are collected; data warehousing, in which huge data sets are stored and linked; and data analysis, where the data are processed and analyzed for meaningful information. Compared to the other two components, data warehousing is a relatively new concept. Since computers were introduced into data-processing centers in the 1960s, operational systems in businesses have gradually been automated. Consequently, data about specific parts of a business are accumulated. Data warehousing is the process that combines data from individual units in an organization (see Bashein & Markus, 2000; Inmon, 1996; Inmon & Hackathorn, 1994 for more information).

Text Mining

Text mining, a relatively new and exciting form of data mining, deals with the estimated 80% of unstructured data used in organizations today (i.e., the data are not organized under variables with known scales, but rather in text which is unstructured, ambiguous, language dependent, heterogeneous, etc.; Shearer & Jouve, 2002). Incident reports, patents, e-mails, or other text-based records are just a few examples. Traditional statistical or probability techniques, which are very powerful in analyzing structured data, often appear ineffective in analyzing text data. Based on linguistics, known as the Natural Language Processing technique, text mining intelligently extracts terms and compounding phrases from text and automatically classifies these into related groups, such as products, organizations, or people using the meaning and context of the text (Hearst, 1999). Text mining can be defined as a process to discover, organize, and anticipate meaningful patterns and relationships from huge volumes of textual information. More important, text mining can provide indices of the extracted concepts, their frequency, and the class to which they belong. This information can then be combined with more traditional statistical and data mining analytic methods.

Web Mining

Web mining is another new and exciting form of data mining that integrates data and text mining together in the context of the World Wide Web. Instead of looking only for meaningful patterns in structured data, as with data mining, or unstructured data, as with text mining; Web mining looks for patterns and their relationships in both structured and unstructured data stored on the Web (Loton, 2002). Like database and text mining, Web mining is a multidisciplinary field. More disciplines, however, such as artificial intelligence, data mining, data warehousing, data visualization, information retrieval, machine learning, pattern recognition, statistics, and Web technology, are involved when other data and approaches are used. Typically, there are three types of Web mining: (a) content mining, (b) usage mining, and (c) Web structure mining (Chang, Healey, McHugh, & Wang, 2001).

Similar to text mining, in which information retrieval and natural language processing techniques search and analyze text, Web-content mining also searches and analyzes graphical content on the Web. Web-usage mining analyzes Web access information available on Web servers, including the automatic discovery of user access patterns from large collections of access logs, which are regularly generated by the servers. This information can help organizations study customers' or visitors' Web-browsing patterns so that a Web site can be better designed and targeted, customized information (e.g., advertisements) can be created, and better long-term strategic decisions can be made relative to Web construction and services provided. Finally, Web-structure mining analyzes the structured information used to describe Web content, including its intra- and inter-page structure. Intra-page structure information refers to the internal document structure of Web documents written in a Web language (e.g., HTML), and inter-page structure information refers to Web linking structure, which is often the external source for a Web structure. A better understanding of Web structures often helps in the design of easy assessable Web sites and enables more efficient searches (Chang et al., 2001; Loton, 2002).

Data Mining Methods

Although the history of data mining is very short, the fact that it is rooted in a variety of disciplines means that there are a multitude of different data mining methods. Based mainly on their functions, data mining methods can be classified into predictive modeling, segmentation, summarization, change and deviation detection, and dependency modeling (Welge & Bushell, 2000). Predictive modeling methods focus on the development of the most accurate model to predict current and future patterns of people, things, or phenomena. Classification, regression, and neural networks are major methods of predictive modeling. Segmentation focuses on determining the interrelationship of a set of variables, and cluster analysis is its major method. Summarization focuses on the relationship among the fields, associations, and information visualization. Change and deviation detection focuses on differences among variables, and dependency modeling focuses on illustrations of the variables. The graphical model is its major technique. It should be noted that the descriptions of the methods are not exclusive and many new methods or procedures (see Riccia, Kruse, & Lenz, 2000) are being developed and frequently employed in practice. The following techniques have received special attention because they function well for their tasks (Berry & Linoff, 1997).

Market Basket Analysis

Market basket analysis takes its name from the idea of a person in a supermarket placing all of their items into a shopping cart (a "market basket"). It functions by examining lists of transactions in order to determine which items are most frequently purchased together. The input to a market basket analysis is normally a list of sales transactions, in which each column represents a product and each row represents either a sale or a customer, depending on whether the goal of the analysis is to learn what items sell together at the same time, or to the same person. With market basket analysis, variables that occur together or in a particular sequence can be identified. The results can be useful to any company that sells products, whether it is in a store, a catalog, or directly to the customer. The information from such analyses has been used for many business purposes, such as planning store layouts (e.g., placing items likely to be purchased together in separate places so that customers must travel from one location to another) and offering coupons for other products when one of them is sold without the others.

Link Analysis

Link analysis follows relationships among records to develop models based on patterns in a relationship and applies concepts from graph theory to data mining. The results of link analyses are displayed as a graph of linked objects, allowing for interesting patterns to be rapidly identified for further investigation. Link analysis has been widely used in business applications. For example, relationships among customers are becoming critical when marketing is targeted to customers, households, or other similar marketing units. In telecommunications marketing, for example, link analysis has been applied to the analysis of phone-call links between a customer and others who are potential customers.

Decision Tree and Rule Induction

Rules are the representation of "cause and effect" relations, but they tend to be simple and intuitive, unstructured and less rigid. There are generally three ways to induce rules in data mining: (a) inductive learning, in which general concepts are determined from specific cause-and-effect examples; (b) generating from training examples; and (c) relying on the fact that classes of examples are known to exist or can be found. Two commonly used rule induction techniques are classification and regression trees (CART; Breiman, Friedman, Olshen, & Stone, 1984) and chi-squared automatic induction (CHAID; Kass, 1980). A **decision tree** is a structure or reasoning process used to divide a large heterogeneous population into a smaller, more homogeneous groups by applying a sequence of simple decision rules. The decision rule can be binary (e.g., "Yes/No" based on presence or absence of a desired attribute) or involve three or more choices (e.g., categorize subjects or data based on a set of predetermined splitting criteria).

Artificial Neural Networks

Created in 1943 by Warren McCulloch, a neurophysiologist, and Walter Pitts, a logician, **artificial neural networks** (ANNs) are one of the most common data mining techniques. ANNs are simple models of neural interconnections in brains, adapted for use on digital computers. In their most common incarnation, they learn from a training set, generalizing patterns inside it for classification or prediction. ANNs can also be applied to undirected data mining (e.g., self-organizing maps and time-series predictions). The major advantage of ANNs is that they can be applied to a variety of data formats and problems. In addition, software programs supporting ANNs analyses are widely available from different vendors and on different platforms (e.g., NeuroDimension, see www.nd.com and NeuroIntelligence, see www.alyuda.com). The limitations of ANNs include (a) the results of ANNs analyses are sometime difficult to interpret and (b) the computations are sensitive to certain data formats.

Genetic Algorithms

Genetic algorithms (GAs) are a part of evolutionary computing, which belongs to the rapidly growing field of artificial intelligence. GAs use selection, crossover, and mutation operations to evolve successive generations of solutions. As the generations evolve, only the most predictive survive, until the functions converge on an optimal solution. It is believed that GAs will become a very useful data mining technique in the near future.

Online Analytical Processing

Online analytical processing (OLAP) integrates a number of data mining techniques and data warehouses. Using a multidimensional view of aggregate data to provide quick access to strategic information for further analysis, OLAP enables analysts, managers, and executives to gain insight into data through fast, consistent, interactive access to a wide variety of possible views of information. The method transforms raw data so that it reflects the real dimensionality of the enterprise as understood by the user. While OLAP systems have the ability to answer "who?" and "what?" questions, it is their ability to answer "what if?" and "why?" questions that sets OLAP apart from data warehouses. OLAP enables decision making about future actions. A typical OLAP calculation is more complex than simply summing data. OLAP and data warehouses are thus complementary. A data warehouse stores and manages data, while OLAP transforms the data from the data warehouse into strategic information. OLAP ranges from basic navigation and browsing, to calculations, to more serious analyses such as time series and complex modeling. As decision-makers exercise more advanced OLAP capabilities, they move from data access to information to knowledge. OLAP applications span a variety of organizational functions. Finance departments in business, for example, use OLAP for applications such as budgeting, activity-based costing (allocations), financial performance analysis, and financial modeling.

Clearly some of the techniques described are not new. For example, the concepts of classification and prediction have been used in exercise science for many years. Many of those "old" techniques are functioning strongly and are associated with machine learning and artificial intelligence. Many new algorithms, such as fuzzy set, genetic algorithms, and genetic programming, which have made machines learn faster and function better, have been developed and applied to data mining. In fact, data analyses, including data mining, based on models with self-improvement ability, have been referred to as "intelligent data analysis" (Berthold & Hand, 1999; Cartwright, 2000).

Requirements for Data Mining Applications

Like other computer-related applications, hardware, software, and a team with disciplinary expertise are all essential to a successful data mining application. In the past, the lack of sufficient computer storage space and slow computation speed were the main hardware concerns. Because of a dramatic decrease in component prices and an increase in hardware performance in the past few years, such concerns have diminished. Progress in software development has also been significant. All major statistical packages have included either a data mining function (e.g., SAS) or added a separate data mining package (e.g., Clementine for SPSS and Insightful Miner for S+). Taken together, it is reported that more than 50 data mining software programs are currently available (see a review at the Web site: www.cs.bham.ac.uk/~anp/software.html) and many new ones are being developed.

Elder and Abbot (1998) compared the strengths and weaknesses of 17 different data mining programs in the following areas: algorithms, multilayer perceptions, decision trees, regression/statistics, usability, visualization, and automation (see www.dataminglab.com for more information). There is also a large variation in software price. Some programs can be downloaded for evaluation without charge, while others can cost $2,000 for a 1-year license agreement. Perhaps the biggest challenge in data-mining applications is building a multidisciplinary team. A variety of expertise is needed to prepare a successful data mining application, and organizing a team with such expertise is often difficult.

Problem Solving Using Data Mining

There is no guarantee that a data mining application will be successful even if the three components are present because many other factors impact the results of a planned data mining application. While there are a number of ways to use data mining to solve real-life problems, the most commonly used process model is known as **Cross-Industry Standard Process for Data Mining** (CRISP-DM). CRISP-DM was conceived in late 1996 by three corporate "veterans" of the young and immature data mining market: DaimlerChrysler, SPSS, and NCR. As illustrated in figure 11.1, there are six phases in this process model (Chapman et al., 2000):

1. *Business understanding*. This initial phase focuses on understanding the project objectives and requirements from a business perspective, and then converting that knowledge into a data mining problem definition, and a preliminary plan designed to achieve the objectives.

2. *Data understanding*. The data understanding phase starts with initial data collection and proceeds with activities to become familiar with the data, to identify data quality problems, to discover first insights into the data, and/or to detect interesting subsets to form hypotheses about hidden information.

3. *Data preparation*. The data preparation phase covers all activities to construct the final dataset (data that will be fed into the modeling tool[s]) from the initial raw data. Data preparation tasks are likely to be performed multiple times and not in any prescribed order. Tasks include tabulating, recording, and attribute selection as well as transformation and cleaning of data for modeling tools.

4. *Modeling*. In this phase, various modeling techniques are selected and applied, and their parameters are calibrated to optimal values. Typically, there are several techniques for the same type of data mining problem. Some techniques have specific requirements for the form of the data. Therefore, revisiting the data preparation phase is often needed.

5. *Evaluation*. At this stage, a model (or models) has been created that appears to be of high quality, from a data analysis perspective. Before proceeding

to the final deployment of the model, it is important to thoroughly evaluate the model, reviewing the steps executed to construct the model, to ensure it properly achieves the program objectives. A key objective is to determine if there is some important issue that has not been sufficiently considered. At the end of this phase, a decision on the use of the data mining results should be reached.

6. *Deployment*. Creation of the model is generally not the end of the project. Even if the purpose of the model is to increase knowledge of the data, the knowledge gained must be organized and presented in a way in which the customer can use it. Depending on the requirements, the deployment phase can be as simple as generating a report or as complex as implementing a repeatable data mining process. In many cases it will be the customer, not the data analyst, who will carry out the deployment steps. If the analyst will not carry out the deployment effort, it is important for the customer to understand, up front, what actions will need to be carried out in order to actually make use of the created model(s).

The purpose of CRISP-DM is to make large data mining projects faster, cheaper, more reliable, and more manageable. Many organizations have adopted CRISP-DM and used it successfully to address varying industry and business problems. CRISP-DM has been referenced by analysts as "the standard" for the industry. A poll of data mining practitioners, conducted by KDnuggets, a data mining Web site, revealed that CRISP-DM was the data mining process most often used; 51% versus 12% using SEMMA, the next most popular standard process (see www.kdnuggets.com/polls/methodlology.html).

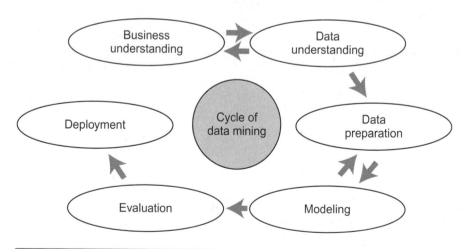

Figure 11.1   Cross-Industry Standard Process for Data Mining (CRISP-DM).

Applications of Data Mining

Data mining has been widely used in many areas, such as customer service, retail, insurance, financial services, health care and medicine, transportation and logistics, energy, and government (see e.g., Berry & Linoff, 1997; Berson, Smith, & Thearling, 2000), and many successful applications have been reported (Delmater & Hancock, 2001). It is expected that data mining and its applications will quickly spread to other disciplines.

Data Mining and Kinesiology Research

The field of kinesiology is replete with problems that may be solved using data-mining techniques. Large data sets of physical activity and health, such as NHNES and BRFSS, are already available. Zhu (2002) recently introduced data mining to the field, and a few initial attempts have been made to apply data mining techniques to problems in kinesiology. For example, using the rule of association, Kang, Zhu, and Ragan (2002) examined the relationships among children's preferred physical activities across gender groups. Boys who participated in individual activities (e.g., walking) were likely to engage in other individual activities, and boys who played team activities (e.g., baseball) were likely to engage in other team activities. This, however, was not true for girls, and no particular pattern of team-activity participation was found. Finally, data mining has been reported in analyzing game data from the National Basketball Association (Bhandari et al., 1997).

Although no research has been reported, both text mining and Web mining have a great application potential to kinesiology. As an example, hundreds of thousands of research abstracts are published and presented annually at research conferences (e.g., about 2,500 in 2005's American College of Sports Medicine annual meeting alone) representing current and future research interests. Few researchers, however, will be able to read these abstracts and absorb the information effectively. With the assistance of text mining, meaningful information can be identified and connections among the different research areas can be established.

Challenges and Issues in Data Mining

While data mining has great potential to help those dealing with very large data sets, many challenges remain. Lack of expertise is perhaps the largest barrier since few domain experts know about data mining and most measurement specialists are trained to handle data using traditional statistical methods. In addition, developing a data mining team is expensive and time-consuming. Lack of usable data is another major barrier as most existing data sets are isolated. For example, although the information from large-scale travel, land use, and air-quality studies may help us understand the roles of environment and policy determinants on people's physical activity participation, the data were often collected using different sampling methods and units making the linking of the data very difficult, if not impossible.

Furthermore, data sets may be unknown (e.g., sporting goods sales, nutrition supplement supplies, etc.), uncollected, and not stored in a systematic way (e.g., fitness club enrollments).

One should also be aware of ethical issues related to data mining. The results of data mining may, for example, lead to a discriminatory decision such as refusing a loan application or denying health insurance, which may have serious ethical implications. A decision based on data mining must be based on careful consideration even if the data mining techniques themselves are appropriately employed and the information discovered is correct. Information privacy is another concern. People should generally be informed how collected information will be used. Finally, even though computers and software are used to collect and discover information, common sense is still critical in interpreting results.

ANALYZING VERY SMALL DATA SETS

Researchers typically use sample data to test statistical hypotheses or statements about a particular population or to establish the nature of the relation between two or more variables within a specific population. In designing a study appropriate to the underlying question of interest, four interrelated aspects of research design need to be considered: sample size, significance or alpha (α) level, the magnitude of the treatment effects within the population or effect size, and statistical power. The last three aspects are critical to researchers operating in domains where sample size is limited.

Ultimately, the ability to correctly refute or support the hypotheses being examined or to accurately estimate the relations among the variables of interest depends on the amount of **statistical power** available. From a hypothesis testing perspective, power is the ability to detect treatment effects and is defined as the probability of correctly rejecting the null hypothesis (H_0). Even in research not intent on hypothesis testing, increased power is related to more precise estimates of relationships, effects, or effect sizes (Venter, Maxwell, & Bolig, 2002). Although we focus on research where increasing sample size to increase statistical power is not possible, consideration of the optimality of the research design may be beneficial even when resources are unlimited. McClelland (1997) noted that nonoptimal designs often produce increased subject costs to compensate for design inefficiencies. Thus, the consideration of design optimality and other factors influencing statistical power may more effectively increase the ability to detect treatment effects.

There are a variety of situations in which small sample sizes are the norm. Kramer and Rosenthal (1999) noted that cases may arise where in addition to limited resources, the phenomenon of interest may be rare, such as a population with a certain disability (e.g., blindness), or the focus of the research might be the completion of a pilot study. Consequently, design or analysis strategies that improve power without increasing sample size or the associated cost of the research are critical. This section provides a conceptual review of the critical issues and options available in cases of low statistical power, specifically within randomized designs focused on group comparisons. Readers are provided with citations for the specific methods of maximizing power and will need to turn to them for the requisite methodological details.

When sample sizes are small, say $n < 30$, the focus is on making the most of the limited number of data points available, making those factors that can enhance statistical power and that are within the researcher's control more critical. Other than sample size, the critical factors are the α level of the test and the effect size. Winer, Brown, and Michels (1991) also identified the variance associated with the dependent measure (σ^2) as critical in increasing statistical power. Before briefly reviewing these factors, notice that these factors are all interrelated and influencing one or more of these factors can enhance power.

Significance Levels

Typically, researchers select in advance the probability level to be considered low enough to enable rejection of H_0. Notice that the logic of hypothesis testing is based upon the assumption that the null hypothesis is true. Because of this logic, the significance level (or α level) selected sets the upper bound for the likelihood of the researcher rejecting H_0 when in fact it is true, in other words, making a **Type I error**. The significance level selected also has some influence on how likely it is that the test will fail to reject H_0 when in fact one of the set of alternative hypotheses (H_1), is true. This is known as a **Type II error** and is designated as β. In this case, however, the actual magnitude of the Type II error depends partly on which one of the possible alternative hypotheses is true. Notice that when a particular H_1 is true, the selection of a smaller significance or α level that reduces the likelihood of a Type I error has the effect of increasing the potential magnitude of a Type II (or β) error.

The level of significance is typically set by convention, and, all else being equal, it follows from the previous discussion that setting a larger α level than usual enables a more powerful test of the H_0. Hays (1994) presents a clear illustration of this relationship and argues that the two critical issues in deciding whether or not to maximize power by selecting a larger α level have to do with the relative costs of Type I versus Type II errors and the fact that power can be increased by other means such as reducing the standard error of the test statistic by reducing variability in the dependent measure (see Franks and Huck [1986] for a discussion of this relationship relative to research in exercise science).

Finally, notice that the probability of rejecting the null hypothesis (H_0) when the alternative hypothesis (H_1) is true is not inherently jeopardized by small sample sizes (Zuckerman, Hodgins, Zuckerman, & Rosenthal, 1993), although this may occur in tests that are asymptotic such as chi-square tests. Small sample size does, however, increase the sensitivity of the tests being used to their underlying assumptions.

Effect Size

Effect sizes provide researchers with an estimate of the magnitude of either the treatment effect or of the relationship between two variables. Notice that **effect size** can be conceptualized in either a standardized format when defined in terms of σ^2 or in an unstandardized fashion, such as (mean$_1$–mean$_2$). The conceptualization is important as it affects the manner by which one attempts to influence effect size in an attempt to enhance statistical power. The remainder of this section views effect size from a standardized perspective. The unstandardized perspective is discussed later.

A number of different types of standardized effect sizes have been defined and different families of effect-size indicators have been identified that indicate the standardized difference between means (Cohen's d, Glass' Δ, and Hedge's g), differences between proportions (Cohen's g and h), and the size of the relationship between variables (Pearson's r, Fisher's Zr, and Cohen's q). Kramer and Rosenthal (1999) provide an excellent review of typically used effect-size indicators, and readers are urged to familiarize themselves with this literature.

Although effect size is not typically thought of when considering factors that researchers can control to increase power, conceptualizing effect size in terms of the underlying noncentrality parameter of the particular statistical test shows that the size of the effect is indeed dependent upon the within-groups variability (σ^2) of the dependent variable (Venter et al., 2002). An example of this conceptualization with additional references is presented in the section below dealing with particular methods of increasing statistical power (see Thomas, Salazar, and Landers [1991] for a discussion of effect-size use in exercise science research).

Variance of the Dependent Measure

Variance is a function of the dependent measure and the reliability of its measurement, the homogeneity of the population from which the sample is being drawn, as well as the nature of the specific experimental design selected. It is central to most, if not all, methods that attempt to maximize statistical power in situations when sample size is small. All else being equal, any reduction in σ, the population standard deviation, produces an increase in statistical power. Framing this in hypothesis testing terms, reducing the "error" variance in the observations leads to reduction in the standard error of the mean and an increase in the power of the test of H_0 against whichever H_1 is true (Hays, 1994). The standard error of the mean $\left(\sigma_{\bar{x}}\right)$ can be expressed as follows:

$$\sigma_{\bar{x}} = \sqrt{\frac{\sigma^2}{N}} = \frac{\sigma}{N} \tag{11.1}$$

making explicit the role that both the variance (σ^2) and sample size (N) have in reducing the standard error of the mean. Notice, therefore, that any reduction in σ^2 (or σ) and any increase in N will have the effect of reducing the standard error of the mean.

Although the actual makeup of the various statistical tests in hypothesis testing may vary, typically the logic remains the same in that any variability that is unaccounted for is treated as "error" variance and is simply added into the denominator term of the particular statistic. Consequently, any reduction in error variance produces a smaller denominator term, and thus a larger test statistic. Hays (1994) notes that experiments in which "error" variance is minimal are characterized by a high degree of precision such that the likelihood of the researcher detecting the presence of an effect of interest is increased. Clearly such precision is of great value to researchers laboring under the constraints of small sample research.

Methods of Increasing Statistical Power in Small Sample Research

The focus of this section is to provide a brief overview of the major methods available for maximizing power in small sample research. In each case we provide readers with the necessary citations so that they can familiarize themselves with the intricacies and details of the relevant methodologies. Interested readers are referred to Shadish, Cook, and Campbell (2002) for an excellent, more detailed overview.

Within- Versus Between-Subjects Designs

Within-subjects designs (see chapter 10 by Ilhyeok Park and Robert Schutz) are the most obvious example of designs that utilize the advantages that multiple measures of subjects' responses over time provide (Venter & Maxwell, 1999). Within-subjects designs are typically more powerful than **between-subjects designs** because the treatment condition constitutes a within-subjects factor. The choice of this design, however, requires some forethought due to the possibility of confounding carryover effects. For example, research in which fatigue, practice, and contamination effects can occur, or research in which the treatment can have permanent effects on the subjects is not suited to this choice of design. In addition, such designs may not always be feasible outside of the formal laboratory setting.

The advantage of within-subjects designs where treatment condition constitutes a within-subjects factor over the traditional between-subjects factor is twofold. In the within-subjects condition, each subject participates in every condition, thus producing as many data points as there are treatment conditions. In the between-subjects design, each subject participates in only a single condition. Clearly, then, given a limited number of participants, more data points are available in the within-subjects design as compared to the between-subjects design. In addition, each subject serves as his or her own control in the within-subjects design, typically producing a reduction in error variance that, at times, can be quite substantial.

Venter and Maxwell (1999) provide a detailed exposition of the statistical model and its requisite assumptions as well as a clear illustration of the difference in the relative power and precision of the within-subjects versus the between-subjects designs. Venter and Maxwell show that using a within-subjects design when scores correlate .80, a researcher interested in achieving a power of .80 to detect a medium effect size would need only 13 participants. In contrast, a pretest–posttest between-subjects design analyzed with difference scores would need 51 participants, the same design analyzed with an ANCOVA (analysis of covariance) would require 46, while a posttest only design would need 128 participants to achieve a power level of .80 to detect a medium effect size. Even when the standardized effect size is only equal to .20, given the same assumptions, a within-subjects design would require 64 subjects as opposed to 256 for a difference-score analysis, and 128 for both the ANCOVA and posttest only design.

A few words of caution are warranted. The analyses of within-subjects designs become complicated when there are more than two levels of the within-subjects' factor because the sphericity assumption may be less tenable. More detail in this regard is provided by Algina (1994), Maxwell and Arvey (1982), and Maxwell and Delaney (1990). In addition, other assumptions may not hold in certain situations, especially with regard to the assumption that there are no carryover effects in the within-subjects design.

Matching, Stratifying, and Blocking

Although the formation of treatment and control groups through the use of matching and stratifying on likely correlates (or concomitant variables) of the posttest is typically done when a pretest is lacking and the design is susceptible to selection biases, the availability of a pretest does not preclude these methods. **Matching** involves the grouping of subjects with similar scores on the concomitant variable so that the treatment and control groups each contain subjects with similar characteristics on the matching variable. Although the terms matching and blocking are most often used interchangeably, some researchers do differentiate between the two, using **blocking** to indicate groups with similar scores and matching to identify groups with identical scores on the matching variable. **Stratifying,** although closely related to matching (and blocking), involves grouping subjects into homogenous sets that contain more subjects than the experiment has conditions. Shadish et al. (2002) used the example of stratifying on gender as a variable upon which matching is not possible—it is impossible to obtain a closer match than "male" among a large group of males. In this case, a stratum will contain many more males than there are experimental conditions. Notice that matches contain the same number of subjects as there are experimental conditions.

These methods all increase the probability that the treatment and control conditions will have similar pretest means and variances on the matching/stratifying variable prior to the experimental manipulation. During the analyses, the variance attributable to the matching/stratifying variable can then be removed from the overall variance, typically reducing the denominator term in the statistical test and thus producing a more powerful test of the hypotheses. Often in the case of very small sample sizes, any single instance of random assignment may not result in equal (or nearly equal) means, variances, and numbers of subjects per condition even though equality will hold as a long-run average. This mechanism becomes critical when planning to do subanalyses comparing particular cells because without matching/stratification, individual cells may have too few data points to enable these analyses. Shadish et al. (2002) recommended randomly assigning from matches or strata whenever possible and a good matching variable is present.

A number of issues raised by Shadish et al. (2002) need to be acknowledged in concluding this section. First, matching consumes degrees of freedom and may actually decrease statistical power when used with variables unrelated to the outcome of interest. Second, the benefits of matching or stratification mostly apply to their use prior to randomization. Finally, care should be taken utilizing these processes within quasi-experimental designs. Shadish et al. provide an excellent review of the various mechanisms available for matching/stratification.

Measuring and Correcting for Covariates

One way of understanding the manner in which covariates can increase statistical power requires reconceptualizing the effect size parameter in terms of the underlying noncentrality parameter of the particular statistical test being used. One example of this method is presented here and readers are directed towards the relevant literature for a more in-depth review.

Taking a two independent group t-test as an example, Venter et al. (2002) showed that the noncentrality parameter (δ^2) associated with the noncentral t distribution can be written as

$$\delta^2 = \frac{n}{2}\left(\frac{\mu_1 - \mu_2}{\sigma}\right)^2 \qquad (11.2)$$

where μ refers to the population mean, which shows explicitly that power can be increased either by (a) increasing n or by (b) increasing $[(\mu_1-\mu_2)/\sigma]^2$. Notice that $[(\mu_1-\mu_2)/\sigma]$ is defined to be the standardized population effect size

$$d = [(\mu_1-\mu_2)/\sigma] \qquad (11.3)$$

If we then substitute Equation 3 into Equation 2, the noncentrality parameter can now be written as

$$\delta^2 = (n/2)\, d^2. \qquad (11.4)$$

Notice then that the noncentrality parameter and thus the statistical power of the test for a set level of significance depend on sample size (n) and effect size (d). From this standardized perspective, ignoring sample size, any increase in the effect size depends upon σ, the standard deviation of the "error term" in the statistical model. Typically, a standard model for comparing the means of two independent groups can be written as

$$Y_{ij} = \mu_j + \varepsilon_{ij} \qquad (11.5)$$

where Y_{ij} is the score on the dependent variable for individual i in group j, μ_j is the population mean on Y for group j, and ε_{ij} is the error for individual i in group *j*. Because ε_{ij} reflects any difference between individuals within a group, σ is the within-group standard deviation of Y. If we vary our statistical model, the meaning and the value of ε_{ij} and thus of σ as well will typically change. For example, the addition of a pretest measure as a covariate in the design produces the following statistical model:

$$Y_{ij} = \mu_j + \beta X_{ij} + \varepsilon_{ij} \qquad (11.6)$$

where β is the regression coefficient for the covariate, and X_{ij} is the score of individual *i* in group *j* on the covariate. Within the constraints of a randomized design, the inclusion of a covariate measured prior to assignment to condition has no effect on $\mu_1 - \mu_2$, but will lower σ in the model to the extent that X and Y correlate with one another. Decreasing the magnitude of σ has the effect of increasing the noncentrality parameter, ultimately producing an increase in power and precision.

Notice that the addition of a covariate to a randomized design is one classic method of increasing power and precision while maintaining a constant sample size, which falls within the broad design and analysis strategy characterized by the utilization of multiple measures of subjects' responses over time. Multiple measures of subjects' responses can occur within either a between-subjects or within-subjects design. For a review of these design options as strategies for maximizing power we suggest the review of the relevant literature provided by Venter and Maxwell (1999).

Increasing the Reliability of the Observations

The effects of unreliable measurements have been well documented and have been shown to be anything but straightforward. Although estimates of actual effects that exist between two variables are attenuated by unreliability, when measuring the relationships among three or more variables the effect of low reliability becomes quite unpredictable (Maxwell & Delaney, 1990). In other words, analyses may indicate the presence of treatment effects when in fact no such effect exists in the population or, conversely, no effects may be found when in fact an actual effect does exist. In addition, often the pattern of differential reliability and the pattern of relationships among the variables influence the actual effect of low reliability within certain correlational designs (Rogosa, 1980).

A number of methods can be used to address the issue of reducing measurement error such as increasing the number of measurements through either additional items or raters, improving the quality of individual items and better training for raters, utilizing techniques such as latent variable modeling, and not dichotomizing continuous variables thus avoiding a restriction of range problem (Shadish et al., 2002). Other options include adding additional waves of measurement (Maxwell, 1994), or allocating more resources to the posttest rather than the pretest (Maxwell, 1998). Once again readers are urged to acquaint themselves with the relevant literature to ascertain how best to deal with any issues of unreliable measurements.

Increasing the Strength, Quality, and Variability of the Treatment

Increasing the strength, variability, or quality of the treatments are methods that can influence effect size when it is conceptualized in the unstandardized form (e.g., $mean_1$–$mean_2$). In some research domains, the ability of the researchers to accurately assess the effect of an intervention is compromised by issues of implementation and attrition. Medical research is, for example, dogged by the problem of nonadherence to the treatment often resulting in subjects not receiving the full treatment. Obviously, any failure in the administration or receipt of the treatment will reduce the size of the effect that the researcher is attempting to detect in the population. Shadish et al. (2002) noted that treatment implementation comprises the processes of treatment delivery, receipt, and adherence and provide an excellent overview of the available methods of addressing these issues.

In terms of increasing the variability of the treatment as a means of increasing power, two related options are typically available. McClelland (1997) used a two-variable linear model as an example to show that in addition to reducing the error variance, statistically increasing the residual variance of the predictor variable can increase the power variable. Given this, McClelland (1997, 2000) noted that this can be achieved either by oversampling from extreme levels of the treatment or by broadening the range of levels of treatment that are implemented and tested.

It should be noted that the methods just described are related to the issues critical to small-sample research. For example, one of the most critical factors involves the variance of the dependent measure as it plays a role in many of the methods discussed here. Not only is σ^2 an integral part of the standardized effect size, it is also the critical factor influenced by selecting a within-versus between-subjects design (or using methods such as matching, stratifying, or blocking) or choosing to use covariates in a between-subjects design. Furthermore, improving the reliability of the dependent measure reduces the effects of any extraneous error variance while increasing the strength, quality, or variability may increase effect size when conceptualized in the unstandardized form (e.g., $mean_1$–$mean_2$).

Randomization/Permutation Tests and Bootstrapping

These methods may also prove to be useful in dealing with small sample sizes. Although space limitations constrain us to simply mentioning these here, readers are encouraged to review the following. For randomization, we suggest Edgington's (1995) presentation of randomization tests, and for bootstrapping, Efron and Tibshirani (1993) and Zhu's (1997) introduction to bootstrapping and Yung and Chan's (1999) review of bootstrapping in small-sample research.

SUMMARY

Dramatically increased information is making traditional statistical query methods less feasible. To identify meaningful patterns or knowledge within large data sets,

Resources: Some Useful Web Links

Introduction to data mining:
www.cs.bham.ac.uk/~anp/dm_docs/dm_intro.html

Kdnuggets News (A data-mining interest group):
www.kdnuggets.com/index.html

Data-mining software:
www.cs.bham.ac.uk/~anp/software.html

Packages for determining statistical power:
www.statsoftinc.com/power_an.html
www.spss.com/spssbi/samplepower
www.statsolusa.com/nquery/review2.htm

For a review of statistical procedures that remain valid for small, sparse, and unbalanced data sets, readers should refer to
www.cytel.com/new.pages/SX.2.html

new query methods are required. By integrating multiple disciplines, data mining appears to be a powerful approach to manage, analyze, and utilize mass data. Successful data mining, however, depends on clearly understood and measurable objects, good quality relevant data, a sponsor for the application (if necessary), and most importantly, a well organized multidisciplinary team. Meanwhile, designs with small sample size ($n < 30$) are commonly employed in laboratory studies. Researchers, who find themselves constrained to working with small sample sizes, should recognize that options to enhance the statistical power are indeed available to them. Even though a well-thought-out, theoretically and methodologically sound design process should be at the foundation of all research, these issues become more critical when operating in the domain of small sample sizes.

PART V

Measurement in Practice

CHAPTER **12**

Current Issues in Physical Education

Terry M. Wood

O ver the past two decades, much has changed in the theory and practice of educational measurement. On the theoretical side,

- item response theory (IRT) (see chapter 4 by Weimo Zhu) has blossomed into a viable alternative to classical test theory as a basis for scaling written tests (see Bock, 1997, for a succinct history of IRT);
- the notion of and debate over consequential validity incorporating social consequences and value implications as evidence for test validity has accompanied the growth of large-scale high-stakes assessments (see Messick, 1989, and the summer 1997 and summer 1998 issues of *Educational Measurement: Issues and Practice*); and
- interest has mushroomed in reliability generalization (see the August 2002 special issue of *Educational and Psychological Measurement* devoted to the topic) and appropriate reporting of test reliability (see Morrow & Jackson, 1993, and Baumgartner & Chung, 2001, for examples in exercise science).

On the practical side, the latest wave of educational reform with its emphasis on **standards-based education** and **performance assessment** has created a renewed interest in criterion-referenced measurement and standard-setting procedures, while raising serious questions about the reliability and validity of performance assessments and how best to communicate student learning to various stakeholders.

Educational measurement in the new millennium can be viewed from two perspectives—**high-stakes assessment** and **classroom assessment**. High-stakes assessment is typically large-scale districtwide or statewide assessment used to inform educational policy, reform curriculum, and monitor accountability (Cooley, 1991). For example, the U.S. Congress passed into law the No Child Left Behind Act of 2001 making states responsible for developing and assessing common learning targets (i.e., content standards) for all students in elementary, middle, and high school (U.S. Department of Education, 2002). The assessment results must be reported in annual state and district report cards. Moreover, parents with a child enrolled in a failing school have the option of transferring their child to another school. In some states mastery of content standards has become a graduation requirement. The stakes are indeed high for teachers, students, schools, and districts! High-stakes assessments are therefore planned and conducted by measurement professionals in state departments of education, often with the assistance of commercial testing agencies such as the Educational Testing Service (Princeton, NJ). In contrast, classroom assessment is developed and conducted by the classroom teacher for monitoring student progress and for grading.

Although many states have adopted **content standards** for physical education (PE), few mandate statewide assessment. Typically, PE teachers are required to teach to state content standards and assess students via teacher-developed classroom assessments. A push for statewide assessment in PE, however, is growing. For example, in Oregon, statewide content standards in PE were adopted in 1996. In 2000 the Oregon legislature mandated statewide testing of PE in grades 3, 5, 8, and 10. Responsibility for developing and administering assessments was given to individual school districts. The task of developing assessments of student learning in PE ultimately fell to PE teachers in each district. Therefore, it is important for PE teachers and those responsible for administering PE programs to understand assessment issues relative to both the high-stakes testing arena and classroom assessment.

This chapter focuses on assessment issues from both the high-stakes and classroom perspectives. To bring readers up-to-date with assessment issues facing today's physical educator, the chapter begins with a brief history of educational reform, the role of assessment in the reform movement, and PE's contributions to educational reform. A description of new assessment practice and terminology is covered next followed by important measurement issues facing physical educators as they begin to implement new assessment practice.

EDUCATIONAL REFORM AND ASSESSMENT PRACTICE

The roots of the latest round of educational reform can be traced to the publication of *A Nation at Risk* (National Commission on Excellence in Education, 1983) and the subsequent Education Summit in 1989 where President George H.W. Bush and state governors established six goals for education. One of the key goals was the development of content standards for five subject areas, of which PE was not included. In 1994 the Goals 2000: Educate America Act formalized federal involvement in education reform by linking federal funding to the voluntary development of content standards and assessments in eight core subject areas. Again, PE was not included among the core subjects. It was clear from the beginning that accountability for learning in the classroom through assessment of common learning targets was a cornerstone of the educational reform movement.

Funding and direction for PE in the reform movement was secured through the National Association for Sport and Physical Education (NASPE). In 1992 NASPE published 20 "Outcomes of Quality Physical Education" as the first step in developing learning outcomes for PE. These outcomes, which formed the basis of the seven national content standards for physical education published by NASPE in

1995, were revised to six content standards in 2004 (NASPE, 1995, 2004). The national standards for physical education provided a framework for states and districts to develop their own PE content standards and assessments at each grade level (K, 2, 4, 6, 8, 10, 12). Since the publication of the national standards, the literature concerning performance assessment and standards-based PE has focused primarily on implementation strategies and practical applications (for example, see the September 1997, October 2001, November/December 2001, and March 2002 issues of the *Journal of Physical Education, Recreation and Dance* and the NASPE *Assessment Series: K-12 Physical Education* edited by Deborah Tannehill). Notably there has been an absence of research into the measurement properties of performance assessments by measurement specialists and researchers interested in sport pedagogy as evidenced by the lack of significant numbers of publications on these topics in research journals such as *Research Quarterly for Exercise and Sport*, *Journal of Teaching in Physical Education*, and *Measurement in Physical Education and Exercise Science* (for an exception, see Joyner & McManis, 1997). It is clear that the development and implementation of content standards and performance assessment in PE continues to be a grassroots movement initiated by a professional teacher organization (NASPE) and sustained by practicing teachers.

UNDERSTANDING
NEW ASSESSMENT PRACTICE

The cornerstones of educational reform are (a) specific and developmentally appropriate common learning targets for all students at all grade levels and (b) criterion-referenced assessment of the extent to which students are mastering learning targets. The purpose of this section is to briefly describe new assessment practice and terminology that has evolved with the practice.

Content standards are broad descriptions of what students should know and be able to do at each grade level. For example, the following list presents the six NASPE (2004) national content standards for PE. These six content standards describe what a physically educated student should know and be able to do upon exiting the public school system. The NASPE standards document outlines developmentally appropriate descriptions of the standards for grades K, 2, 4, 6, 8, 10, and 12. Standards-based physical education uses content standards as the framework for planning and delivering instruction.

NASPE (1995) defined a **benchmark** as "behavior that indicates progress toward a performance standard" (p. vi) and "which describe[s] developmentally appropriate behaviors representative of progress toward achieving the standard" (p. viii). Benchmarks are more directly measurable than content standards and can be viewed as specific signposts indicating whether students are meeting a specific content standard. Table 12.1 gives sample benchmarks identified by the Oregon Department of Education (2001) for the state's fourth content standard, *Provide evidence of engaging in a physically active lifestyle.*

NASPE (2004) National Content Standards for Physical Education

A physically educated person:

1. Demonstrates competency in motor skills and movement patterns needed to perform a variety of physical activities.
2. Demonstrates understanding of movement concepts, principles, strategies, and tactics as they apply to the learning and performance of physical activities.
3. Participates regularly in physical activity.
4. Achieves and maintains a health-enhancing level of physical fitness.
5. Exhibits responsible personal and social behavior that respects self and others in physical activity settings.
6. Values physical activity for health, enjoyment, challenge, self-expression, and/or social interaction.

Alternative assessment strategies have evolved to assess student mastery of benchmarks. These strategies differ from more traditional assessment strategies in that they integrate higher-order cognitive, affective, and psychomotor processes in more real-life contexts (Wood, 1996). Traditional assessment strategies in PE include multiple-choice tests and sport-skill tests, whereas alternative assessments include such methods as portfolios (Danielson & Abrutyn, 1997; Melograno, 1994, 2000), journals (Cutforth & Parker, 1996), and student logs and demonstrations (NASPE, 1995).

Although alternative assessments can take many forms, two types have attracted the attention of educators. Performance assessments are those in which "the student completes or demonstrates the same behavior the assessor desires" (Meyer, 1992, p. 40), whereas authentic assessments are performance assessments administered in real-life settings (Meyer, 1992). Kane, Crooks, and Cohen (1999) further noted that "the defining characteristic of a performance assessment is the close similarity between the type of performance that is actually observed and the type of performance that is of interest" (p. 7). For example, in a middle school basketball unit, a PE teacher desires to assess student ability to referee a game. A traditional approach might be to administer a written test targeting basketball rules and techniques of officiating. This test provides an indication of students' knowledge of rules but does not indicate their ability to apply and interpret those rules in a real-life context. A performance assessment consisting of students' ability to demonstrate their officiating technique and knowledge of rules in a controlled environment (e.g., ask

Table 12.1   Sample Benchmarks From the Oregon Department of Education (2001) Physical Education Content Standards

3rd Grade	5th Grade	8th Grade	10th–12th Grade
Identify changes in his/her body during moderate to vigorous exercise	Identify changes in his/her body before, during, and after moderate to vigorous exercise (e.g., perspiration, increased heart and breathing rates)	Develop personal activity goals, and describe benefits that result from regular participation in physical education.	Participate in physical activities, and evaluate personal factors that impact participation.
3rd Grade	**5th Grade**	**8th Grade**	**10th–12th Grade**
		Analyze and categorize physical activities according to potential fitness benefits	Through physical activity, understand ways in which personal characteristics, performance styles, and activity preferences will change over the life span

Reprinted, by permission, from Oregon Department of Education, 2001, *Benchmarks 1 through CIM* (Salem, Oregon), 3.

students to demonstrate the charging infraction) more directly assesses required behavior. Assessing students while they officiate a basketball game represents a more authentic assessment because students' mastery of rules and officiating technique are assessed directly in a real-life context.

Scoring alternative assessments has received significant attention in the educational literature (see Arter & McTighe, 2001; Lund, 2000; and Taggart, Phifer, Nixon, & Wood, 1998) because such assessments often require subjective evaluation of student performance. Of critical importance is the development of valid and reliable scoring guides or **rubrics**. Scoring guides explicitly outline **performance criteria** or the behavior(s) to be assessed and a **performance scale** that assigns a numerical value to the performance criteria. It is not uncommon for the performance scale to consist of descriptors such as *novice, basic, proficient,* and *distinguished* instead of numbers. For each level of a performance scale we can identify the

performance cutscore as the numerical value associated with that level and the **performance standard** as the qualitative description of the performance criteria expected at that level (Kane, 2001). Figure 12.1 presents a scoring guide for the NASPE (1995) national content standard 6, *Demonstrates understanding and respect for differences among people in physical activity settings,* constructed by the Wichita Public Schools (1999). Note that this scoring guide has seven levels ranging from NA (no attempt) to 6. At each level the score represents the performance cutscore, and the descriptor beside the score represents the performance standard. The performance standards reflect the performance criteria deemed important at each level.

A traditional model of instruction begins with setting unit objectives, developing instructional materials and strategies, pretesting students, delivering instruction, and then assessing student progress in a formative framework. Assessment often involves testing discrete behaviors in the cognitive, psychomotor, and affective domains and is usually viewed as different, and conducted separately, from instruction. Historically in PE, the affective domain has been emphasized in evaluation plans.

Assessment under the educational reform movement has assumed a more prominent role in the instructional process. **Assessment-driven instruction** refers to "teaching and planning for teaching that are based on, derived from, and focused on performance assessment" (Glatthorn, Bragaw, Dawkins, & Parker, 1998, p. 7). Performance assessments are designed to assess student mastery of content standards and benchmarks. Therefore the teacher's focus is on developing effective instructional strategies to

---

6—Shows an appreciation of differences and helps others understand the importance of cooperating, respecting, and sharing in a physical activity setting

5—Cooperates, respects, and shares with others in a physical activity setting, regardless of the differences among them

4—Participates with, and shows respect for persons of like and different skill levels

3—Respects and accepts others regardless of similarities and differences

2—Can identify the similarities and differences between themselves and a partner

1—Shares and takes turns during physical activity

NA—No attempt

Figure 12.1 A sample scoring guide or rubric developed by the Wichita Public Schools (1999).

Adapted, by permission, from Wichita Public Schools, 1999, *Wichita Public Schools Physical Education Program Standards: Curriculum Alignment and Assessment Rubrics* (Wichita, KS: Wichita Public Schools).

assist students in mastering performance assessments, thus creating a seamless integration of assessment and instruction. Although this may seem like "teaching to the test," it is argued that valid performance assessments measure the behaviors that students are expected to know and be able to do in real-life contexts. Mastery of the assessments, therefore, is presumed to indicate mastery of the behavior.

MEASUREMENT ISSUES

Performance assessments combined with standards-based education offer great promise for increased accountability in education. Frechtling (1991) outlined the following advantages of performance assessment over more traditional assessments such as multiple-choice tests:

> First the new assessments go beyond simplistic, multiple-choice questions, requiring students to perform in situations that are both more lifelike and more complex. Both process and product can be examined, providing a far richer picture of what students do and do not understand. Second, because the assessments look more like familiar instructional situations, the results from them are more likely to be perceived as valid by teachers. Third, because the results are judged to be more valid, they may also have a greater impact on the improvement of instructional programs. The distance between testing and instruction should be reduced, and the value of test data for program improvement should be increased. (p. 23)

Kane and colleagues (1999) further noted that "another factor favoring a shift from objective testing to performance assessment has been the realization that assessment practices can have a profound influence on how students study and on how teachers teach . . . . This has led some reformers to advocate using assessment programs to shape what goes on in schools, rather than simply to monitor the outcomes of education" (p. 5).

Although the potential of performance assessment is great, realizing the promise may not be easy. Logistically, performance assessments tend to be more time consuming, personnel intensive, and costly than more traditional test formats (see Solano-Flores & Shavelson, 1997, for a discussion of practical and logistical constraints associated with developing and administering performance assessments). In addition, measurement issues regarding the validity and reliability of performance assessments have yet to be resolved.

Evidence for Validity of Performance Assessments

The majority of research into validity issues of performance assessment is aimed at high-stakes testing. The types of evidence for validity of high-stakes tests are necessarily more complex than those for classroom assessments; however, the principles and assumptions of validity are important for both levels of assessments because important decisions are being made about students and programs based on test scores.

Linn, Baker, and Dunbar (1991) stated that questions of validity should focus on the extent to which assessments represent the desired long-range objectives of instruction. In other words, if performance assessments are designed to measure students' mastery of content and their ability to solve problems relative to content in real-life scenarios, then test developers must present evidence for the adequacy of test scores for making such interpretations.

Because performance assessment tasks have high fidelity (i.e., appear closely related to what students should know and be able to do), many test developers are lured into relying on "face validity" (Safrit & Wood, 1995) as the primary evidence for test validity. Face validity; however, is not sufficient evidence for validity (Mehrens, 1992; Wood, 1989). For example Messick (1995) suggested six aspects of validity to consider when evaluating performance assessments:

- *Content relevance and representativeness.* Both content standards and performance assessments should be relevant and representative of the content domain being assessed. **Content relevance** defines and evaluates the worth of the universe of content and behaviors being assessed (Wood, 1989), whereas **content representativeness** evaluates the extent to which assessment tasks are representative of the defined universe. Both relevance and representativeness are evaluated via professional judgment (Crocker, 1997).

- *Substantive theories, process models, and process engagement.* These aspects of validity include data-based studies examining whether the desired processes are being used by respondents during the assessment. For example, what proof can a test developer provide to conclude that critical thinking or problem-solving skills are actually being employed in an assessment task? Burger and Burger (1994) provided one of the few examples of how this type of evidence can be achieved in their examination of the degree of relationship between locally developed performance assessments of writing and student performance on a valid and reliable standardized test of reading and writing.

- *Scoring models as reflective of task and domain structure.* Scoring guides should be logically consistent with the underlying content and behavioral domains being assessed so that scores can be compared across different tasks and settings.

- *Generalizability and the boundaries of score meaning.* **Generalizability** examines the extent that inferences from test scores can be generalized across different assessment tasks, contexts, and raters (Dunbar, Koretz, & Hoover, 1991). This is particularly important for high-stakes assessments because they involve different raters and, given the uneven allocation of resources among schools, different tasks. The often subjective nature of performance assessment scoring has resulted in rather low reported interrater reliability (Burger & Burger, 1994; Mehrens, 1992). Even more problematic is evidence that performance assessments do not generalize across different tasks (Mehrens, 1992), which "limits the extent to which results from small samples of tasks can be generalized to broader domains" (Kane et al., 1999, p. 5). Linn and Burton (1994) pointed out, however, that focusing on generalizability of tasks rather than the size of standard errors and the effect of standard errors on decision making is misguided. Resolving this issue will not be easy because performance assessments typically are time-consuming to administer and therefore are limited to few assessment tasks at the expense of generalizability to a larger behavioral domain.

• *Convergent and discriminant correlations with external variables.* These types of evidence indicate the degree to which scores from performance assessments are related to predictable outcomes such as selection and placement in applied settings. For example, in PE we might claim that students who score well on a performance assessment of movement skill in a team sport are likely to participate in such activities outside of class (convergent correlation).

• *Consequential validity.* Consequential validity focuses on the positive and negative, intended and unintended consequences of performance assessments. Messick (1996) emphasized that "the primary measurement concern with respect to adverse consequences is that any negative impact on individuals or groups should not derive from any source of test invalidity such as construct underrepresentation or construct-irrelevant variance" (p. 13). Frequently mentioned negative consequences of performance assessments include teaching to the test, narrowing the curriculum, and inflating test scores as a result of overemphasis on test preparation (Chudowsky & Behuniak, 1998). For teachers of PE, awareness of the potential consequences of performance assessments is a necessary first step in the evaluation process. Taleporos (1998) suggested that managing the potential consequences of tests also involves soliciting from various stakeholders, such as students, parents, and administrators, input regarding their needs, wants, and desired outcomes of testing (see Chudowsky & Behuniak, 1998, for an example of using teacher focus groups for this purpose; chapter 5 by Pat Patterson provides further insight into consequential validity).

It is clear that evaluating the validity of performance assessments is a rather complex task that demands consideration of many factors. Messick's (1995) list of six considerations just outlined can be extended. For example, Linn and colleagues (1991) described two additional considerations: (a) cost and efficiency and (b) meaningfulness of the tasks to students.

Kane and colleagues (1999) offered an insightful look into the validation of performance assessments, reducing validation to a question of how we can generalize from the results of observations during assessment to the underlying abilities being measured. Kane and colleagues described a chain of inference beginning with observations of performance linked to observed scores, which are in turn linked to a universe score representing a subset of tasks from the target domain (see figure 12.2). Lastly, the universe score must be linked to the target score representing the expected score over the larger domain of interest. The three inferential links represent three convincing arguments that must be made when validating performance assessments. For example, if scoring criteria are appropriate and scoring guides are used properly, there is support for the first link between observations and observed scores. Therefore, as assessments become more complex and subjective, particular attention must be paid to constructing valid and reliable scoring guides. The second inferential link between observed scores and the universe score provides evidence that the assessment can be generalized beyond the specific tasks used in the assessment and across different raters and occasions. Reliability and generalizability studies (see chapter 3 by Ted Baumgartner) are typically used to provide such evidence. The third and final inferential link between the universe score and the expected score over the target domain is typically accomplished using theoretical rationale, although criterion-related validity studies (Burger & Burger, 1994) have been proposed.

It is clear that "face validity" is desirable but not sufficient evidence for validating performance assessments. It is also clear that inservice PE teachers do not have the time or the technical expertise to include all the validity evidence outlined earlier. Teachers do, however, have the expertise to

- examine the content relevance and content representativeness (see Wood, 1989) of their classroom assessments;

- construct scoring criteria that are logically consistent with the abilities being assessed (see Quellmalz, 1991, for six characteristics of effective performance criteria) and scoring guides and test administration procedures that maximize

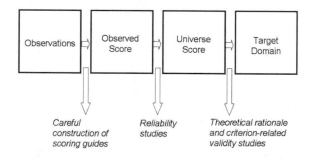

Figure 12.2   The inferential process of validating performance assessments (Kane et al., 1999).

Adapted, by permission, from M. Kane, T. Crooks and A. Cohen, 1999, "Validating measures of performance," *Issues and Practice* 18(2): 5-17.

generalizability across raters and occasions (see Arter & McTighe, 2001; Lund, 2000; and Taggart et al., 1998); and

- consider the positive and negative, intended and unintended consequences of the assessment (for example, pressure from districts to increase test scores may result in teaching to the test and excluding other worthwhile curricula; Miller & Legg, 1993).

Opportunity to Learn

As state-mandated assessment of PE content standards becomes a reality, an issue of some importance is one of **opportunity to learn** (OTL) (also referred to as *opportunity bias* or *opportunity for success;* Phillips, 1996). That is, can schools provide the appropriate instruction, curricula, and resources so that all students can master what is expected of them in the classroom (Herman, Klein, & Abedi, 2000)? For example, if a school plans to assess critical thinking skills in PE, will the curriculum and instructional planning include critical thinking skills and will students be prepared to answer the types of open-ended questions demanded in assessments of critical thinking? Additionally, do PE teachers have the resources and expertise to foster critical thinking skills in their classes? Phillips (1996) provided several interesting illustrations of how OTL can be compromised through differential equipment, using group work to evaluate individual students, bias as a result of outside assistance, and procedural differences in classrooms. Given the large class sizes, relatively short class periods, large numbers of classes per PE teacher, and small PE budgets, OTL in PE is a potential problem and should be monitored constantly. Valid assessment of OTL in low-stakes environments can be accomplished through teacher interviews and student survey instruments (Herman et al., 2000).

Setting Valid Performance Mastery Cutoff Scores

At the core of standards-based education are the notions that (a) content experts can delineate what students should know and be able to do for each content area at each grade level and that (b) student progress can be monitored with performance assessments. A question of major importance is, How do we know when students have reached content mastery? Or, How good is good enough? What performance cutscore on a performance assessment indicates student mastery of the benchmark being assessed? For example, in figure 12.1, which performance cutscore (NA through 6) represents mastery of the content assessed by the test? "Considering that the decisions about students resulting from using mastery cutoffs (e.g., high school graduation) can have serious ramifications, it is little wonder that students, teachers, parents, and administrators are looking carefully at how these indicators of student mastery are determined" (Wood, 2003).

Although the genesis of many standard-setting methods can be traced back to the explosion of criterion-referenced test methodologies in the 1970s and 1980s (see Berk, 1980, 1984; Safrit & Wood, 1989), the recent emphasis on standards-based education has caused a renaissance in criterion-referenced measurement and standard setting (see Cizek, 2001). Jaeger (1989) categorized **performance mastery cutoff** (PMC) setting methods as examinee centered or test centered. Methods using a test-centered approach examine the test alone and not actual examinee responses to the test when deciding on the most valid PMC. For example, the popular Angoff procedure (Angoff, 1971) uses a panel of content experts to determine the probability that a student with mastery of the content will answer a question correctly. The probabilities for all questions on a test are summed to compute the PMC. The primary weakness of test-centered methods is their failure to take into account student performance on the test. Examinee-centered methods, in contrast, use students' test performance to determine a cutoff score. For example, in the body-of-work method, a panel of content experts reviews students' test scores and student work to identify responses demonstrating content mastery. The test scores on these responses are used to determine the PMC (see Kingston, Kahl, Sweeney, & Bay, 2001, and Plake & Hambleton, 2001, for detailed descriptions of examinee-centered methods). The contrasting-groups method is a more statistically oriented examinee-centered approach described in the physical education literature (see Safrit, 1989, and Safrit & Wood, 1995). This method can be used to determine the PMC for tests of motor skill. It employs a decision-validity approach to compare the mastery ratings of content experts to mastery ratings predicted by the test.

More recent contributions to the standard-setting literature include cluster analysis and the bookmark procedure. Cluster analysis is a statistical technique that uses test scores to group examinees into identifiable groups or clusters and is recommended as supplemental evidence for test-centered methods (Sireci, 2001). The bookmark procedure is a test-centered method that employs IRT to sequence test items in order of difficulty. Panelists are instructed to identify the most difficult and easiest items a borderline student would be able to answer (Mitzel, Lewis, Patz, & Green, 2001). A "bookmark" representing "judgment of the divide between items that a student at the threshold of a performance level (the minimally qualified student) should master from those items that are not necessary to master" (Mitzel et al., p. 254) is placed between the most difficult and easiest items to denote a PMC.

Setting valid PMCs is not merely a matter of choosing and implementing a particular standard-setting method. Complicating the process are policy decisions arising from the types of decisions that will be made based on the PMC and the degree of real and perceived abitrariness in the process. Kane (1994) outlined three forms of evidence for validating PMCs. Procedural evidence refers to the adequacy of the procedures used in the standard-setting process and includes the particular standard-setting method chosen and how the method was implemented (e.g., selection and training of content experts; see the summer 1991 issue of *Educational Measurement: Issues and Practice* devoted to the topic and Raymond & Reid, 2001). Documenting criteria for selection of content-expert panels, selection of a psychometrically defensible standard-setting procedure, and detailed documentation of the standard-setting process are legally defensible procedural evidence

(Phillips, 2001). Estimating the standard error due to intrajudge, interjudge, and across-occasions variability, along with analysis of the performance of borderline examinees, provides validity checks based on internal criteria. Lastly, validity checks based on external criteria involve (a) examination of how decisions based on the PMC relate to examinee readiness for a subsequent event such as employment or higher education, (b) comparison of the PMC to the PMC computed from another standard-setting method, or (c) logical comparisons of decisions made with the PMC to decisions made from other assessments. Kane concluded: "None of these sources of evidence makes it possible to fine tune the passing score. Rather, these methods provide reality checks which could detect major flaws . . . . Therefore, even if all available checks on the validity of the standard are implemented, the best that researchers are likely to be able to do is to show that the proposed standard is reasonable" (p. 457).

Clearly, in spite of the recent proliferation of standard-setting procedures, many issues remain in setting valid standards. As Brennan (2001) succinctly put it,

> Standards are not anyone's truths, and just about every aspect of standard setting is a value-laden activity that is subject to disagreement. . . . Many contentious arguments center on the choice of standard-setting methodology . . . . There is little logical or empirical justification for assuming that different methods will or should converge to the same result. Surely, some methods are better than others in specific cases, but there is no Holy Grail. (p. 13)

Does this mean that PE teachers should abandon setting PMCs for their performance assessments? The answer lies in the types of decisions made using such scores. PMCs for classroom assessments used to monitor progress in PE can be developed using modified standard-setting procedures and attending to factors that maximize reliability. As the stakes become higher (e.g., high school graduation), however, greater attention must be given to the process to successfully defend the PMCs to various stakeholders.

Evaluating Student Learning

Unlike the current time-based paradigm in which achievement and success are comparative and competitive, outcomes-based education (OBE) assumes that most, if not all, students can demonstrate mastery on specified educational goals, given enough time (Burger & Burger, 1994).

The emphasis on accountability for student learning through standards-based education and performance assessments is changing teachers' instructional strategies and the way they evaluate students. It is not surprising, then, that new perspectives on assessing student performance have been accompanied by a burgeoning literature on evaluating student learning (see Guskey, 1996a; Guskey & Bailey, 2001; Marzano, 2000; Smith, Smith, & De Lisi, 2001).

Traditionally, "physical education teachers, in general, are basing their grades on questionable criteria and in fact are showing little evidence of using the information normally provided in the undergraduate measurement and evaluation class" (Hensley & East, 1989, p. 310). These practices differ little from classroom teachers who use a hodgepodge of factors when grading and "use a variety of assessment techniques, even if established measurement principles are often violated" (McMillan, 2001). Although it is recommended that PE teachers base student evaluation on cognitive, affective, and psychomotor objectives measured in a formative framework using criterion-referenced standards (Morrow, Jackson, Disch, & Mood, 2000), it is not uncommon for PE teachers to evaluate students based on affective components such as participation and effort using a norm-referenced standard.

In contrast, standards-based education is based on the assumptions that (a) content experts can specify in advance what students should know and be able to do and (b) with appropriate instruction and resources all students can master the standards. Given the tenability of these assumptions, the evaluation process is criterion referenced and formative in nature. Student progress toward mastery instead of student achievement becomes the focus of attention. For example, in Oregon, student progress in meeting the PE content standards is monitored in grades 3, 5, 8 and 10. The focus in grades 3, 5, and 8 is progress toward mastering the exit requirements in grade 10. It is the mastery of 10th grade standards that determines whether a student has met the state's PE requirement. A student who is not meeting the PE benchmarks in grades 3, 5, and 8 presumably will receive the additional instruction necessary to master the all-important exit requirements in the 10th grade. Students who do no meet the exit requirements in grade 10 will be given additional opportunities to meet the requirements prior to graduation.

It is informative to consider how these new perspectives on student evaluation provide clear answers to the following four commonly asked questions concerning evaluation:

- *What criteria should PE teachers use when evaluating students?* Traditionally, measurement specialists have advocated that teachers formulate curricular objectives from the psychomotor, cognitive, and affective domains. It follows that evaluation criteria would come from measurement of these domains. Over the past three decades, PE teachers have relied primarily on assessments in the affective domain (e.g., participation and effort) to evaluate student achievement in PE (Hensley & East, 1989). A much clearer picture of grading criteria is provided using a standards-based approach because the grading criteria are simply the common learning targets (i.e., benchmarks and content standards) developed for PE. Under a standards-based approach, the benchmarks for each content standard at each grade level along with their associated performance assessments are specified in advance, so there is little question concerning grading criteria. For readers who insist that student behavior remains an important grading criterion, I suggest that either a separate content standard be developed for such behavior or that assessments of the affective domain be used to formulate a grade separate from the content standards (e.g., a citizenship grade).

- *Should students fail PE?* A traditional approach to grading in PE employs letter grades to assess achievement of course objectives. Such an evaluation system does not preclude students achieving a failing grade. This issue has less relevance under a standards-based approach to PE. Under a standards-based approach emphasis is placed more on student progress toward meeting benchmarks at each grade level than on evaluating student achievement. Report cards are more of a report on the progress toward meeting benchmarks. If a student does not meet the grade-level benchmark, then his or her progress is reported to the next grade level and appropriate remedial action is taken. Only in high school, when students are required to demonstrate mastery of the exit benchmarks for each content standard, does the risk of failing become real. The risk, however, can be minimized with careful planning. For example, if students are tested for mastery of exit benchmarks in the 10th or 11th grade instead of during their senior year, it is possible for them to be reexamined several times prior to graduation.

- *What method(s) should teachers use in formulating grades?* Although many PE teachers rely on a norm-referenced approach to grading (Hensley & East, 1989), a more defensible approach to grading is the criterion-referenced approach (NASPE, 1995; Safrit & Wood, 1995). Because standards-based PE specifies common learning targets for students and uses performance assessments with PMCs to evaluate the degree to which students have mastered the targets, it is clear that the evaluation approach is criterion referenced.

- *How should grades be reported?* Typically, PE teachers report a single letter grade reflecting students' overall achievement. Guskey (1996b) argued for more extensive reporting formats to adequately communicate student learning to students and parents. Specifically, the following learning criteria should be included when reporting student evaluations:
  - **Product criteria** are the results of performance assessments and indicate students' mastery of benchmarks.
  - **Process criteria** include work habits and affective components that indicate "not just the final results, but how students got there" (Guskey, 1996b).
  - **Progress criteria** evaluate how far students have come and indicate whether students are on track to meet the exit benchmarks.

Moreover, reporting formats should evaluate each criterion separately and delineate the assessments used to measure each criterion. For an example of such a reporting format, see Wood (2003). Readers will note that no overall grade is advocated here because "a single letter grade or a percentage score is not a good way to report student achievement in any subject area because it simply cannot present the level of detailed feedback necessary for effective learning" (Marzano, 2000, p. 106).

Measurement Classes for Teacher Training Programs: A Plea for Authenticity

A plethora of research from the 1960s through the 1990s on teachers' grading practices has identified a theory–practice gap between measurement practices taught in teacher-education measurement classes and the measurement practices of in service PE teachers and classroom teachers (Wood, 1996). As Stiggins (2001) lamented,

> It is as if someone somewhere in the distant past decided that teachers would teach, and they would need to know nothing about accurate assessment. On the other hand, measurement experts would develop and conduct our assessments and would need to know little about day-to-day life in the classroom or the connection of assessment to instruction. As disconnected entities, each facet of the schooling process has developed over the decades with little regard for the other. . . . We enter the new millennium under the same classroom assessment pall that troubled us in 1950. We still cannot guarantee the accuracy of the assessments developed and used day to day by three million teachers in the U.S. (pp. 5-6)

Over the past two decades several authors (e.g., Stiggins, 2001; Veal, 1992; Whittington, 1999; and Wood, 1990) have challenged the measurement community to develop educational measurement courses that have greater authenticity—that is, greater relevance to the environment in which teachers work. Stiggins offered four solutions to the theory–practice gap, including redirecting research resources away from high-stakes testing programs and toward (a) the interactions of assessments and students and (b) the role of assessment in instruction and learning; taking stock of how school assessments are treated in other countries; exploring how assessment practice manifests in teachers' work environment; and using inservice teachers and authentic assessment experiences in educational measurement classes. Whittington offered five suggestions for the curriculum and delivery of instruction in teacher education programs:

1. Clearly address what teachers do and how to do it well.

2. Acknowledge and work with the value systems of preservice and inservice teachers.

3. Incorporate the underlying values associated with various assessment activities.

4. Build on students' entering skills and expectations.

5. Recognize that learning assessment is continuous; no one course, or even one program, can cover it all.

Recommended content for educational measurement classes can be found in the *Standards for Teacher Competence in Educational Assessment of Students* (American Federation of Teachers, National Council on Measurement in Education, and National Education Association, 1990) and in sources such as Schafer (1991) and Arter (1999). In particular, Arter offers a useful summary of performance assessment knowledge and skills based on the 1990 *Standards*.

SUMMARY

Educational reform over the past two decades is rapidly changing the way we view assessment in PE. Emphasis on accountability in the classroom through common learning targets for all students accompanied by authentic performance assessments evaluated within a criterion-referenced framework is creating new perspectives of assessment and its role in the educational process. The seamless integration of assessment into the teaching–learning process has, in its extreme, made assessment the hub around which curriculum and instruction revolve. The promise of the new assessment practice, however, will not be realized unless advocates come to grips with several measurement issues. Of foremost concern are (a) the lack of attention to appropriate validation of classroom assessment devices and (b) validation of performance mastery cutoff scores. In addition, physical educators need to become aware of issues relating to opportunity to learn and the consequences of test score interpretation. Lastly, if a new generation of PE teachers is to be successful in implementing educational reform in PE, measurement courses in teacher training programs must focus on teaching the new assessment practice in more authentic environments integrated throughout the program.

CHAPTER 13

Measuring Physical Activity

Michael J. LaMonte, Barbara E. Ainsworth, and Jared P. Reis

Accurate measurement of physical activity and its related attributes is central to several professional disciplines including physical education (Hastad & Lacy, 1998), exercise science (Maud & Foster, 1995), and public health (LaMonte & Ainsworth, 2001). Physical activity is, however, a complex behavior, and its assessment is difficult, particularly in free-living populations. Lack of a feasible gold standard measure and inconsistent use of physical activity terminology have been long-standing challenges to physical activity measurement. In this chapter we provide (a) definitions for several terms related to physical activity measurement, (b) a conceptual framework for measuring physical activity, (c) an overview of physical activity assessment methods, and (d) suggestions for choosing a method of measurement.

TERMINOLOGY

Different terminology pertaining to physical activity has often been used interchangeably by educators, researchers, and practitioners. This has resulted in confusion, inconsistent study designs, limitations to interstudy comparisons, and lack of standardized measurement practices (Caspersen, Powell, & Christenson, 1985; LaMonte & Ainsworth, 2001; Pate, 1988). Attempts to standardize terminology, however, have been made (Caspersen et al.,1985; Corbin, Pangrazi, & Franks, 2000; Howley, 2001; Pate, 1988). These efforts aimed to develop a universal framework from which definitions could be drawn to aid in operationalizing constructs into measurable study variables, and within which more precise interpretation and comparison of data could be made among studies that relate physical activity to health or physical performance. Several definitions pertaining to physical activity measurement are presented next. Although each of these terms will not be discussed in detail, the list provides a standardized guide for use in various settings.

It is important to recognize that *physical activity* and *energy expenditure* are not synonymous terms. **Physical activity** is a behavioral process in which body movement is produced by skeletal muscle contraction, whereas **energy expenditure** is a direct result of physical activity and reflects the net transfer of energy required to support the skeletal muscle contraction (LaMonte & Ainsworth, 2001). Physical activity is traditionally defined in terms of **mode** (e.g., walking, jumping, dishwashing, chopping wood), **frequency** (number of sessions), **duration** (minutes

Definitions of Terms Related to the Measurement of Physical Activity

Physical activity—Bodily movement that is produced by the contraction of skeletal muscle and that substantially increases energy expenditure.

Exercise—Planned, structured, and repetitive bodily movement done to improve or maintain one or more components of physical fitness. Exercise is a specific subcategory of physical activity.

Physical fitness—A set of attributes (e.g., muscle strength and endurance, cardiorespiratory fitness, flexibility) that people have or achieve that relate to the ability to perform physical activity.

Energy—A product of substrate metabolism that facilitates biological work.

Energy expenditure—The total exchange of energy required to perform a specific type of biological work. It is often used to express the *volume of physical activity* performed during a defined time frame. Energy expenditure can be expressed as a *gross* or *net* term.

Gross energy expenditure—The total amount of energy expended for a specific activity including the resting energy expenditure. Gross energy expenditure is typically used for between-persons comparisons.

Net energy expenditure—The energy expenditure associated exclusively with the activity itself. It is computed as gross energy expenditure minus the individual's resting energy expenditure. This term is used to compare the energy costs of specific activities.

Calorimetry—Methods used to calculate the rate and quantity of energy expenditure when the body is at rest and during physical activity.

Calorie—A unit of energy that reflects the amount of heat required to raise the temperature of 1 g of water by 1 °C.

Kilocalories (kcal)—1,000 calories, 4.184 kilojoules. Used to express the energy expended during physical activity.

Kilojoules (kJ)—The unit of energy in the International System of Units; 1,000 joules, 0.238 kcal.

Metabolic equivalent (MET)—A unit used to estimate the metabolic cost (oxygen consumption) of physical activity. One MET equals the resting metabolic rate of approximately 3.5 ml $O_2 \cdot kg^{-1} \cdot min^{-1}$, or, 1 $kcal \cdot kg^{-1} \cdot hr^{-1}$.

MET-minutes—The rate of energy expenditure expressed as METs per minute, which is calculated by multiplying the minutes a specific activity is performed by the corresponding energy cost (METs) of the activity.

MET-hours—The rate of energy expenditure expressed as METs per hour, which is calculated by multiplying the hours a specific activity is performed by the corresponding energy cost (METs) of the activity.

Mode—The dimension of physical activity that identifies the specific type of activity being performed (e.g., walking, bicycling, jumping, weightlifting, bowling).

Frequency—The dimension of physical activity referring to how often an activity is performed.

Duration—The dimension of physical activity referring to the amount of time (e.g., minutes, hours, days) an activity is performed.

Intensity—The dimension of physical activity referring to the level of effort or physiological demand required to perform the activity. Intensity can be expressed as an *absolute* or *relative* term.

Absolute intensity—A standard or actual rate of energy expenditure (e.g., L O_2 uptake · min^{-1}, METs, kcal · min^{-1}) assumed for a specific activity.

Relative intensity—The rate of energy expenditure for a specific activity expressed as a percentage of the person's maximal capacity to do physical work (e.g., % maximal oxygen uptake, % heart rate reserve).

Hours, minutes—Typical units of time used in quantifying the rate of energy expenditure or the period of physical activity measurement (e.g., kcal per minute or kcal · min^{-1}).

Unitless indices—A unitless number that is computed as an ordinal measure of physical activity (e.g., 10 to 50, low to high activity level).

Dose-response—A relationship in which increasing levels, or "doses," of physical activity result in corresponding changes in the expected levels of a defined physical performance or health parameter.

Adapted from: Caspersen et al., 1985; Corbin et al., 2000; Howley, 2001.

per session), and **intensity** (*absolute* energy cost [e.g., METs] or *relative* effort as a percentage of maximal work capacity [e.g., % heart rate reserve]).

Several types or categories of physical activity exist as outlined in figure 13.1, and they likely overlap to some extent depending on a person's purpose for performing the activity. For example, a brisk walk to and from the store may be a form of transportation for one person, whereas the same brisk walk may be part of a planned exercise program aimed at weight loss for another person. **Exercise** training and competitive sport are subcategories of physical activities that are systematically structured to enhance one or more dimension of **physical fitness** or sport-specific skill to optimize physical performance. Because these types of activities are intentional and often strenuous, they tend to be more easily quantified,

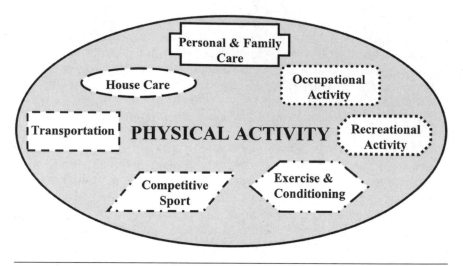

Figure 13.1 Physical activity and related subcategories.

particularly when self-reported questionnaires or motion sensors are the measurement tools (Durante & Ainsworth, 1996; Jacobs, Ainsworth, Hartman, & Leon, 1993; Matthews et al., 2000). In contrast, household, family, and self-care activities are routine in daily life and may contribute considerably to a person's physical activity level. Because of their incidental nature and low exertion level, these types of activity are more difficult to quantify. The degree of overlap among these subcategories of physical activity likely varies from person to person; therefore, measuring independent categories of activity is particularly challenging among large heterogeneous populations.

Additional categorization of physical activities can be based on the intensity or rate of energy expenditure attributed to a specific activity (Ainsworth, Haskell, et al., 2000; Howley, 2001). Activities can be self-rated as *light, moderate,* or *vigorous* intensity (Wilcox et al., 2001), or described according to published intensity categories (Ainsworth, Haskell, et al., 2000; Howley, 2001; Pate et al., 1995). Hence, physical activity may be classified by purpose, such as sports, occupation, and home care; by mode, such as walking, mopping, ice skating; or by intensity, as in light, moderate, and vigorous.

The term *energy expenditure* is used to quantify the total amount, or *volume,* of physical activity performed during a specified period of time. Energy expenditure for individual activities can be estimated by multiplying the frequency, duration, and **absolute intensity** (e.g., METs) of the activity. Total energy expenditure for combined activities can be estimated by summing the energy expenditure estimates of each activity.

To standardize the quantification of energy expenditure and reduce sources of extraneous variation in physical activity measurement, Ainsworth and colleagues (Ainsworth, Haskell, et al., 1993, 2000) have published a systematic approach for assigning MET levels of energy expenditure to specific physical activities. A **metabolic equivalent (MET)** represents the ratio of work to resting metabolic rate (Howley, 2001). It is accepted that resting energy expenditure is approximately 1 MET, which is equivalent to 3.5 ml $O_2 \cdot kg^{-1} \cdot min^{-1}$, or about 1 kcal $\cdot kg^{-1} \cdot hr^{-1}$ (Howley, 2001). Multiplying the MET level of a given physical activity by the duration (e.g., minutes) the activity was performed results in the MET-minute (MET-min). This index quantifies the rate of energy expenditure for the duration an activity is performed, standardized for body size and resting metabolism. The *Compendium of Physical Activities* (Ainsworth, Haskell, et al., 1993, 2000) provides researchers and practitioners with linkages among specific activities, their purpose, and their estimated absolute energy cost expressed in METs. A sample entry from the Compendium is listed in table 13.1.

Column 1 shows a five-digit code that indexes the general class or purpose of the activity. In this example, 12 refers to running and 120 refers specifically to running a 6 min/mi (4 min/km) pace. Column 2 shows the absolute energy cost of the activity in METs. Columns 3 and 4 show the type of activity (running) and a specific example with the related activity code. If a 55 kg (121 lb) female runner completes a 45 min run at a 6 min/mi pace (4 min/km), her energy expenditure for this activity would be 720 MET-min based on the following computation: Frequency (1) × Duration (45 min) × Energy cost of running a 6 min/mi pace (16 METs).

Much of the original work to standardize the energy cost of physical activity was calibrated for a 60 kg (132 lb) person (Ainsworth, Haskell, et al., 1993). Therefore, the conversion between MET-min and **kilocalorie** (kcal) of energy expenditure is approximated by multiplying MET-min by the quotient of a person's body mass divided by 60 (Ainsworth, Haskell, et al., 1993). For a 60 kg person, MET-min is equivalent to kcal; for persons who weigh more than 60 kg, the caloric equivalent will be slightly higher; and for those weighing less than 60 kg, the caloric equivalent will be slightly lower than the MET-min value. The caloric equivalent of 150 MET-min of walking for a 70 kg (154 lb) person is about 175 kcal; for a 60 kg (132 lb) person the caloric equivalent is about 150 kcal; and for a 50 kg (110 lb) person, about 125 kcal. Returning to the 55 kg runner, completing the 45 min run at a 6 min/mi (4 min/km) pace would result in 720 MET-min (16 MET activity × 45 min) or 660 kcal (720 MET-min × [55 / 60]).

Table 13.1   Sample Entry From *Compendium of Physical Activities*

Code	MET	Activity	Example
12120	16	Running	Running 10 mph (6 min/mi)

The previous calculation determined the **gross energy expenditure** for running 45 min at a 6 min/mi pace (4 min/km). This value, however, reflects both the activity-related and resting energy expenditure (Howley, 2001). To account for only the energy expended during the running activity, one must compute the **net energy expenditure** by subtracting from the gross energy expenditure the amount of energy assumed to be expended to sustain resting metabolic functions within the specified activity duration. A 55 kg (121 lb) person has a resting energy expenditure of approximately 55 kcal · hr^{-1} or 0.92 kcal · min^{-1}. Therefore, the amount of energy expended during the 45 min running bout that is attributed to resting metabolism is approximately 41.2 kcal (0.92 kcal · min^{-1} × 45 min). After subtracting this value from the previously computed gross energy expenditure of 660 kcal, a net energy expenditure of about 619 kcal is attributed to the 45 min of running activity. Net energy expenditure should be used when comparing the energy cost of one activity to another, whereas gross energy expenditure should be used when comparing activity-related energy expenditures among individuals.

Another important consideration pertaining to quantifying activity-related energy expenditure is the use of **absolute intensity** versus **relative intensity** scales to index the energy cost of specific activities. Although several factors may influence energy expenditure on a relative scale (e.g., age, body size, fitness level, health status), if one assumes that human mechanical efficiency to perform physical work is fairly constant (~25%), then absolute energy expenditure is generally constant for a given activity. It is, therefore, possible to standardize methods of assigning energy costs to specific activities for the purpose of assessing activity-related energy expenditure among large populations of free-living individuals. Factors such as age, gender, fitness level, and health status will undoubtedly influence the precision by which a standardized activity-specific absolute energy cost reflects a given person's relative intensity level (Arroll & Beaglehole, 1991; Howley, 2001). For example, a "brisk" 3.5 mph (5.6 km/hr) walk on level ground has an absolute energy cost of 4 METs (Ainsworth, Haskell, et al., 1993). For a young, healthy person with a maximal cardiorespiratory capacity of 12 METs, the relative intensity is 33% of maximal capacity; whereas for an older person with a maximal capacity of 6 METs, the relative intensity is 67%. The issue of absolute versus relative intensity is probably most important when prescribing exercise or when categorizing people into intensity-specific activity levels (e.g., moderately versus vigorously active). Furthermore, feasibility considerations related to individualized measures of relative energy expenditure limit assessment methods in large, free-living populations to the use of absolute energy cost.

Because energy expenditure is closely related to body size, it is essential to account for this factor when comparing activity-related energy expenditure among individuals (Gretebeck & Montoye, 1992; Howley, 2001). Energy expenditure should, therefore, be expressed relative to body mass, for example, as METs or kcal per kilogram of body mass per minute (kcal · kg^{-1} · min^{-1}). Returning to the 55 kg (121 lb) female runner who completes a 45 min run at a 6 min/mi pace (4 min/km), the absolute net energy expenditure was 619 kcal, whereas the net energy expenditure relative to this woman's body mass would be about 11.3 kcal · kg^{-1} during the 45 min run. A 75 kg (164 lb) person who completes the same running task would have an absolute net energy expenditure of 844.75 kcal, but when expressed per kg of body weight, the energy expenditure is the same (11.3 kcal · kg^{-1}) as that computed for the lighter runner.

Defining and standardizing terms associated with physical activity measurement is a critical step in reducing unwanted sources of variation and producing unbiased estimates of activity-related energy expenditure for use in research and applied settings (Ainsworth, Haskell, et al., 1993; Arroll & Beaglehole, 1991; Jacobs et al., 1993; Montoye, Kemper, Saris, & Washburn, 1996). Although the use of a standardized compendium of physical activities potentially results in large differences among people's absolute net energy expenditure, after accounting for body size, differences in energy expenditure for a given activity are quite small. The *Compendium of Physical Activities* may not resolve every issue related to individual versus population-based assessment of activity-related energy expenditure. It does, however, provide a standardized method for use in research and practical settings, which should facilitate more consistency, precision, and reproducibility in measuring activity-related energy expenditure. Seasonal and day-to-day intraindividual variation in physical activity patterns (Gretebeck & Montoye, 1992; Levin, Jacobs, Ainsworth, Richardson, & Leon, 1999), and discordance between self-rated and actual activity intensity (Robertson et al., 1982; Wilcox et al., 2001), influence the precision of measuring activity and energy expenditure. Further, because subcategories of physical activity have different meanings according to gender, race, ethnicity, and cultural perspectives (Ainsworth, Richardson, Jacobs, & Leon, 1993; Warnecke et al., 1997), self-report activity instruments must reflect the specific demographics and lifestyle of the target population. Accordingly, these issues should be considered when choosing a method of assessing physical activity and its related energy expenditure.

CONCEPTUAL FRAMEWORK FOR PHYSICAL ACTIVITY MEASUREMENT

Measurement begins with identifying the construct of interest (e.g., energy expenditure), which is operationalized into a measurable variable (e.g., caloric expenditure in units of kcal · kg^{-1} · day^{-1}). An appropriate measurement method (e.g., doubly labeled water) is then chosen based on considerations related to characteristics of the target population, measurement costs, and administrative burden. To incorporate the terminology described previously into a framework that can guide the measurement of physical activity and energy expenditure under laboratory and field conditions, we could argue that the construct of interest might best be defined as "movement." Figure 13.2 shows that movement can be operationalized into two measurable variables: physical activity (e.g., min · d^{-1} of "brisk" [2.5 to 4.0 mph or 4 to 6 km/hr] walking) and energy expenditure (MET-min · d^{-1} of "brisk" [2.5 to 4.0 mph or 4 to 6 km/hr] walking). Direct and indirect measures of physical

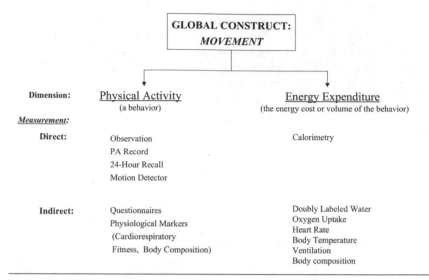

Figure 13.2   Conceptual framework for defining and assessing physical activity and energy expenditure.

Reprinted, by permission, from M.H. LaMonte and B.E. Ainsworth, 2002, "Quantifying energy expenditure and physical activity in the context of dose response," *Medicine and Science in Sports and Exercise 33*(6 Suppl), S370-S378.

activity and energy expenditure are available. In some settings (e.g., public health research) physical activity measures may be extrapolated to units of energy expenditure prior to evaluating associations between energy expenditure and specific health parameters (e.g., cardiovascular disease mortality). Measures of physical activity may provide acceptable estimates of activity-related energy expenditure depending on the degree of concordance between the activity measure and direct measures of energy expenditure. Table 13.2 summarizes several methods of assessing physical activity and energy expenditure that will be described in the next sections. Comprehensive reviews of free-living and laboratory measures of physical activity and energy expenditure for children (Freedson & Melanson, 1996; Kohl, Fulton & Caspersen, 2000; Welk, Corbin, & Dale, 2000) and adults (Ainsworth, Montoye, & Leon, 1994; LaMonte & Ainsworth, 2001; Maud & Foster, 1995; Montoye et al., 1996; Washburn, 2000; Welk, 2002) have recently been published.

MEASURES OF ENERGY EXPENDITURE

A complex series of biochemical processes results in the transfer of metabolic energy to drive skeletal muscle contraction during physical activity (Brooks, Fahey, & White, 1996). In addition to body movement, a large amount of heat energy is produced during the coupling of energy transfer and skeletal muscle contraction. The rate of heat production is directly proportional to the net activity-related energy

Table 13.2 Common Methods of Assessing Physical Activity and Energy Expenditure

	Dimension measured (M) or estimated (E)	Units†	Adminis-trative/ technical burden††	Advantage (A)/ disadvantage (D) §
Direct measures				
Observation	(M): Physical activity	F, D, T	Moderate	A: 1, 3, 4, 9
				D: 3, 10
Room calorimetry	(M): Energy expenditure	kcal	High	A: 2, 3
				D: 1, 10
Doubly labeled water	(E): Energy expenditures	kcal	High	A: 2, 7
				D: 1, 2, 4, 5, 6, 10
Accelerometer	(M): Physical activity	F, D, I	High	A: 1, 3, 8, 9
		kcal, METs		D: 1, 3, 5, 8, 9
	(E): Energy expenditure			
Pedometer	(M): Physical activity	Steps	Low	A: 1, 3, 5, 6, 10
				D: 3, 4, 5, 6, 7, 8, 9
Activity records, diaries, logs	(M): Physical activity	F, D, T	High	A: 1, 3, 4, 5, 8, 9
		kcal, METs		D: 3, 10
	(E): Energy expenditures			
Indirect measures				
Indirect calorimetry	(E): Energy expenditure	kcal, METs	High	A: 2, 3, 10
				D: 1, 3, 4, 5, 7, 8, 10
Physiologic measures (e.g., heart rate)	(E): Energy expenditure	kcal, METs	High	A: 2, 8, 9, 10
				D: 1, 3, 5

(continued)

Table 13.2 *(continued)*

	Dimension measured (M) or estimated (E)	Units†	Adminis-trative/ technical burden††	Advantage (A)/ disadvantage (D) §
Indirect measures				
Physical activity questionnaires	(M): Physical activity (E): Energy expenditure	F, D, T kcal, METs	Low	A: 4, 5, 6, 7, 9 D: 6, 11, 12

† F = frequency; D = duration; I = intensity; T = type

†† Administrative/technical burden refers to the measurement cost, administration time, examiner and subject burden, technical aspects of measurement instrumentation, and data processing.

§ Advantages (A): 1 = objective indicator of body movement; 2 = objective measure of energy expenditure; 3 = noninvasive; 4 = provides quantitative and qualitative information; 5 = inexpensive/can use in large samples; 6 = quick administration; 7 = not likely to alter behavior; 8 = stores data for extended observation; 9 = allows for assessment of activity bouts (e.g., specific time intervals of activity [i.e., 10 min]); 10 = can use to prescribe activity

§ Disadvantages (D): 1 = cost prohibitive for large-sample applications; 2 = invasive; 3 = likely to alter behavior; 4 = lacks temporal dimension for activity; 5 = lacks information on type of activity; 6 = lacks information on intensity of activity; 7 = lacks information on frequency of activity; 8 = limited range of activity assessment; 9 = lack of standard methods for data application; 10 = time intensive; 11 = recall bias; 12 = sensitivity/meaning varies by sex, age, race, ethnicity, culture

expenditure; therefore, energy expenditure can be precisely quantified by measuring body heat at rest or during exercise (Jequier, Acheson, & Schutz, 1987; Montoye et al., 1996). The oxidation of food substrate is a primary source of energy production at rest and during and following physical activity. Based on assumptions about the energy cost of substrate oxidation, activity-related energy expenditure can be estimated by measuring the fractional concentrations of expired carbon dioxide and oxygen at rest or during physical activity (Brooks et al., 1996; Jequier et al., 1987; Montoye et al., 1996). Laboratory and field methods exist for direct measures of heat production (e.g., room calorimetry) and ventilatory gas exchange (e.g., indirect calorimetry). Following is an overview of these methods.

DIRECT MEASURES OF ENERGY EXPENDITURE

Direct measures of activity-related heat production require sophisticated instrumentation, which precludes such measures in large free-living populations. While providing highly precise measures of energy expenditure, laboratory methods generally restrict the type and pattern of physical activity that can be studied. However, these measures are often used in small controlled laboratory studies aimed at developing cost-effective field measures of activity-related energy expenditure.

Direct Calorimetry

The measurement of body heat production is known as direct calorimetry and is the most precise measure of energy expenditure (Brooks et al., 1996; Jequier et al., 1987; Horton, 1983; Montoye et al., 1996). Calorimetry is performed under laboratory conditions with the person in a small airtight chamber that contains insulated pipes used to circulate water through the calorimeter. Based on the temperature and flow rate of water as well as heat loss from the chamber, a subject's heat production, and thus energy expenditure, can be measured at rest to within 1 or 2% error (Horton, 1983; Jequier et al., 1987). Measurements are typically made over a 24 hr period and follow a 10 to 12 hr fast so that resting metabolic rate (RMR) can be accurately assessed. Room calorimeters are typically small and confined, and are not practical for assessing energy expenditure related to free-living activity patterns. A phase delay between body heat release during exercise and calorimeter sampling may require extended observation periods. Webb, Annis, and Troutman (1980) used a water-cooled suit to measure energy expenditure with direct calorimetry during controlled bouts of exercise. Although this technique may be more conducive to studying energy expenditure for a wider variety of physical activities, its accuracy may be lower by 3 to 23% than room calorimetry (Webb et al., 1980) because of issues with dissipating perspiration and altered mechanical efficiency during exercise. Cost and technical limitations make direct calorimetry generally infeasible for use outside laboratory settings.

Indirect Measures of Energy Expenditure

Direct measures of activity-related heat production may not be practical in many settings. Activity-related energy expenditure, however, can be estimated by measuring the rate of oxygen uptake ($\dot{V}O_2$) and carbon dioxide production ($\dot{V}CO_2$) associated with the energy transfer of substrate oxidation (Ferrannini, 1988; Jequier et al., 1987). $\dot{V}O_2$ and $\dot{V}CO_2$ measurements can be based on expired ventilatory gas analysis or radioisotope concentrations obtained from serial urine samples. These methods of assessing energy expenditure are referred to as **indirect calorimetry** (Ferrannini, 1988; Jequier et al., 1987).

Doubly Labeled Water

Doubly labeled water (DLW) provides a precise measure of activity-related energy expenditure. Because energy expenditure is estimated from the rate of metabolic CO_2 production (Lifson, Gordon, & McClintock, 1955; Speakman, 1998), DLW should be considered a form of indirect calorimetry (Ferrannini, 1988; Jequier et al., 1987). DLW consists of the stable water isotopes ^2H_2O and $H_2{}^{18}O$ dosed according to body weight. Urinary isotope excretion is tracked over several days using an isotope-ratio mass spectrometer. Labeled hydrogen (^2H_2O) is excreted as water, whereas labeled oxygen ($H_2{}^{18}O$) is lost as water and CO_2 ($C^{18}O_2$) produced by the carbonic anhydrase system. The difference in isotope turnover rates provides a measure of metabolic $\dot{V}CO_2$. Oxygen uptake and total body energy expenditure is extrapolated from CO_2 and estimated respiratory quotient (RQ) obtained from published equations (Black, Coward, Cole, & Prentice, 1986). Under steady-state conditions, RQ reflects the relative percentage of carbohydrate and fat oxidation and is calculated as $\dot{V}CO_2/\dot{V}O_2$. Inherent error will exist in DLW energy expenditure measures when RQ is estimated and when measurements are made under non-steady-state conditions such as exercise (Montoye et al., 1996).

The DLW method has been used to assess activity-related energy expenditure in laboratory and field settings (Black et al., 1996; Bouten et al., 1996; Conway, Seale, Jacobs, Irwin, & Ainsworth, 2002; Seale, Rumpler, Conway, & Miles, 1990; Westerterp, Brouns, Saris, & Ten-Hoor, 1988). DLW estimates of energy expenditure have ranged from 1 to 20% of those derived from measured oxygen uptake at rest and during a variety of physical activity intensities (Conway et al., 2002; Seale, Rumpler, Conway, & Miles, 1990; Westerterp et al., 1988). Discrepancies between free-living DLW studies and laboratory-based indirect calorimetry likely reflect a greater amount of free-living activity-related energy expenditure than can be simulated and measured under controlled laboratory conditions. Although DLW provides precise estimates of free-living energy expenditure over prolonged periods (e.g., a week), a major shortcoming of this technique is the fact that it quantifies only total energy expenditure. DLW does not differentiate the duration, frequency, or intensity of activity-related energy expenditure, which results in a lack of information about individual or population physical activity patterns, and precludes examining **dose-response** issues pertaining to specific intensities, durations, or frequencies of activity in relationship with health parameters or physical performance. A "physical activity level" index (PAL) has been computed as total daily energy expenditure measured with DLW divided by measured or estimated resting metabolic rate (Black et al., 1996). Lack of qualitative information and potential errors associated with estimated rather than measured resting metabolic rate limit the application and precision of the PAL.

Although DLW may be considered the gold standard field measure of total energy expenditure, it has very little value in settings in which quantification of intensity-specific energy expenditure (e.g., 150 kcal in activities ≥ 3 METs [Pate et al., 1995]) is desired. The cost of the isotopes and mass spectrometry assays precludes DLW measures in large studies of free-living individuals, and it generally limits this technique to small studies aimed at validating more feasible methods of assessing total daily energy expenditure.

Isotope-Labeled Bicarbonate

The labeled bicarbonate ($NaH^{14}CO_3$) method is very similar to DLW and has been used to measure free-living total daily energy expenditure over shorter observation periods (e.g., days) than in studies of DLW (Elia, Fuller, & Murgatroyd, 1992). A specific amount of isotope is infused at a constant rate and eventually diluted by the body's CO_2 pool. Labeled carbons are recovered from expired air, blood, urine, or saliva, and metabolic $\dot{V}CO_2$ is determined from an isotope dilution curve. Total energy expenditure is estimated from $\dot{V}CO_2$ and assumptions made about RQ. Experimental studies of this method have shown energy expenditure estimates to be within 6% of that measured in a respiratory chamber (Elia et al., 1992, 1995; el-Khoury, Sanchez, Fukagawa, Gleason, & Young, 1994). Practical concerns similar to those discussed for DLW apply to the labeled bicarbonate method.

Oxygen Uptake

The heat energy released from substrate oxidation during physical activity can be estimated from measured **oxygen uptake** ($\dot{V}O_2$). Energy expenditure estimates are based on assumed relations between $\dot{V}O_2$ and the caloric cost of substrate oxidation (Brooks et al., 1996; Ferrannini, 1988; Jequier et al., 1987). Similar to the room calorimeter (see the section on direct calorimetry), the respiratory chamber is an airtight, insulated temperature- and humidity-controlled room (Brooks et al., 1996; Jequier & Schutz, 1983; Jequier et al., 1987; Montoye et al., 1996). Specific concentrations of O_2 and CO_2 are introduced to the room at a controlled rate of airflow. Fractional concentrations of O_2 and CO_2 are measured as the air leaves the system. Based on the gas concentrations and flow rate of expired air, $\dot{V}O_2$ and $\dot{V}CO_2$ can be determined. Together, $\dot{V}O_2$ and RQ ($\dot{V}CO_2/\dot{V}O_2$) can be used to estimate energy expenditure in kcal $\cdot$ min^{-1} according to Weir's equation (Weir, 1949):

$$\text{Energy expenditure (kcal} \cdot \text{min}^{-1}) = \dot{V}O_2 \, (3.9 + 1.1 \, \text{RQ}) \quad (13.1)$$

RQ is typically estimated from the respiratory exchange ratio (RER) computed as $\dot{V}CO_2/\dot{V}O_2$ from expired ventilatory gas concentrations. Because the quantity of CO_2 produced relative to that of O_2 consumed varies with the substrate mixture being metabolized, error can be introduced into the estimation of energy expenditure when RQ is not directly measured. Therefore, measured RQ, requiring urinary nitrogen, is recommended for precise estimates of activity-related energy expenditure from indirect calorimetry (Ferrannini, 1988; Jequier et al., 1987). It is, however, likely that the error introduced from RQ estimates is minimal because the contribution of protein to energy metabolism during physical activity is typically very small

(Brooks et al., 1996). Consequently, a simplified approach to estimating energy expenditure from $\dot{V}O_2$ is based on an assumed 5 kcal of energy expenditure for every liter of oxygen consumed per minute (L · min⁻¹) (Brooks et al., 1996):

$$\text{Energy expenditure (kcal · min}^{-1}) = \dot{V}O_2 \text{ (L · min}^{-1}) \times 5 \quad (13.2)$$

Correction factors have been published that match the measured RER with a specific energy cost of oxidizing the substrate mixture represented by the RER in order to improve the estimation of energy expenditure from $\dot{V}O_2$ (Brooks et al., 1996). Other sources of error in quantifying activity-related energy expenditure from $\dot{V}O_2$ can result because of bicarbonate buffering of metabolic CO_2 during exercise, and postexercise oxygen consumption kinetics.

Respiratory chambers are expensive and require a large space and extensive upkeep. Nevertheless, they have been used to study activity-related energy expenditure in a few instances (Jequier & Schutz, 1983; Schulz, Nyomsa, Alger, Anderson, & Ravussin, 1991). It is likely, however, that free-living physical activity patterns and energy expenditure are altered in a respiratory chamber. Therefore, methods for performing oxygen-uptake-based indirect calorimetry outside the chamber have been developed. These techniques are based on the same principles described previously and use an integrated measurement system comprised of an O_2 and CO_2 analyzer, a ventilation flow-volume meter, and a microcomputer to process expired air collected through a fitted hood, face mask, or mouthpiece (Davis, 1996). Small portable indirect calorimeters have been used to measure $\dot{V}O_2$ and energy expenditure in field settings (Bassett et al., 2000; King, McLaughlin, Howley, Bassett, & Ainsworth, 1999; Strath, Bassett, Thompson, & Swartz, 2002). Cost issues, and the necessity of wearing instrumentation that likely alters usual patterns of physical activity limit the utility of portable indirect calorimeters for quantifying activity-related energy expenditure among free-living individuals.

Heart Rate (HR) Monitoring

Because of the administrative and technical difficulties associated with indirect calorimetry, methods of monitoring physiological parameters that are closely associated with oxygen uptake have been used in field settings. HR has been used to estimate activity-related energy expenditure based on the assumption of a linear relationship between HR and $\dot{V}O_2$ (Wilmore & Haskell, 1971). Because the HR–$\dot{V}O_2$ relationship is somewhat attenuated during low and very high intensity activities (Acheson, Campbell, Edholm, Miller, & Stock, 1980), and because of considerable between-persons and day-to-day HR–$\dot{V}O_2$ variability (Li, Deurenberg, & Hautvast, 1993; McCroy, Mole, Nommsen-Rivers, & Dewey, 1997), individual HR–$\dot{V}O_2$ calibration curves are necessary for precise estimation of activity-related energy expenditure (Haskell, Yee, Evans, & Irby, 1993). One method requires establishing a threshold HR, known as the "FLEX HR," prior to estimating activity-related energy expenditure from the HR–$\dot{V}O_2$ calibration curve (Livingstone et al., 1990). The FLEX HR is determined in the laboratory with indirect calorimetry studies of various types and intensities of physical activity. Free-living activities eliciting an HR below the FLEX HR are assigned an activity-related energy cost based on resting energy expenditure (e.g., 1 MET). Activities eliciting an HR above the FLEX HR are assigned an activity-related energy cost based on the person's HR–$\dot{V}O_2$ regression.

Moderate to high correlations (r = .53 to .73) have been reported between total energy expenditure estimated from DLW and HR values based on individual HR–$\dot{V}O_2$ calibrations (Shultz, Westersterp, & Bruck, 1989). Individual variability has, however, been high (Li et al., 1993; Livingstone et al., 1990). Livingstone and colleagues reported difference scores of –22% to +52% between total energy expenditure estimated from DLW and FLEX HR. Because of the potential for large individual errors in activity-related energy expenditure based on HR–$\dot{V}O_2$ calibrations, some researchers recommend against using this method (Washburn & Montoye, 1986). Recently, however, Strath and colleagues (2000) showed a strong correlation between activity-related energy expenditure estimated from HR reserve and indirect calorimetry (r = .87, standard error of estimate [SEE] = .76 METs) among adults performing moderate-intensity lifestyle activities. The authors emphasized the potential value of basing activity-related energy expenditure on relative versus absolute HR measures to reduce sources of between-persons variance such as age, gender, fitness, and health status.

Simultaneous measurement of heart rate and body motion has also been used to improve assessment of free-living activity-related energy expenditure (Haskell et al., 1993; Strath et al., 2000). Strath and colleagues described the following method: Individual HR–$\dot{V}O_2$ calibrations are established from oxygen uptake during upper body (e.g., arm ergometry) and lower body (e.g., treadmill exercise) graded exercise. Motion sensors (e.g., accelerometer) are worn on upper and lower body segments to determine whether activities are primarily upper or lower body movements, and to screen out non-activity-related elevations in HR (e.g., emotional or temperature stimuli). A threshold level of motion sensor output is used to differentiate low-intensity from moderate- and vigorous-intensity activities. Activities below the motion sensor threshold are weighted with a constant for resting metabolism (e.g., 1 MET), whereas activities above the motion sensor threshold are assigned an intensity weight based on the measured HR and either the upper or lower body HR–$\dot{V}O_2$ regression. A ratio technique is used to determine which HR–$\dot{V}O_2$ regression should be applied for activities that exceed the motion sensor threshold for both arms and legs. Using oxygen uptake as the criterion, Strath and colleagues (2002) showed that the simultaneous HR-motion sensor method was significantly better at estimating total energy expenditure and time spent in different intensities of physical activity than the FLEX HR during 6 hr of field activities.

Several factors influence HR without having substantial effects on oxygen uptake, which may result in imprecise estimates of activity-related energy expenditure. Such factors include body temperature, size of the active muscle mass (e.g., upper vs. lower body), type of exercise (static vs. dynamic), stress, and medication

(Acheson et al., 1980; Montoye et al. 1996). Day-to-day HR variability reduces the reliability of energy expenditure estimates (Washburn & Montoye, 1986). The combination of HR and motion monitoring may overcome some limitations; however, the need to develop individual HR–$\dot{V}O_2$ calibration curves and instrumentation costs ($\geq$\$150 per HR unit, >\$400 per accelerometer) makes this method less suitable for routine use in free-living settings.

MEASURES OF PHYSICAL ACTIVITY

The complexity of physical activity behaviors and the lack of a universally accepted criterion measure have resulted in the development of several assessment methods for quantifying individual and group activity levels. Physical activity measures range in precision from crude categorization of activity status (e.g., sedentary vs. active) to detailed descriptions of activities (e.g., type, duration, frequency) and their estimated energy cost, to monitoring body motion or physiological parameters known to vary with individual levels of physical activity. Following is an overview of some direct and indirect methods used to assess physical activity in various settings.

Direct Measures of Physical Activity

Direct measures of physical activity are used to characterize the actual type, frequency, duration, and intensity of movement during specific observation periods. Physical activity records and motion sensors are used to collect information while a person engages in various activities. Activity logs and checklists collect information at the end of a defined observation period (e.g., the end of a day, every 60 min). None of these methods is adequate to completely characterize physical activity patterns. They are, however, considered objective quantitative measures of activity exposure. Therefore, direct measures of physical activity are often used as validation criteria for indirect or subjective indexes of physical activity.

Physical Activity Records

Physical activity records are ongoing accounts of activity patterns kept in diary format during a defined observation period (Ainsworth, Irwin, Addy, Whitt, & Stolarczyk, 1999; Ainsworth, Montoye, & Leon, 1994; Conway et al., 2002; Eason et al., 2002). A sample physical activity record is shown in figure 13.3(a). Typically, respondents record information about the type (e.g., sleep, walking, digging), purpose (e.g., transportation, occupation, exercise), duration (e.g., minutes), self-rated intensity (light, moderate, vigorous), and body position (reclining, sitting, standing, walking) for every activity completed during the specified observation period (e.g., 24 hr). Physical activity records are scored using the *Compendium of Physical Activities* (Ainsworth, Haskell, et al., 1993, 2000) to link a MET intensity score with the type and purpose of each activity entered. Once scored, the physi-

Time began	Position (circle one)	Description (What are you doing?)	How hard? (circle one)	Activity group (circle one)	(leave blank) Code	Mins
8:05 (AM) PM	Recline Sit Stand *(Walk)*	*Walk in house to fix breakfast*	*(Light)* Moderate Vigorous	SC HH PAR TRANS OCC (WALK) INAC LG EC MISC	17150	1
8:06 (AM) PM	Recline Sit *(Stand)* Walk	*Fix breakfast*	*(Light)* Moderate Vigorous	SC (HH) PAR TRANS OCC WALK INAC LG EC MISC	05050	4
8:10 (AM) PM	Recline (Sit) Stand Walk	*Eat breakfast*	*(Light)* Moderate Vigorous	(SC) HH PAR TRANS OCC WALK INAC LG EC MISC	13030	5
8:15 (AM) PM	Recline Sit Stand *(Walk)*	*Gather things to leave home*	Light *(Moderate)* Vigorous	SC HH PAR TRANS OCC WALK INAC LG EC (MISC)	09071	3
8:18 (AM) PM	Recline Sit Stand *(Walk)*	*Walk to car*	Light *(Moderate)* Vigorous	SC HH PAR TRANS OCC (WALK) INAC LG EC MISC	17161	2
8:20 (AM) PM	Recline (Sit) Stand Walk	*Drive car to work*	*(Light)* Moderate Vigorous	SC HH PAR (TRANS) OCC WALK INAC LG EC MISC	16010	

Figure 13.3(a) Direct and indirect methods of physical activity assessment: physical activity record.

cal activity record provides a detailed account of the time (e.g., minutes) spent in various types, intensities, and patterns of physical activity. Activity-related energy expenditure (e.g., MET-min · d^{-1}) can then be quantified from these physical activity dimensions. The physical activity record is an accurate (Conway et al., 2002) and reproducible (LaMonte, Durstine, Addy, Irwin, & Ainsworth, 2001) method of tracking the time spent in specific types and patterns (e.g., single continuous bouts, sporadic intermittent bouts) of physical activities that accounts for individual or population activity-related energy expenditure. Physical activity records can be kept seasonally (e.g., winter, summer) to obtain information on seasonal variations in habitual activity patterns (Levin et al., 1999). Instructing participants to put entries into the activity record at the time a behavior is executed reduces the influence of recall bias (Durante & Ainsworth, 1996) on the precision of quantifying physical activity and related energy expenditure.

Physical activity records have been used to obtain comprehensive accounts of free-living physical activities and energy expenditure (Ainsworth et al., 1999; Ainsworth, LaMonte, et al., 2000; Bouchard et al., 1983; Conway et al., 2002; Eason et al., 2002; Richardson et al., 2001; Wilcox et al., 2001). Conway and colleagues reported a difference of only 7.9 ± 3.2% between a seven-day physical activity record and DLW estimates of free-living activity-related energy expenditure in men. Moderate to strong age-adjusted correlations ($r = .35$ to.68) have been reported between total MET-min per day of energy expenditure from physical activity records and an accelerometer in free-living men and women (Richardson et al., 1995).

Habitual physical activity patterns can be characterized in free-living populations using physical activity records. Ainsworth and colleagues (1999) described the types and patterns of physical activity and related energy expenditure among Caucasian, African American, and Native American women living in the southeastern and southwestern regions of the United States as part of the Women's Health Initiative (Riley & Finnegan, 1997). Physical activity records were administered during three 4-day observation periods separated by one month to better quantify usual activity patterns. Focus groups and individual debriefings were used to enhance the richness and interpretation of data gleaned from the physical activity records (Henderson, Ainsworth, Stolarzcyk, Hootman, & Levin, 1999). Physical activity surveys were then developed for field assessment of habitual daily physical activity and its related energy expenditure among these specific groups of women (Ainsworth, LaMonte, et al., 2000).

Physical activity records provide a very detailed and comprehensive method of assessing free-living physical activity patterns. Feasibility is limited by cost, the potential for altered behavior, the need for multiple and extended observation periods, and administrative burden on the participant and investigator. For these reasons, physical activity records may be most useful for individual activity assessments or as a criterion measure for validating simple physical activity measurements that are more practical for field use.

Physical Activity Logs and 24 Hr Recalls

Physical activity logs, as shown in figure 13.3(b), are simplified physical activity records that are structured as a checklist of activities specific to the administrator's interests or the target population's usual activity pattern. The Bouchard Physical Activity Log (Bouchard et al., 1983) requires respondents to record an item from nine intensity- and type-specific activity categories every 15 min during three 24 hr sampling frames (e.g., two weekdays, one weekend day). The Ainsworth Physical Activity Log, shown in figure 13.3(b), is a modifiable list consisting of 20 to 50 activities originally developed for use in ethnically diverse populations of women (Ainsworth, Bassett, et al., 2000; Ainsworth, LaMonte, et al., 2000; Henderson et

ID # _____ **Day of the week** _____ **Day #** ___ **Date** ___ / ___ / ___

HOUSEHOLD			Amount of time Hours Minutes					Amount of time Hours Minutes

HOUSEHOLD

Indoors

Cooking	Yes	No	___	___
Cleaning up	Yes	No	___	___
Laundry	Yes	No	___	___
Shopping	Yes	No	___	___
Dusting	Yes	No	___	___
Scrubbing	Yes	No	___	___
Vacuuming	Yes	No	___	___
Home repair	Yes	No	___	___
Mopping	Yes	No	___	___
Washing car	Yes	No	___	___
Other	Yes	No	___	___

Outdoors

Mowing lawn	Yes	No	___	___
Raking lawn	Yes	No	___	___
Weeding	Yes	No	___	___
Sweeping	Yes	No	___	___
Shoveling	Yes	No	___	___
Pruning	Yes	No	___	___
Chopping wood	Yes	No	___	___
Other	Yes	No	___	___

CARE OF OTHERS

Bathing	Yes	No	___	___
Feeding	Yes	No	___	___
Playing	Yes	No	___	___
Lifting	Yes	No	___	___
Pushing wheelchair	Yes	No	___	___

TRANSPORTATION

Drive car	Yes	No	___	___
Ride in car	Yes	No	___	___

WALKING

To get places	Yes	No	___	___
For exercise	Yes	No	___	___
With the dog	Yes	No	___	___
Work breaks	Yes	No	___	___

DANCING

In church	Yes	No	___	___
For pleasure	Yes	No	___	___
Other	Yes	No	___	___

SPORTS

Golf	Yes	No	___	___
Team sports	Yes	No	___	___
Other	Yes	No	___	___

CONDITIONING

Aerobics	Yes	No	___	___
Jogging	Yes	No	___	___
Swimming	Yes	No	___	___
Bicycling	Yes	No	___	___
Stretching	Yes	No	___	___
Weightlifting	Yes	No	___	___
Other	Yes	No	___	___

INACTIVITY

Watching TV	Yes	No	___	___
Reading or sewing	Yes	No	___	___
Using computer	Yes	No	___	___
Other	Yes	No	___	___

OCCUPATION

Sitting tasks	Yes	No	___	___
Standing tasks	Yes	No	___	___
Walking tasks	Yes	No	___	___
Heavy labor	Yes	No	___	___

VOLUNTEER

Sitting tasks	Yes	No	___	___
Standing tasks	Yes	No	___	___
Walking tasks	Yes	No	___	___
Heavy labor	Yes	No	___	___

MONITORS

Time put ON	_____ a.m.
Time taken OFF	_____ p.m.
Steps taken	_____

Figure 13.3(b) Direct and indirect methods of physical activity assessment: physical activity log.

Copyright © 2002 from Field assessment of physical activity and energy expenditure among athletes. In *Nutritional assessment of athletes* by LaMonte, M.J., & Ainsworth, B.E., edited by J.A. Driskell & I. Wolinsky, 2002. Adapted by permission of Routledge/Taylor & Francis Group. LLC.

al., 1999). At the end of a specified observation cycle (e.g., every 8 hr, the end of the day), respondents identify the type and duration of activities performed during that time period. Activity-related energy expenditure (e.g., MET-min · d^{-1}) is computed by assigning intensity values from the *Compendium of Physical Activities* to each activity selected. The log takes only a few minutes to complete and can be scored quickly to provide information on time spent in specific types (e.g., walking, computer work) or categories (e.g., exercise, house care) of physical activity and related energy expenditures.

Physical activity logs may be more convenient to complete and process than physical activity records. Activity logs, however, may underestimate actual physical activity levels and related energy expenditure if participants engage in activities other than those listed on the log. Because the log is completed at the end of the day, the degree of recall bias associated with this method is likely to be somewhat higher than with the physical activity record.

A **24 hr physical activity recall** has been conducted as a telephone interview to acquire detailed information on activity levels during the past 24 hr (Matthews et al., 2000, Matthews, DuBose, LaMonte, Tudor-Locke, & Ainsworth, 2002). Activity recalls take 20 to 50 min to complete and are similar to physical activity records and logs in that they assess the type, purpose, duration, and related energy expenditure of reported activities. Test–retest reliability coefficients have ranged from $r = .22$ to $r = .58$, and criterion-related validity correlations between total MET-hr · d^{-1} from the 24 hr recalls and an accelerometer were $r = .74$ and $r = .32$ for men and women, respectively (Matthews et al., 2000). The 24 hr recall may reduce administrator and participant burden, alterations in physical activity patterns during assessment, and the potential for recall bias. This method may not be suitable in populations with limited telephone access or with those unwilling to complete the phone interview, and it may use a time frame (e.g., past 24 hr) that does not capture a person's true habitual activity level.

Motion Detectors

Electronic motion detectors have become increasingly popular direct measures of free-living physical activity (Freedson & Miller, 2000; Tudor-Locke & Myers, 2001; Welk, Blair, Wood, Jones, & Thompson, 2000). Energy expenditure is often extrapolated from the activity data under the assumption that movement (or acceleration) of the limbs and torso is closely related to whole-body activity-related energy expenditure (Freedson & Miller, 2000). Figure 13.3, c through d, shows several types of motion detectors that differ in cost, technology, and data output.

Pedometers

Pedometers are inexpensive devices (~ $20) used to quantify ambulatory activity as steps per unit time of observation (e.g., per day) (Tudor-Locke & Myers, 2001). Electronic pedometers are small battery-operated devices worn at the waist. The vertical forces of footstrike cause movement of a spring-suspended lever arm to open and close an electrical circuit, which registers a "step." In theory, step reg-

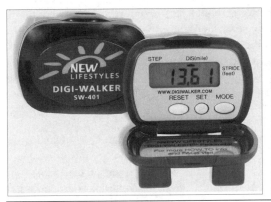 

Figure 13.3(c-d) Direct and indirect methods of physical activity assessment: (c) Yamax Digiwalker; (d) CSA accelerometer.

Copyright © 2002 from Field assessment of physical activity and energy expenditure among athletes. In *Nutritional assessment of athletes* by LaMonte, M.J., & Ainsworth, B.E., edited by J.A. Driskell & I. Wolinsky, 2002. Adapted by permission of Routledge/Taylor & Francis Group. LLC.

istration should reflect only the vertical forces of footstrike and hence walking activity. However, any vertical force through the hip area (e.g., sitting down hard onto a chair or riding on a bike or in a car over rough terrain) can trigger the device. In regards to normal vehicle travel, a recent study suggested that any erroneous steps measured with the pedometer likely amount to less than 1% of total daily ambulatory activity (Le Masurier & Tudor-Locke, 2003). An estimate of distance walked is obtained by calibrating pedometer steps to a person's stride length at a usual walking pace over a known distance (e.g., 1/4 mile). Welk, Corbin, and Dale (2000) reported mean step counts of 1,875 and 1,996 for average-sized men and women, respectively, during 1 mi (1.6 km) of walking at 4 mph (6.4 km/hr) (107 m · min^{-1}) on an outdoor track.

Pedometers have demonstrated reasonable precision for use in research and clinical settings in which walking is the primary type of activity (Tudor-Locke & Myers, 2001). Correlations of $r = .48$ to .93 have been reported between pedometer steps per day and activity data from accelerometers worn on the hip (Bassett et al., 2000; Leenders, Sherman, & Nagarja, 2000). Correlations of $r = .21$ to .49 have been shown between steps per day and self-reported total daily activity (Ainsworth, LaMonte, et al., 2000; Welk, Differding, et al., 2000). Additional studies reporting the convergent and construct validity of pedometers have been reviewed in detail elsewhere (Tudor-Locke, Williams, Reis, & Pluto, 2002, 2004). The precision of pedometers has varied considerably according to manufacture brand (Bassett et al., 1996; Schneider, Crouter, & Bassett, 2004), and has been lower at slower walking speeds versus more moderate-paced walking and running activities (Bassett et al., 1996; Welk, Differding, et al., 2000). Another major limitation to pedometers as an objective field measure of physical activity is a lack of temporal information on the type, duration, and intensity of activities performed while steps were being recorded. This precludes accurate quantification of activity-related energy expenditure, time spent in type- or intensity-specific activities, and patterns of activity (e.g., short vs. continuous bouts). Furthermore, activities that require upper body involvement are not measured. Pedometers may, however, be a reasonable method for promoting and assessing changes in walking activity (Tudor-Locke & Myers, 2001).

Accelerometers

Accelerometers are small battery-operated devices that measure the rate and magnitude of displacement in the body's center of mass or a body limb. Movement is measured in a single plane (uniaxial; Caltrac, MTI [formerly called CSA]) or multiple planes (triaxial; Tritrac) by way of piezoelectric signaling. Microcomputer technology integrates and sums the absolute value and frequency of acceleration forces over a defined observation period (e.g., every minute). Physical activity data is output as an activity "count." Regression equations have been developed from controlled laboratory experiments to allow for the estimation of activity-related energy expenditure from the activity counts.

The Caltrac accelerometer (Muscle Dynamics Fitness Network, Torrance, CA) was first described in the early 1980s (Montoye et al., 1983). The device estimates resting energy expenditure (kcal $\cdot$ min^{-1}) with activity counts recorded during body movement. Activity-related energy expenditure is continuously added to the resting energy expenditure value and summed across the specified observation period (e.g., 24 hr) to yield a value of total, or *gross,* energy expenditure. Net activity-related energy expenditure can be approximated by dividing gross energy expenditure by the product of resting metabolic rate multiplied by the number of minutes the Caltrac was worn. Physical activity can be quantified in terms of accumulated activity counts rather than kcal of energy expenditure (Melanson & Freedson, 1995). Caltrac cannot store data; therefore, data must be recorded by hand at the end of each sampling frame. Furthermore, because it cannot be programmed for time-interval sampling, only total daily activity counts or energy expenditure can be measured.

The MTI Actigraph (Manufacturing Technology, Inc., Fort Walton Beach, FL [formerly called CSA accelerometer]) is popular for use in field studies of physical activity and energy expenditure because of its small size and ability for time-interval sampling and data storage. The Actigraph must be computer initialized and downloaded. Activity data are presented as counts per unit sampling time (e.g., per minute). Regression equations have been developed from controlled studies of treadmill exercise or limited simulations of lifestyle physical activities to estimate energy expenditure from activity count data (Freedson, Melanson, & Sirard, 1998; Hendelman, Miller, Bagget, Debold, & Freedson, 2000; Melanson & Freedson, 1995; Swartz et al., 2000). Regression equations ($R^2 = .82$, $SEE = 1.12$ METs and $R^2 = .82$, $SEE = 1.4$ kcal $\cdot$ min^{-1}) derived from treadmill exercise are described in Melanson and Freedson (1995) and Freedson and colleagues (1998).

The Tritrac (Hemokinetics Inc., Madison, WI) is a triaxial accelerometer that provides activity count data for the anterior–posterior, medial–lateral, and vertical planes, as well as an integrated vector magnitude (Vmag) of counts for all three planes combined. A regression equation ($R^2 = .90$, SEE = 0.014 kcal $\cdot$ kg^{-1} $\cdot$ min^{-1}) derived from treadmill walking and running allows for activity-related energy expenditure estimates that account for body mass and resting energy expenditure (Nichols, Morgan, Sarkin, Sallis, & Calfas, 1999).

Several studies have examined the precision of accelerometry-based assessment of physical activity and energy expenditure (Ainsworth, Bassett, et al., 2000; Bassett et al., 2000; Campbell, Crocker, & McKenzie, 2002; Hendelman et al., 2000; King, Torres, Potter, Brooks, & Coleman, 2004; Leenders et al., 2000; Leenders, Nelson, & Sherman, 2003; Matthews et al., 2000; Swartz et al., 2000; Welk, Blair, et al., 2000; Welk, Almeida, & Morss, 2003). Caltrac and Actigraph data have correlated strongly ($r \sim .80$) with $\dot{V}O_2$ during laboratory studies of treadmill walking and running (Freedson et al., 1998; Melanson & Freedson, 1995; Montoye et al., 1983; Welk, Blair, et al., 2000). Lack of association ($r \sim .03$) between Caltrac and Actigraph output with treadmill grade (Melanson & Freedson, 1995) illustrates the inability of accelerometers to detect changes in the energy cost of activities as a result of increased resistance to body movement (Freedson & Miller, 2000; Haskell et al., 1993). Actigraph counts have varied significantly with monitor placement at three different ipsilateral hip locations (Welk, Blair, et al., 2000). Correlations between $\dot{V}O_2$ and accelerometer activity data have been higher during controlled laboratory activities such as treadmill exercise (e.g., $r = .80$ to .95) (Bassett et al., 2000; Melanson & Freedson, 1995; Welk, Blair, et al., 2000) than lifestyle activities (e.g., $r = .40$ to .60) (Ainsworth, Bassett, et al., 2000; Bassett et al., 2000; Welk, Blair, et al., 2000). Similarly, the precision of energy expenditure estimates from accelerometry-based regression equations have been lower for lifestyle activities ($R^2 = .32$ to .35, SEE = 0.96 to 1.2 METs [Hendelman et al., 2000; Swartz et al., 2000]) compared with controlled laboratory activities ($R^2 = .82$ to .89, SEE = 1.1 METs [Freedson et al., 1998]). Large discrepancies have been shown for time spent (e.g., min $\cdot$ d^{-1}) in specific intensity categories of activity from detailed physical activity logs and Actigraph data from three different regression equations (Ainsworth, Bassett, et al., 2000; Strath, Bassett, & Swartz, 2003). Taken together, these observations suggest that accelerometry-based assessment of physical activity and related energy expenditure may be of limited use in situations aimed at measuring habitual daily activities other than walking or running.

Measurement in three planes should account for more sources of body movement and therefore provide more precise estimates of activity-related energy expenditure, particularly under lifestyle conditions. Studies have shown that triaxial devices (e.g., Tritrac) have only slightly better correlations with both laboratory ($r_{\text{Triaxial}} = .84$ to .93 versus $r_{\text{Uniaxial}} = .76$ to .85) and lifestyle ($r_{\text{Triaxial}} = .59$ to .62 versus $r_{\text{Uniaxial}} = .48$ to .59) activity-related energy expenditure (Hendelman et al., 2000; Welk, Blair, et al., 2000). Activity-specific equations have been suggested to improve the precision of Tritrac measures of free-living activity-related energy expenditure (Campbell et al., 2002). Small improvements in the precision of estimating energy expenditure from accelerometer output have been reported when uniaxial wrist and hip accelerometers and a heart rate monitor were used simultaneously (Strath et al., 2002; Swartz et al., 2000).

Accelerometers provide information about the frequency, duration, intensity, and patterns of physical activity. The specific type of physical activity, however, is unknown. Accelerometers tend to overestimate walking-related energy expenditure and underestimate lifestyle-activity-related energy expenditure, and they do not account for energy expenditure owed to upper body involvement or uphill walking. These limitations likely reflect the accuracy of the regression equations used to predict activity-related energy expenditure rather than imprecision of motion detection by the accelerometer. Subject compliance issues, potentially altered physical activity patterns, and the cost of the more sophisticated instruments (uniaxial ~ $300, triaxial ~ $550 per unit) are additional limitations to the practicality of accelerometers as measures of activity-related energy expenditure among free-living populations.

Emerging Technologies

In recent years, technological advancements have contributed to the development of several innovative electronic motion detectors. One such device is the Sense-Wear Pro Armband (SWA) (Body Media, Pittsburgh, PA). The SWA is a relatively small device worn around the upper arm that incorporates a variety of measured parameters (two-axis accelerometry, heat flux, galvanic skin response, skin temperature, near-body temperature) and demographic characteristics (gender, age, height, weight) into algorithms developed by the manufacturer to estimate energy expenditure. The Intelligent Device for Energy Expenditure and Activity (IDEEA) (MiniSun, CA) is another more recent invention. The IDEEA consists of one small recorder worn at the waist and five sensors placed on the chest, the frontal part of each thigh, and under each foot, which measure the acceleration and angle of each body segment. The sensors are connected to the recorder by thin, flexible wires. A unique aspect of the IDEEA is its ability to determine the subject's body position (sitting, standing, walking, running) and then incorporate the details (intensity from the speed of walking or the rate of stepping up and its associated duration) of that activity into relevant proprietary equations to estimate energy expenditure at that moment. Data from the SWA and the IDEEA are uploaded to manufacturer-provided software for analysis. To date, the SWA and the IDEEA have been subjected to only a small number of laboratory-based evaluations under controlled conditions, and no study has assessed their feasibility or accuracy in free-living populations (Fruin & Rankin 2004; Jakicic et al., 2004; Zhang, Pi-Sunyer, & Boozer, 2004; Zhang, Werner, Sun, Pi-Sunyer, & Boozer, 2003).

One preliminary study showed that the generalized proprietary equation of the SWA significantly underestimated energy expenditure during treadmill walking ($6.9 \pm 8.5\%$), cycling ($28.9 \pm 13.5\%$), and stepping exercise ($17.7 \pm 11.8\%$) and overestimated energy expenditure during arm ergometry exercise ($29.3 \pm 13.8\%$) when compared to indirect calorimetry (Jakicic et al., 2004). Under controlled conditions the IDEEA was capable of detecting the type, onset, duration, and intensity of several physical activities with an accuracy of more than 98% (Zhang et al., 2003) and was able to estimate energy expenditure within 95% of that measured with indirect calorimetry (Zhang et al., 2004). The future is promising for motion detectors such as the SWA and IDEEA. Prototypes of other devices that have the capacity to combine tiny sensors and cameras embedded into clothing and accessories to conveniently monitor free-living physical activity behaviors will ensure the advancement of activity assessment well into the next decade (Healey, 2000).

Emerging interest in environment–individual interaction has focused attention on using geographic information systems (GIS) to identify individual movement patterns in geographic spaces. Global positioning systems (GPS) capable of recording the location and distance traveled have been built into small wearable devices (e.g., wristwatches, cell phones, shoulder bags) to provide a relatively effortless direct assessment of activity patterns (Schutz & Herren, 1999; Terrier, Ladetto, Merminod, & Schutz, 2000; Terrier & Schutz, 2003). Preliminary data evaluating the accuracy of these GPS devices in small, well-defined geographic areas show acceptable accuracy ($r = .95$); however, additional studies are necessary to determine their accuracy among free-living populations.

Indirect Measures of Physical Activity

Indirect assessment methods provide surrogate measures of physical activity among free-living individuals when direct methods would be cost prohibitive or impractical. Indirect measures are typically validated against objective direct measures of activity in controlled small-sample studies. Activity-related energy expenditure can be estimated from indirect measures with reasonable precision. The most common indirect method is the self-report physical activity questionnaire.

Physical Activity Questionnaires

Self-report questionnaires are the most frequently used method of assessing physical activity levels among free-living individuals. Questionnaires are generally classified as global, recall, and quantitative history instruments based on their level of detail (Ainsworth et al., 1994; LaMonte & Ainsworth, 2001). Table 13.3 provides a list of questionnaires that have been used to estimate physical activity and energy expenditure in several settings.

Global Questionnaires

Global activity questionnaires are short (e.g., one to four items) surveys that are self-administered (Siscovick et al., 1988; Sternfeld, Cauley, Harlow, Liu, & Lee, 2000) or given as a phone interview (Macera et al., 2001) to obtain a general index of physical activity, as seen in figure 13.3(e). Because global questionnaires provide little detail on specific types and patterns of physical activity, they provide only simple classifications of activity status (e.g., active vs. inactive) and do not allow for precise assessment of activity-related energy expenditure (LaMonte & Ainsworth, 2001). Global questionnaires are preferred in physical activity surveillance systems in which sample

Table 13.3   Self-Report Questionnaires Used to Quantify Levels of Physical Activity and Energy Expenditure in Free-Living Populations

Method/ questionnaire[a]	Type of activity[b]	Recall time frame	Burden[c]	Expression of PA score[d,e]	Author
Global					
NSPHPC	TOTAL (relative to peers)	General	Low	5-point qualitative scale	Sternfeld et al., 2000
				2-point qualitative scale	Belloc et al., 1972
Lipid Research Clinics	JOB, EX	Usual day	Low	2-point qualitative scale	Siscovick et al., 1988
	JOB, NON-JOB, EX			4-point qualitative scale	Ainsworth et al., 1993
BRFSS Occupation	OCC	General	Low	1-point nominal scale	Yore et al., 2006
Recall questionnaires					
Baecke	JOB, SP, LEIS	General	Moderate	5-point ordinal scale	Baecke et al., 1982
Seven-day recall	EX, LEIS	Past 7 days	Moderate	$kcal \cdot kg^{-1} \cdot d^{-1}$	Blair et al., 1985
College alumnus	EX, SP, LEIS	Past year	Moderate	$kcal \cdot wk^{-1}$	Paffenbarger, 1986
CARDIA	JOB, EX, SP, LEIS, HH, YRD	Past year	Moderate	Unitless index	Jacobs et al., 1989
Typical week survey	JOB, EX, SP, LEIS, Tran, HH, YRD, CARE, VOL	Typical year in past month	Moderate	$MET\text{-}min \cdot d^{-1}$	Ainsworth, LaMonte et al. 2000
IPAQ	EX, SP, LEIS, HH, OCC, TRAN	Past 7 days	Moderate	3-point qualitative scale	Craig et al., 2003
BRFSS	EX, SP, LEIS, HH	Usual week	Moderate	3-point qualitative scale	Macera et al., 2001

Occupational Physical Activity (OPAQ)	OCC	Usual week	Moderate	MET-min · d^{-1}	Reis et al., 2005
Quantitative history					
MN LTPA	EX, SP, LEIS, HH	Past year	High	AMI · d^{-1}	Taylor et al., 1978
Tecumseh occupation	JOB, TRAN	Past year	Moderate	MET-hr · wk^{-1}	Montoye, 1971
Historical PA	SP, LEIS	Lifetime	High	Ordinal Scale in hr · wk^{-1} and kcal · wk^{-1}	Kriska et al., 1988
LTPAQ	JOB, EX, SP, HH	Lifetime	High	Hr · wk^{-1}	Friedenreich et al., 1998
Modified HLPAQ	EX, SP, HH, YRD, CARE	Lifetime	High	MET-hr · wk^{-1}	Chasan-Taber et al., 2002

[a] NSPHPC = National Survey of Personal Health Practices & Consequences; BRFSS = Behavioral Risk Factor Surveillance System; CARDIA = Coronary Artery Risk Development in Young Adults; MN LTPA = Minnesota LTPA; Tecumseh LTPA = Tecumseh, Michigan, LTPA; LTPA = leisure-time physical activity; LTPAQ = Lifetime Physical Activity Questionnaire; HLPAQ = Historical Lifetime Physical Activity Questionnaire. IPAQ = International Physical Activity Questionnaire.

[b] JOB = occupational; EX = exercise; SP = sport; TRAN = transportation; LEIS = leisure; HH = household; YRD = yardwork; CARE = caregiving; VOL = volunteer

[c] Administrative burden including cost, time to administer or respond, data management or processing time

[d] MET = metabolic equivalent; AMI = activity metabolic index; kcal = kilocalorie; kg = kilogram

[e] Qualitative scale refers to categories such as more active versus less active, or sedentary versus active. Ordinal scale refers to an ordered range of numbers (e.g., 1 through 5) used to rank activity status (e.g., low to high).

Adapted from Ainsworth, Montoye, & Leon, 1994, and LaMonrte & Ainsworth, 2001.

Reprinted, by permission, from M.H. LaMonte and B.E. Ainsworth, 2002, "Quantifying energy expenditure and physical activity in the context of dose response," *Medicine and Science in Sports and Exercise* 33(Suppl 6): S370-S378.

When you are at work, which of the following best describes what you do?

1. Mostly sitting or standing
2. Mostly walking

   or

3. Mostly heavy labor or physically demanding work
4. Do not work

   [*Interviewer note:* If respondent has multiple jobs, include all jobs]

Figure 13.3(e)   Direct and indirect methods of physical activity assessment: global activity questionnaire.

Copyright © 2002 from Field assessment of physical activity and energy expenditure among athletes. In Nutritional assessment of athletes by LaMonte, M.J., & Ainsworth, B.E., edited by J.A. Driskell & I. Wolinsky, 2002. Adapted by permission of Routledge/Taylor & Francis Group. LLC.

sizes are very large, administrative time is limited, and the assessment goal is to merely track long-term population behavioral trends (Macera et al., 2001; Macera & Pratt, 2000). Global reports might also provide enough information to control for physical activity status (e.g., as a covariate) in multivariable analyses that might be confounded by physical activity level (Stevens et al., 1998). Although the precision and reproducibility of global activity questionnaires have been reasonably good (Ainsworth et al., 1993; Jacobs et al., 1993), misclassification on activity status is likely to occur (Macera et al., 2001), and lack of detail precludes comprehensive summaries of physical activity dimensions and related energy expenditure.

Recall Questionnaires

Recall questionnaires are typically 10- to 20-item instruments that detail the frequency, duration, and types of activities performed during a defined recall period (e.g., past month or year) (see figure 13.3f). They can be interview based (Ainsworth, LaMonte, et al., 2000; Blair et al., 1985; Jacobs et al., 1989) or self-administered (Baecke, Burema, & Frijters, 1982; Paffenbarger et al., 1986; Wolf et al., 1994). Scoring systems can be a simple ordinal scale (e.g., 1 to 5 representing low to high activity levels [Baecke et al., 1982]), a unitless activity index (e.g., higher number reflects higher activity level [Jacobs et al., 1989]), or a summary score of continuous data (e.g., frequency or minutes of activity per week, MET-min $\cdot$ d^{-1} [Ainsworth, LaMonte, et al., 2000; Blair et al., 1985; Paffenbarger et al., 1986; Wolf et al., 1994]). The latter measure allows for quantification of time spent (e.g., min $\cdot$ d^{-1}) performing specific physical activities and their related energy expenditure, and therefore allows for the most precise interpretation of physical activity patterns and effects for application in various settings (LaMonte & Ainsworth, 2001).

Recall surveys have demonstrated acceptable levels of accuracy and reproducibility (Ainsworth, LaMonte, et al., 2000; Jacobs et al., 1989, 1993; Wolf et al.,

Think about the types of activities you did in a typical week in the past month (see calendar provided). For each activity, note which of these activities you did by checking a box for Yes or No. Then, for each item you marked as Yes, write the number of days you did the activity Monday to Friday and Saturday to Sunday, and the average time in hours and minutes that you did these activities. Refer to the following intensity levels before responding to each question:

Intensity levels:
Light → easy or no effort at all
Moderate → harder than light, some increase in breathing or heart rate
Vigorous → all-out effort, large increase in breathing or heart rate

Example:

CONDITIONING ACTIVITIES		Monday–Friday			Saturday–Sunday		
Moderate Effort		# of days Hours/day Minutes/day			# of days Hours/day Minutes/day		
Low-impact aerobics,	❏ 1. Yes→						
health club machines,	❏ 2. No	3	0	30	0	0	0
bicycling, tai chi							

During a typical week in _____, did you do:

Month

HOUSEHOLD ACTIVITIES		Monday–Friday			Saturday–Sunday		
1 Light Effort		# of days Hours/day Minutes/day			# of days Hours/day Minutes/day		
Cooking, cleaning up,	❏ 1. Yes→						
laundry, shopping,	❏ 2. No						
dusting	CAP301	CAP301A	CAP301B	CAP301C	CAP301D	CAP301E	CAP301F
2 Moderate or Vigorous Effort							
Scrubbing, vacuuming,	❏ 1. Yes→						
repairs, mopping,	❏ 2. No						
washing car	CAP301	CAP301A	CAP301B	CAP301C	CAP301D	CAP301E	CAP301F
LAWN/YARD/GARDEN/FARM							
3 Moderate Effort							
Weeding, sweeping,	❏ 1. Yes→						
mowing, raking	❏ 2. No						
	CAP301	CAP301A	CAP301B	CAP301C	CAP301D	CAP301E	CAP301F

Figure 13.3(f) Direct and indirect methods of physical activity assessment: recall activity questionnaire.

Copyright © 2002 from Field assessment of physical activity and energy expenditure among athletes. In Nutritional assessment of athletes by LaMonte, M.J., & Ainsworth, B.E., edited by J.A. Driskell & I. Wolinsky, 2002. Adapted by permission of Routledge/Taylor & Francis Group. LLC.

1994). One-month test–retest correlations for activity-related energy expenditure computed from self-reported walking, stair climbing, and sport activity were $r = .61$ and $r = .75$ for men and women, respectively, who completed the College Alumnus Questionnaire (Ainsworth, Leon, et al., 1993). Criterion-related validity correlation between activity-related energy expenditure and Caltrac accelerometer METs · d⁻¹ was $r = .29$ for all participants. Age-adjusted test–retest correlations of $r = .60$ and $r = .36$ for total MET-min · d⁻¹ of energy expenditure were reported for men and women, respectively, who completed the 7-Day Activity Recall (Richardson et al., 2001). Age-adjusted criterion validity correlations for total daily energy expenditure between the seven-day recall and the Caltrac accelerometer were $r = .54$ and $r = .20$, and, between the seven-day recall and 48 hr physical activity records were $r = .58$ and $r = .32$ for men and women, respectively.

Recall surveys typically have not assessed nonoccupational, nonleisure, or nonsport activities, which may be particularly relevant sources of physical activity and energy expenditure among women and minorities (LaMonte & Ainsworth, 2001; Jacobs et al., 1993). This may partly explain the lower correlations between self-reported and objectively measured activity levels among women (Jacobs et al., 1993). Ainsworth and colleagues (2000) developed a comprehensive Typical Week Physical Activity Survey for use among middle-aged minority women. This interview-based survey, shown in figure 13.3(f), is very detailed and requires respondents to recall the frequency and duration of several types of activities including items for house and family care, exercise and sport, and occupation. Moderate to high intraclass reliability correlations (e.g., $r = .57$ to .77) and age-adjusted criterion validity correlations (e.g., $r = .45$ to .54 with physical activity records and logs) were observed for total, light, moderate, and vigorous MET-min · d⁻¹ summary scores. Despite being one of the most detailed and comprehensive recall questionnaires, validity and reliability characteristics of the Typical Week Physical Activity Survey were similar to other frequently used recall questionnaires (Jacobs et al., 1993).

The International Physical Activity Questionnaire (IPAQ) is a recall questionnaire designed to compare population levels of physical activity across countries (Craig et al., 2003). The IPAQ has gained considerable popularity because it has been evaluated in 12 countries and is easily accessible in various languages. The short version of the IPAQ contains seven items to collectively assess the daily duration and weekly frequency of moderate- and vigorous-intensity activities and walking. One-week test–retest reliability for the short IPAQ ranged from $r = .66$ to .88, and validity compared against the MTI Actigraph worn concurrently for seven days ranged from $r = .02$ to .47 (Craig et al., 2003). A longer version of the IPAQ was also created to provide a more detailed description of physical activity participation in various domains (household, transportation, occupation, walking, sport, and leisure). Test–retest reliability for the long IPAQ ranged from $r = .70$ to .91 and $r = .05$ to .52 for validity compared to the MTI Actigraph (Craig et al., 2003). Telephone and in-person formatted versions of the short and long IPAQ can be downloaded from www.ipaq.ki.se.

Several limitations apply to physical activity recall instruments; however, three are particularly important. First, activity information is subject to response bias (e.g., recall bias, social desirability), which may influence the precision in measures of physical activity and related energy expenditure (Coughlin, 1990; Durante &

Ainsworth, 1996). Bias appears to be highest for light- and moderate-intensity physical activities that are habitual behaviors (e.g., walking, housework) compared with vigorous sports and conditioning activities that are planned and intentional behaviors (Chasan-Taber, Erickson, Nasca, Chasan-Taber, & Freedson, 2002; Hayden-Wade, Coleman, Sallis, & Armstrong, 2003; Strath, Bassett, & Swartz, 2004). Second, questionnaires with ordinal scales and summary indexes may be inappropriately used to examine physical activity effects that are based on interval or ratio level data (Zhu, 1996, 2002; Zhu, Timm, & Ainsworth, 2001). Third, the structure of the instrument, including the question order, the domains of activity assessed, and the exclusion of relevant population-specific activities, may lead to a misrepresentation of a person's actual physical activity level and related energy expenditure (Ham, Macera, Jones, Ainsworth, & Turczyn, 2004; LaMonte & Ainsworth, 2001; Warnecke et al., 1997). This issue may be particularly problematic when a questionnaire that has been validated for use in a population with unique age, gender, race, ethnicity, and sociocultural characteristics is used (or modified for use) in a population with distinct differences in these parameters. As with most measurements, practitioners should take great care when considering the population within which a specific questionnaire was validated and the population for whom the measurement will be used or to whom the findings will be generalized (Masse, 2000). Notwithstanding, recall questionnaires are the most frequently used method to assess free-living physical activity and related energy expenditure. Another major measurement problem is that so many questionnaires have been developed that we have little information on their equivalence (Zhu, 2000). A new development framework should be proposed in which new development effort is based on the previous questionnaires, and every effort should be made to link the data collected by the new questionnaires to those collected by previous ones (see chapter 6 by Weimo Zhu).

Quantitative Histories

Quantitative histories, as shown in figure 13.3(g), record the frequency and duration of occupational, leisure, and other physical activities over the past year (Craig et al., 2003; Montoye, 1971; Taylor et al., 1978) or lifetime (Chasan-Taber et al., 2002; Friedenreich, Courtneya, & Bryant, 1998; Kriska et al., 1988). These surveys generally include more than 20 items and may be interview based (Friedenreich et al., 1998; Kriska et al., 1988; Taylor et al., 1978) or self-administered (Craig et al, 2003.; Chasan-Taber et al., 2002; Montoye, 1971). Activity scores are usually expressed as a continuous variable (e.g., kcal $\cdot$ kg$^{-1}\cdot$ wk^{-1}; MET-hr $\cdot$ wk^{-1}). One-year test–retest correlations have been high (e.g., $r = .61$ to .86) for interview-based (Richardson, Leon, Jacobs, Ainsworth, & Serfass, 1994) and self-administered (Chasan-Taber et al., 2002) methods. Compared with scores from activity diaries, criterion validity correlations for total, moderate, and vigorous activity have been stronger for an interview-based survey ($r = .70$ to .75 [Richardson et al., 1994]) than a self-administered survey ($r = .15$ to .52 [Chasan-Taber et al., 2002]).

Quantitative activity histories are useful for investigators and practitioners who are interested in physical activity patterns and related energy expenditure over long

PART IV—LEISURE TIME PHYSICAL ACTIVITIES

Listed below are a series of leisure time activities. Related activities are grouped under general headings. Please read the list and check Yes in column 3 for those activities you have performed in the last 12 months and No in column 2 for those you have not. Do not complete any of the other columns.

To be completed by participant ACTIVITY (1)	Did you perform this activity?		For clinic personnel use only														Average number of times per month	Time per occasion		Do not write in this space 25
	No (2)	Yes (3)	Jan	Feb	Mar	Apr	May	Jun	Jul	Aug	Sep	Oct	Nov	Dec				Hrs.	Min.	
SECTION A: Walking and miscellaneous																	27	30		24
Walkin for pleasure																				010
Walking to work																				020
Using stairs when elevator is available																				030
Cross country hiking																				040
Back packing																				050
Mountain climbing																				060
Bicycling to work and/or for pleasure																				115
Dancing: Ballroom and/or square																				125
SECTION B: Conditioning exercise																				
Home exercise																				150
Health club exercise																				160
Job/walk combination																				180
Running																				200
Weight lifting																				210
SECTION C: Water activities																				
Water skiing																				220
Sailing in competition																				235
Canoeing or rowing for pleasure																				250
Canoeing or rowing in competition																				260
Canoeing on a camping trip																				270
Swimming (at least 50 ft.) at a pool																				280
Swimming at the beach																				295
Scuba diving																				310
Snorkeling																				320
SECTION D: Winter activities																				
Snow skiing, downhill																				340
Snow skiing, cross country																				350
Ice (or roller) skating																				360
Sledding or tobogganing																				370
SECTION E: Sports																				
Bowling																				390
Volley ball																				400
Table tennis																				410
Table, singles																				420
Table, doubles																				430
Softball																				440
Badminton																				450
Paddle ball																				460

Figure 13.3(g) Direct and indirect methods of physical activity assessment: quantitative history questionnaire.

Copyright © 2002 from Field assessment of physical activity and energy expenditure among athletes. In *Nutritional assessment of athletes* by LaMonte, M.J., & Ainsworth, B.E., edited by J.A. Driskell & I. Wolinsky, 2002. Adapted by permission of Routledge/Taylor & Francis Group. LLC.

periods of time (e.g., ≥12 months) because of the presumed influence on developmental (Malina & Bouchard, 1991) or disease issues (Chasan-Taber et al., 2002; Friedenreich et al., 1998). The intensive administrative burden and recall effort, however, limit the feasibility and practicality of these instruments in many settings.

Physiological Markers

Cardiorespiratory fitness ("fitness") has been used as a surrogate of physical activity status (Paffenbarger, Blair, Lee, & Hyde, 1993; Stofan, DiPiettro, Davis, Kohl, & Blair, 1998) and as a validation measure for physical activity questionnaires (Jacobs et al., 1993). Although several factors such as age, gender, genetics, and health status influence individual fitness levels, physical activity, particularly of moderate to vigorous intensities, is the principal determinant of changes in measured $\dot{V}O_2$ or fitness (Bouchard, Malina, & Perusse, 1997; Saltin & Rowell, 1980). Differences in mechanical and metabolic efficiency among adults tend to be small during habitual activities such as walking, jogging, and stationary cycling. Therefore, fitness can be estimated from maximal or submaximal exercise work rate (e.g., treadmill speed and grade, bicycle resistance) without the additional burden of indirect calorimetry (American College of Sports Medicine, 1995). Fitness can also be estimated from field tests (Cooper, 1968) and equations that do not require exercise test data (Ainsworth, Richardson, Jacobs, & Leon, 1992); however, these methods are generally less precise than exercise testing measures of individual fitness levels. Exercise testing is safe and provides a highly objective and reproducible method of assessing the physiological effect of recent (e.g., past month) physical activity (American College of Sports Medicine, 1995; Paffenbarger et al., 1993; Saltin & Rowell, 1980). Fitness assessment may, therefore, be the most accurate assessment of sedentary or irregularly active lifestyles (Saltin & Rowell, 1980). An advantage of measuring fitness as an index of habitual physical activity is the ability to prescribe individualized activity interventions and verify changes in activity behavior. Conversely, fitness measures are not influenced by light-intensity activity and provide no information about the type or pattern of physical activity that contributes to a person's measured fitness level. Feasibility considerations challenge the practicality of assessing fitness as an indirect measure of physical activity among free-living populations.

Body composition varies with physical activity and related energy expenditure (Ching et al., 1996; Hill, Melby, Johnson, & Peters, 1995). Therefore, body composition measures have been used as validation criteria for physical activity questionnaires (Jacobs et al., 1993). Typical measures of body composition include body weight, percent body fat from skinfolds or hydrodensitometry, and girth measurements (Roche, Heymsfield, & Lohman, 1996). Higher levels of physical activity are generally associated with more favorable body composition measures (e.g., lower percent body fat or BMI); however, the association is typically strongest for vigorous (>6 METs or 7.5 kcal · min^{-1}) activities and weak for light- and moderate-intensity activity (Jacobs et al., 1993; Slattery & Jacobs, 1987). Body composition measures are at best a crude surrogate of physical activity and related energy expenditure, providing no information about the type, frequency, duration, or intensity of individual or group activity patterns.

SELECTING A METHOD OF MEASUREMENT

After considering the extensive list of possible methods for measuring physical activity and energy expenditure, choosing the method best suited for a given application can be challenging. The following considerations are useful in this regard. First, consider the *objective of the measurement* and the *dimension of movement* being addressed. Some investigators may be interested in measuring the prevalence of physical inactivity within a defined free-living population, for which a global physical activity questionnaire may be adequate. Another situation may aim to quantify time spent in walking activity (e.g., for exercise, for transportation), for which one may use a physical activity record or recall questionnaire that contains items specific to various types of walking behavior and records both the frequency and duration of activity. Yet another situation may involve quantifying the activity-related energy expenditure during a weight loss intervention, in which case DLW or heart rate–accelerometer measures may be the methods of choice. A related consideration is the *level of detail* required from the measurement. Although a global activity survey and DLW provide adequate information to determine gross activity status or total energy expenditure, respectively, neither would be useful to precisely quantify dose-response issues related to frequency, duration, or volume of moderate-intensity occupational activity in relationship to cardiovascular risk. In this case detailed physical activity questionnaires, an activity log or record, or heart rate–accelerometer methods combined with an activity log might provide the needed level of detail to address this question.

The *units of measurement* should allow for direct application or easy extrapolation to the desired metric. If kcal $\cdot$ wk^{-1} of activity-related energy expenditure is the metric of interest, using a physical activity survey that is scored to a unitless activity index or an ordinal activity scale would not be appropriate. Rather, DLW, accelerometry, or a physical activity questionnaire that quantifies the frequency and duration of several activity categories and that can be weighted with standard energy costs should be used. Similarly, if time spent in a specific type or intensity of activity is desired, then heart rate monitoring, accelerometry, physical activity log, or some combination should be used.

The *measurement setting* is also an important consideration. Although laboratory-based studies of physical activity and energy expenditure permit the use of complex measurement methods, free-living assessment must be nonobtrusive and practical with respect to a person's daily lifestyle. Characteristics of the *participants* being measured are a major concern. This is particularly true when using physical activity questionnaires, but also applies to exercise testing methods and equations used to extrapolate accelerometer data to activity-related energy expenditure. Characteristics such as age, gender, education level, sociocultural and race or ethnic status, and motivation to comply with measurement methodology must be considered. It would make little sense to survey habitual activity levels among end-stage heart-failure patients using a physical activity questionnaire comprised primarily of occupational and sport activities. Similarly, using a physical activity survey or accelerometry regression equation that was validated for populations of college-aged white men to describe activity patterns and quantify activity-related energy expenditure among elderly black women would be inappropriate. When possible, choose a measure that has demonstrated adequate psychometric properties among people similar to those in the target-specific population. If such a measure does not exist, consider a pilot study to examine these issues prior to widespread use of the measurement within the intended population.

Finally, but certainly of extreme importance, is consideration of the *size of the population* to be measured and the *measurement burden*. DLW and heart rate-accelerometry methods may provide highly precise measures of activity-related energy expenditure among individuals or small groups, but these are impractical for use in large free-living populations. Financial costs related to instrumentation, data processing, and participant motivation play a large role in identifying a realistic measurement technique. Methods that impose a substantial time burden on the administrator or participant may result in poor data quality. The most costly and burdensome measures are direct methods, and the least costly and burdensome methods are self-administered paper-and-pencil questionnaires. The tradeoff is in the precision of measurement.

SUMMARY

Physical activity and *energy expenditure*, two terms that have often been used interchangeably in the physical activity literature, are distinct constructs with a variety of direct and indirect assessment methods. Measures of energy expenditure are generally reserved for smaller, laboratory-based studies, whereas measures of physical activity are often used among larger, free-living populations and provide an estimate of energy expended. With an array of assessment methods available, the most appropriate measurement instrument for a given application should be based on a comprehensive list of considerations. The measurement construct, cost, measurement burden, dimension of movement, level of detail, measurement unit, and study population should all be considered when selecting a physical activity assessment method. A critical review of these factors will help to ensure that a suitable instrument is chosen to meet the needs of a given application.

CHAPTER 14

Epidemiology and Physical Activity

Richard A. Washburn, Rod K. Dishman, and Gregory Heath

Epidemiology, in simple terms, is the study of the distribution of a disease in a population. Epidemiology has three distinct goals: (a) to describe the distribution of disease (e.g., who gets the disease and when and where the disease occurs); (b) to analyze this descriptive information in order to identify risk factors that are associated with an increased probability of disease occurrence; and (c) to prevent disease occurrence by modifying the identified risk factors. Physical activity epidemiology studies factors associated with participation in a specific behavior—that is, physical activity—and how this behavior relates to the probability of disease or injury. Examples include a description of the level of physical activity in a population, a comparison of levels of physical activity among populations, a determination of factors that are associated with participation in physical activity, and an investigation of the association between physical activity and the risk for chronic diseases such as coronary heart disease (CHD), stroke, diabetes, osteoporosis, and cancer. This chapter provides an introduction to the methods that physical activity epidemiologists use to scientifically confirm that the lack of physical activity is a burden on public health.

PHYSICAL ACTIVITY EPIDEMIOLOGY— A HISTORICAL PERSPECTIVE

Interest in the association between physical activity and health is not new. As early as the ninth century BC both active and passive exercise were recommended for the preservation and restoration of health. For example, Hippocrates wrote:

> Exercise should be many and of all kinds, running on the double track increased gradually, . . . sharp walks after exercises, short walks in the sun after dinner, many walks in the early morning, quiet to begin with, increasing till they are violent and then gently finishing. (Dolan & Adams-Smith, 1978)

In contrast, the modern history of physical activity epidemiology is relatively short. In 1975 Milton Terris, former president of the American Public Health Association, stated in a keynote address at the sixth annual meeting of the Society for Epidemiologic Research that

physical fitness and physical education have no respected place in the
American public health movement. . . . On the subject of physical fitness
I speak with no authority. Having spent a large portion of my life seated
at a desk, I have no personal acquaintance with the concept. On a more
intellectual level, I have been far too bound by the philosophical rigidi-
ties of the American public health movement to become knowledgeable
in the literature of this field, and am therefore in no position to judge
the relation of physical exercise and physical fitness to performance
of activities of daily living and to "physical, mental and social well-
being," that is, to "positive health," vitality, and joy of life. These are
issues which are eminently worth studying. (Terris, 1975)

Therefore, as late as the mid-1970s the importance of physical activity as
a serious preventive health behavior was not clearly established in the United
States. It is thus quite remarkable that over the next 20 years, interest and evidence
regarding the role of physical activity and health progressed to the point that in
1996, the surgeon general of the United States and United States Public Health
Service published a detailed report on physical activity and health (United States
Department of Health and Human Services, 1996). The surgeon general's report
outlined the scientific consensus regarding the association of physical activity
with total mortality, cardiovascular disease, cancer, non-insulin-dependent diabetes,
osteoarthritis, osteoporosis, obesity, mental health, health-related quality of life, as
well as risks of musculoskeletal injury and sudden death. The foundations for the
report were established by the 1984 Workshop on Epidemiologic and Public Health
Aspects of Physical Activity and Exercise sponsored by the Centers for Disease
Control (CDC) in Atlanta (Powell & Paffenbarger, 1985); the establishment of
the Behavioral Epidemiology and Evaluation Branch at CDC; the first and second
International Conferences on Physical Activity, Fitness, and Health organized by
the Canadian Association of Sport Sciences and the Ontario Ministry of Tourism
and Recreation held in Toronto in 1988 and 1992; a 1992 position statement by the
American Heart Association recognizing physical inactivity as an independent risk
factor for CHD (Fletcher et al., 1992); and Recommendations on Physical Activity
and Public Health prepared jointly by the CDC and the American College of Sports
Medicine in 1993 (Pate et al., 1995).

Physical activity epidemiology can trace its roots to the work of two remarkable
scientists: Jeremy Morris, emeritus professor of public health, University of London
at the London School of Hygiene and Tropical Medicine; and Ralph Paffenbarger,
emeritus professor of medicine at Stanford University and adjunct emeritus professor at Harvard University. Understanding the association of physical activity
and physical fitness with reduced risk of chronic diseases did not have a scientific
basis in the field of public health until Morris and his colleagues began to study
CHD in the late 1940s. In the early 1950s Morris formulated the hypothesis that
"men in physically active jobs suffer less coronary – ischaemic heart disease than
comparable men in sedentary jobs, such disease as the active do develop is less

severe and strikes at later ages" (Morris, Heady, Raffee, Roberts, & Parks, 1953). The hypothesis evolved when Morris observed what appeared to be a protective effect of occupational physical activity against CHD, observations that were viewed skeptically by the scientific community at the time. In 1953 Morris published the famous bus driver study in which he observed that the physically active conductors on London's double-decker buses were at lower risk of CHD than the drivers, who sat through their shifts at steering wheels. If conductors did develop CHD, it was less severe and occurred at later ages. The London bus driver study began the modern era of physical activity epidemiology (Morris et al., 1953).

Though several important studies of occupational physical activity and disease were begun internationally following publication of the findings by Morris, the most sustained and compelling studies were conducted in the United States by Paffenbarger. Paffenbarger is most recognized for his seminal reports from the San Francisco Longshoremen Study (Paffenbarger & Hale, 1975) and the College (Harvard College and the University of Pennsylvania) Health Study (Paffenbarger, Wing, & Hyde, 1978) begun in the 1960s and 1970s. Those large-cohort studies helped fuel scientific and public interest in physical activity as an important component of health promotion and focused the broader fields of preventive medicine and public health on the importance of physical inactivity as a public health problem.

CONCEPTS AND METHODS
IN PHYSICAL ACTIVITY EPIDEMIOLOGY

The term **epidemiology** is derived from the Latin root *epi* ("upon") and *demo* ("the community"). In simple terms, epidemiology is the study of the distribution of disease in a population. More specifically, the term is defined as the application of the scientific method to the study of the distribution and dynamics of disease in a population for the purposes of identifying factors that affect this distribution and then modifying risk factors to reduce the frequency of morbidity and mortality from the disease (Jekel, Elmore, & Katz, 1996).

The following sections describe the methods used by epidemiologists to assess the distribution of disease or behavior (e.g., physical activity) in a population and to identify factors that affect this distribution.

Epidemiologic Measures

A fundamental measurement in epidemiology is the frequency with which an event occurs, usually an injury, disease, or death in a population under study. **Incident cases** are the new occurrences of these events in the study population during the time period of interest. In other words, incident cases are those cases in which the health status changes (i.e, alive/dead, not injured/injured, or not sick/sick) during the period of observation. This is in contrast to **prevalent cases,** which represent the number of persons in the population who have a particular disease or condition at some specific point in time. Prevalence of a disease is a function of both incidence and duration. Thus, the prevalence of a disease could increase as a result of either an increase in the number of new cases (incidence) or an increase in the length of time people with the disease survive before death or recovery.

Rates

If the incidence or prevalence of a condition is known, the incidence rates and prevalence rates can be calculated. A rate is the frequency or number of events that occur over some defined time period (numerator) divided by the average size of the population at risk (denominator). The usual estimate of the average number of people at risk is the population at the midpoint of the time interval under study. The general formula for a calculating a rate is

$$Rate = \frac{Numerator}{Denominator} \times Constant \qquad (14.1)$$

Because rates are usually less than 1, they are generally multiplied by a constant (100, 1,000, or 10,000) for ease of discussion. Therefore, if the death rate in the United States were calculated to be .0090 deaths per year, this could be multiplied by 1,000 and expressed as nine deaths per 1,000 individuals in the population per year.

Incidence rates are the number of incident cases over a defined time period divided by the population at risk over that time period. Incidence rates provide a measure of the rate at which people without disease develop disease over a specified time interval. Likewise, prevalence rates are calculated as the number of prevalent cases divided by the size of the population at a particular time. **Prevalence rates** indicate how many people have a particular disease or engage in a behavior such as physical activity or smoking at a certain time.

Prevalence rates are useful for planning purposes. For example, if a survey of a city finds that the prevalence of people with CHD is particularly high, it may be economically feasible for a local hospital to consider opening a cardiac rehabilitation program. Prevalence data, however, are not useful when trying to determine which factors may be related to an increased probability of disease. High prevalence does not necessarily indicate high risk but could be reflective of increased survival. The high prevalence of CHD in a city survey may not necessarily indicate that people in that city are at an increased risk of getting coronary heat disease but could be reflective of high-quality emergency services and medical care that increase the rate of survival. In contrast, low prevalence may simply reflect rapid death or rapid cure, not low incidence. The problem of having prevalence data only is that one does not know which of these possibilities are true.

It is particularly important that the information used to make comparisons among groups actually represents rates. This may seem obvious, but this prerequisite is often overlooked. For example, a sports medicine physician reports 100 cases of ruptured patellar tendons in runners over the past year. Does this indicate that running is the cause of this problem and that indeed it is a large problem that needs to be dealt with? With only information on the number of cases (numerator) and no information regarding the denominator (the number of people at risk), it is impossible to tell. To properly make these assessments, one must know how many runners visited the clinic over the course of the year. If 100 runners were seen and 100 cases of ruptured patellar tendons were diagnosed, then the incidence would be 100%—a potentially serious problem! Conversely, if 1,000 runners were seen, the rate would be reduced to 10%, and one could make a completely different interpretation. Using numerator data without considering the size of the population at risk should be avoided; however, data of this nature can often be found in the sports medicine literature.

Crude, Specific, and Standardized Rates

Three general categories of rates are common in epidemiology: crude, specific, and standardized. Rates that are based on a total population, without considerations of any of the population characteristics such as the distribution of age, gender, and ethnicity, are referred to as **crude rates.** When rates are calculated separately for population subgroups (typically age, gender, and ethnicity), they are called **specific rates** (age specific, gender specific, and so on). **Standardized rates** are crude rates that have been standardized (adjusted) for some population characteristic such as age or gender to allow for valid comparisons of rates among populations in which the distribution of age or gender may be quite different.

Crude rates, because they are dependent on the characteristics of the population from which they are calculated, can be misleading. For example, the crude prevalence of participation in vigorous physical activity in Boulder, Colorado (demographically a relatively young population), would be expected to be higher than that in a community such as Sun City, Arizona (demographically a relatively older population), simply because of the difference in the age distribution of residents in those communities. Likewise, a comparison of breast cancer rates in two populations in which the gender distribution varies greatly could be misleading. There are two solutions to this problem. First, valid comparisons among populations can be made if specific rates are used. In the preceding examples it would be reasonable to compare the rates of participation in vigorous physical activity between Boulder and Sun City by five-year age groups, or the rates of breast cancer for men and women, separately. Although the use of specific rates provides a valid comparison, the procedure can become cumbersome, particularly when numerous categories, such as five-year age groups over a large age range, need to be compared. To simplify the issue, epidemiologists often standardize crude rates.

Standardized rates, also referred to as adjusted rates, are simply crude rates that have been adjusted to control for the effect of some population characteristic such as age or gender. The most common method for the adjustment is called direct standardization. For example, the data in table 14.1 (part A) represent the death rates from two different populations. The crude death rate in population A is 4.51% and in population B is 3.08%. This is curious, and misleading, when you

Table 14.1 Illustration of the Principle of Direct Standardization of Crude Rates From Two Hypothetical Populations

A. Calculation of crude rate

Age group	Number	Population A: age-specific death rate	Expected	Number	Population B: age-specific death rate	Expected
20–49	2,000	.001	2	8,000	.002	16
50–79	10,000	.01	100	10,000	.02	200
80 and over	8,000	.1	800	2,000	.2	400
Total	20,000		902	20,000		616
Crude death rate	902 / 20,000 = 4.51%			616 / 20,000 = 3.08%		

B. Calculation of standardized rates using the combined population to form the standard population

Age group	Number	Population A: age-specific death rate	Expected	Number	Population B: age-specific death rate	Expected
20–49	10,000	.001	10	10,000	.002	20
50–79	20,000	.01	200	20,000	.02	400
80 and over	10,000	.1	1,000	10,000	.2	2,000
Total	40,000		1,210	40,000		2,420
Standardized death rate	1,210 / 40,000 = 3.03%			2,420 / 40,000 = 6.05%		

consider that the age-specific death rates in population B are twice that of population A. An inspection of the age distributions in these populations will suggest the problem. Population A has a high proportion of people in the older age group, where the age-specific death rate is highest, compared with population B. To make a valid comparison of the death rates between these two populations, death rates need to be adjusted to account for the difference in age distribution. The direct standardization method involves applying the age-specific rates of the populations to be compared to a single standardized population. The standard population can be any reasonable or realistic population. In this example the standard population is simply the combination of populations A and B. In practice, the population of a particular state or of the entire United States is often used. Because the age distribution in the standard population is the same for all the age-specific death rates that are applied to it, the effect of the different age distribution in the two actual populations being compared is eliminated. This procedure allows for the overall death rates in the two populations to be compared without the bias introduced by differences in the age distribution.

As illustrated in table 14.1 part B, after adjustment for age, the overall death rate in population B (6.05 %) is higher than that in population A (3.03%), accurately reflecting the fact that the age-specific death rates in population B are twice as high as those in population A. The same principles of direct standardization can also be used to compare incidence rates of disease or injury in populations differing by gender, health status, cholesterol or blood pressure level, or any other characteristic that may bias the rate comparison. Though standardized rates are useful for making valid companions across populations, it must be remembered that they are fictional rates. The actual magnitudes of the adjusted rates will vary depending on the standard population that is used in the adjustment process.

Research Design in Epidemiologic Studies

A research design is the way participants are grouped and compared according to a behavior or attribute (e.g., physical activity or fitness), the health-related events being studied, time, and factors other than physical activity or fitness that could explain the occurrence of health-related events. The goal of a design in physical activity epidemiology is to make sure that the comparisons of groups based on differences in physical activity or fitness are not biased by other factors. In a true research design, passage of time is needed between the change in the independent variable and the subsequent change in the dependent variable (i.e., temporal sequence). When the change occurs as the result of natural history (i.e., it is self-initiated by the people being studied), the design is **observational.** When change in the independent variable is manipulated by the investigator, the design is **experimental.** When the independent and dependent variables are observed or manipulated across a period of time, the design is longitudinal or **prospective.** When the study looks back in time after the occurrence of injury, disease, or death in an attempt to reconstruct physical activity habits, the design is **retrospective.**

Several types of research designs are commonly used in epidemiologic research; cross-sectional surveys, case-control studies, cohort studies, and randomized controlled trials. The design employed in any particular study will depend on the questions to be answered, the time and financial resources available, and the availability of data. The major advantages and disadvantages of the commonly used epidemiologic study designs are summarized in table 14.2.

Cross-Sectional Surveys

Cross-sectional surveys, sometimes called prevalence studies, measure both risk factors and the presence or absence of disease at the same point in time. Although this approach is expedient and relatively inexpensive, it is not possible to determine the temporal sequence of a potential cause-and-effect relationship. For example, the Iowa Farmers Study (Pomrehn, Wallace, & Burmeister, 1982) used a cross-sectional design to examine the association of physical activity and all-cause mortality. Sixty-two thousand all-cause deaths occurring from 1962 to

Table 14.2 Study Design in Epidemiologic Research—Advantages and Disadvantages

Design	Advantages	Disadvantages
Cross-sectional surveys	Quick and easy to conduct. Hypothesis generation.	No temporal relationship between risk factors and disease. Not appropriate for hypothesis testing.
Case-control studies	Appropriate for the study of rare events. Can study multiple risk factors. Inexpensive and quick to perform.	Cannot determine absolute risk. Subject to recall bias. Can study only one disease at a time. Temporal relationships may be uncertain.
Cohort studies	Provide an absolute measure of risk. Allow for the study of multiple disease outcomes.	Expensive and time-consuming to conduct. Not applicable for studying rare outcomes. Results may be affected by loss to follow-up. Can only assess the effect of risk factors obtained at baseline.
Randomized controlled trials	Investigator has control over the research process. Are the gold standard for the evaluation of interventions.	Expensive and time-consuming to conduct. Generalizability is often limited. Problems with compliance and dropouts.

1978 in male residents of Iowa aged 20 to 64 years were examined. A randomly selected group of 95 farmers was compared to 158 men who lived in a city. Farmers had a 10% lower rate of death, and they were twice as likely to participate in strenuous physical activity. They also were more physically fit as determined by lower exercise heart rate and longer endurance time on a treadmill test. Because this was a cross-sectional survey, it is possible that healthy and fit individuals self-selected to be farmers (or that men living in the city were less healthy and fit to begin with) and that the activity level of the farmers had very little to do with the observed reduction in mortality rate.

Cross-sectional surveys can be useful for generating hypotheses regarding potential risk-factor–disease associations and also for the assessment of the prevalence of risk factors or behaviors in a defined population. For example, the United States Centers for Disease Control and Prevention, in cooperation with state health departments, conducts a Behavioral Risk Factor Surveillance Survey each year to determine the prevalence of several disease risk factors including smoking and sedentary behavior. A specific type of cross-sectional survey, called an **ecologic study,** is used to compare the frequency of some risk factor of interest (e.g., sedentary behavior) with an outcome measure such as obesity in the same geographic region. For example, surveys conducted in a particular state may find high rates of sedentary behavior as well as high rates of obesity. Though this type of information may suggest the hypothesis that sedentary behavior results in obesity, drawing this conclusion is unjustified. Data from this type of survey should never be used to make any conclusion regarding cause and effect because there is no information as to whether the people who are sedentary are the same people who are obese. This problem is referred to as the ecologic fallacy.

Case-Control Studies

In a **case-control study,** subjects are selected based on the presence or absence of a disease. After cases and controls are selected, a comparison of the two groups is made relative to the frequency of past exposure to potential risk factors for the disease. In essence, the risk of having the risk factors between the case and control groups is compared. Risk factor information is typically obtained by personal interview or a review of medical records. As will be discussed later in this chapter, in the section on risk assessment, the case-control design does not allow for a direct determination of the absolute risk of the disease. This is because the incidence rates are not available because a group of people is not followed over time. However, an estimate of the risk of disease in those exposed to the risk factor compared to those not exposed can be calculated. Additional disadvantages of the case-control design include the difficulty in obtaining a truly representative control group, the inability to study more than one disease outcome at a time, and the potential for recall bias. To obtain a representative group of controls that are generally matched with cases on age, gender, and race, controls are often obtained from the same setting as the cases (i.e., from the hospital where the cases were diagnosed or from the same neighborhood where the cases reside). Investigators often use multiple control groups to increase the probability of obtaining a representative comparison group.

One limitation of case-control studies is **recall bias,** which may result in a spurious association between a risk factor and disease. Recall bias is the notion that people who have experienced an adverse event (cancer, heart attack) may think more about why they had this problem than healthy people and thus might be more likely to recall exposure to potential risk factors. Errors in recall can also be a problem in mortality studies because records must be available in archives or must be obtained from a witness to past behavior (e.g., a spouse).

Case-control studies are relatively quick and inexpensive to conduct, are useful for studying rare disease outcomes, require a relatively small number of subjects, and allow the study of multiple risk factors. Case-control studies are particularly useful for the initial development and testing of hypotheses to determine whether moving forward to conduct a more time-consuming and expensive cohort study or randomized trial is warranted.

A number of case-control studies are available in the physical activity epidemiology literature, particularly in the area of physical activity and cancer. The case-control methodology is ideal for the study of diseases such as cancer, which occur rather infrequently and have a long latent period between exposure to a risk factor and actual manifestation of the disease. For example, a group of investigators in the Netherlands (Verloop, Rookus, van der Kooy, & van Leeuwen, 2000) reported results from a case-control study on lack of physical activity as a risk for breast cancer in women age 20 to 54 years. A sample of 918 women diagnosed with invasive breast cancer between 1986 and 1989 was selected from a cancer registry. Each patient was matched with a control subject based on age and region of residence. Both cases and controls were interviewed in their home to collect information on lifetime physical activity and other risk factors including reproductive and contraceptive history, family history of breast cancer, smoking, and alcohol use, as well as premenstrual and menstrual complaints. Control subjects were assigned a date of pseudo-diagnosis that corresponded to the date on which they was the same age as their matched case subjects at diagnosis. The analyses were restricted to risk events that occurred prior to actual or pseudo-diagnoses. Results indicated that women who were more active than their peers at age 10 to 12 years were at significantly reduced risk of breast cancer. Also, women who had engaged in recreational physical activity at any point prior to diagnosis were also at significantly reduced risk for breast cancer. These data support the hypothesis that recreational physical activity decreases the risk of breast cancer in women, but the results must be interpreted in the light of potential problems associated with the case-control study design, including recall bias and nonrepresentativeness of the control group.

Prospective Cohort Studies

The term **cohort** in epidemiology defines a generational or demographic group large enough to permit the statistical control of a number of confounding variables and, thus, permits inference to a population. **Prospective cohort studies,** sometimes referred to as incidence or longitudinal follow-up studies, involve the selection of a group of people at random from some defined population or the selection of groups exposed or not exposed to a risk factor of interest. After the cohort is selected, baseline information concerning potential risk factors is collected and people are followed over time to track the incidence of disease. The prospective approach is more costly and time-consuming than either the cross-sectional survey or the case-control study, cannot be used to study diseases that occur infrequently, and can only assess the effect of risk factors that were measured at baseline (i.e., at the beginning of the study).

The major advantage of the prospective cohort study is that the risk profile is established before the outcome is assessed. Therefore, any recall information assessed at baseline cannot be biased by the knowledge of results. Prospective studies also allow the investigator to have control over the data collection process. The investigator also has the ability to assess changes in risk factors over time and correctly classify the disease endpoints (i.e., CHD, diabetes, osteoporosis), and the ability to study multiple disease outcomes, some of which may not have been planned at the study outset. Perhaps most important, a prospective design allows for the estimation of the true absolute risk for developing disease. Issues relative to definition and measurement of risk will be discussed later in this chapter.

A number of prospective cohort studies—for example, the Nurse's Health Study (Hu, Li, Colditz, Willett, & Manson, 2003), the Framingham Heart Study (Kiely, Wolf, Cupples, Beiser, & Kannel, 1994), the Harvard Alumni Study (Sesso, Paffenbarger, & Lee, 2000), the Honolulu Heart Study (Hakim et al., 1999), the Physician's Health Study (Manson et al., 1992), and the Aerobics Center Longitudinal Study (Blair et al.,1989)— generated valuable information regarding the associations among physical activity, physical fitness, and health outcomes. For example, the Aerobics Center Longitudinal Study measured physical fitness, defined as endurance time on a treadmill test, in over 10,000 men and 3,000 women when they visited the Cooper Clinic in Dallas, Texas, for preventive medical examination and followed them for eight years for total mortality. During the period of observation, 240 deaths among men and 43 deaths among women occurred after about 110,000 person-years of exposure. Age-adjusted death rates (per 10,000 person-years of exposure) from all causes were lower with each successive level of fitness in men from the least fit (64) to the most fit (19) and similarly in women from the least fit (40) to the most fit (9). The effects of higher fitness were independent of age, smoking, cholesterol level, systolic blood pressure, blood sugar, and parental history of CHD. Much of the decrease in rates of total mortality in the more fit subjects was explainable by reduced rates of cardiovascular disease and all-site cancers.

Randomized Controlled Trial

The **randomized controlled trial** is the gold standard design for testing a research hypothesis. With this design the researcher has more control over the research process than with any of the other epidemiologic research design options. The validity of a controlled trial depends on having a sample representative of the population and matching treatment and control groups on characteristics thought to affect the health outcome of interest. In a randomized controlled trial, participants are selected for study and randomly assigned to receive either an experimental manipulation or a control condition. Measurements are taken before and after the intervention period in both groups to assess the degree of change in the outcomes between the intervention and control conditions. The key to this approach is randomization, which assures that the experimental and control groups are comparable at baseline with respect to all factors, known or unknown, except for the factor being studied by the experimental intervention.

The randomized controlled trial is a powerful research design; however, the actual conduct of these trials poses a number of interesting problems. For example, to enroll in a randomized trial, the potential participant must agree to participate without knowing whether he or she will be assigned to the intervention or control group. This can be particularly problematic in an exercise intervention trial in which the motivation to participate is to receive the intervention, not to be assigned to a control condition. If possible, it is best to conduct a randomized trial in a **double-blind** manner (i.e., neither the participant nor the observers who collect the data are aware of group assignment). The double-blind approach is obviously not possible in exercise interventions. Only a **single-blind** trial, in which data collection personnel are unaware of group assignment, is feasible. Failure to maintain compliance with the intervention and differential dropout from both the exercise and control groups can introduce bias. Given the difficulty in recruiting participants for large randomized trials, these trials are often conducted using highly select samples. This reduces external validity, which is the ability to generalize the study results to other populations. For example, the Physician's Health Study, which was conducted to determine the effectiveness of aspirin on cardiovascular disease and beta carotene on cancer, was conducted on mostly white, healthy, middle-aged, male physicians. The generalizability of these results to other groups such as young men, women, minorities, and nonphysicians is questionable.

Although the randomized controlled trial is the research design of choice for hypothesis testing, many important questions in physical activity epidemiology cannot be answered with this approach. For example, to learn what type, frequency, intensity, and duration of exercise is most beneficial for reducing the incidence of CHD, an experiment could be conducted in which people are randomly assigned to a specific exercise regimen or a control condition and followed for incidence of heart disease. To obtain enough cases of disease requires that large numbers of people need to adhere to both the exercise and control conditions over a long period of time. Such a study is exceedingly expensive and a logistical nightmare to complete. Estimates of the number of participants needed for a randomized trial to adequately test the efficacy of exercise for primary prevention of CHD among middle-aged men and women have ranged from several thousand to several hundred thousand depending on the amount of reduction in the risk of heart disease that was to be detected. Needless to say, a trial of this type has never been, and most likely will never be, undertaken. The randomized trial has a place in physical activity epidemiology. Smaller, more manageable studies on the effects of different levels of physical activity or exercise training on muscle performance; balance and gait; and CHD risk factors such as lipids, obesity, blood pressure, and insulin can be and have been conducted.

The ultimate goal of a research design is to permit the conclusion that changes in the independent variable (i.e., physical activity or fitness) is the only reasonable explanation for changes seen in the dependent variable (i.e., injury, disease, or death). Factors other than the independent variable can vary simultaneously and might also cause a change in the dependent variable. Researchers must control such extraneous or confounding factors before they can reach a conclusion that the presumed effect of the independent variable was indeed independent. The types of research designs used in physical activity epidemiology differ in their ability to control extraneous factors. It is important to remember the inherent strengths and weaknesses of the study design options when evaluating the evidence for physical activity in reducing the risk of chronic disease.

EVALUATING ASSOCIATIONS IN EPIDEMIOLOGIC STUDIES

Conceptually, epidemiologic research is a comparison of (a) groups formed based on the presence or absence of a risk factor under study, the independent variable (i.e., physical activity, smoking, obesity), and (b) the presence or absence of a disease or condition of interest, the dependent variable (i.e., CHD, diabetes, osteoporosis). The goal is to identify the risk factors for a particular disease and to determine how much of an impact those factors have on the disease probability. A **risk factor** is a characteristic that, if present, increases the probability of disease in a group of people who have that characteristic compared with a group of people who do not have that characteristic. In practice, the situation can easily become much more complex because it is often of interest to study several independent variables at the same time and to determine how those variables interact to affect disease risk. For example, it is of interest to study the association of physical activity and the incidence of type 2 diabetes while considering the influence of age, gender, body fat, family history of diabetes, diet, smoking, and so forth. It may also be of interest to assess the effect on disease outcome of different levels (i.e., low-, moderate-, or high-intensity exercise) and durations of exposure to a risk factor.

Regardless of how complex the issues under study become, most epidemiologic research can conceptually be considered as a standard two-by-two table as shown in figure 14.1. There are slight differences in the interpretation of the two-by-two table depending on whether data were obtained from a prospective cohort or a case-control study. Both study designs are discussed in further detail in the following sections.

Disease Status

		Present	Absent	Total
Risk factor	Present	a	b	a + b
status	Absent	c	d	c + d
	Total	a + c	b + d	a + b + c + d

a = subjects with risk factor and disease
b = subjects with risk factor and no disease
c = subjects with no risk factor and disease
d = subjects with no risk factor and no disease
a + b = all subjects with risk factor
c + d = all subjects with no risk factor
a + c = all subjects with disease
b + d = all subjects with no disease
a + b + c + d = all subjects

Figure 14.1 Two-by-two table for assessing the association between a risk factor and disease.

Prospective Cohort Study

The interpretation of the two-by-two table for a prospective study is presented in figure 14.2. Recall that in a prospective cohort study a population is selected, baseline measures are obtained, and the population is followed forward in time to document the development of disease.

Figure 14.3 provides data from a hypothetical prospective cohort study on the association between physical activity at baseline and the incidence of CHD to illustrate the process of assessing the strength of the association between a risk factor and disease. In this example, 500 men in the total sample of 10,000 developed CHD. The overall incidence rate can be calculated as 500 cases per 10,000 population or .05 (5%). The question of interest is: What impact does risk factor status (i.e., level of physical activity) have on the incidence of disease? To answer this question, we need to calculate the incidence of disease in both the active and sedentary groups. The incidence rate for CHD in the sedentary group is $[a / (a + b)] = [400 / (400 + 5600)] = 400 / 6000 = 0.067$ or 6.7%. The incidence rate of disease in the active group is $[c / (c + d)] = [100 / (100 + 3,900)] = 100 / 4,000 = .025$ or 2.5%. The risk of sedentary behavior on CHD can be evaluated with this information.

The **risk difference** is simply the risk of disease in the group exposed to the risk factor minus the risk of disease in the unexposed group. The risk in the exposed group (sedentary) is 6.7%, and the risk in the unexposed group (active) is 2.5%, so the risk difference is 6.7% − 2.5% = 4.2%. If the level of risk in the exposed and unexposed groups were the same, the risk difference would be zero. If the exposure to the risk factor were harmful, as is the case for sedentary behavior in the example, then the risk difference would be greater than zero. If the exposure were protective (for example, exposure to a drug that lowers cholesterol level), then the risk difference would be less than zero.

The risk difference is also called the **attributable risk** because it is an estimate of the amount of risk attributable to the risk factor. In the example, 4.2% of the risk for the development of CHD in this population is attributable to exposure to the risk factor of sedentary behavior. The relative risk (RR), or risk ratio, is the ratio of the risk in the exposed group compared with that in the unexposed group. In the example, the RR of CHD in the exposed group (sedentary) compared with the unexposed group (active) is

$$RR = [a / (a + b)] / [c / (c + d)]$$
$$= [400 / (400 + 5,600)] / [100 / (100 + 3,900)] \qquad (14.2)$$
$$= 6.7 / 2.5 = 2.68$$

If the risk of disease in both the exposed and unexposed groups is the same, the RR is 1.0. If the risks in the two groups are not the same, calculation of the RR provides an easily interpretable method of demonstrating, in relative terms, how much different (greater or smaller) the risks are. In the example, the risk of CHD in the sedentary group is 2.68 times higher than that in the active group. When considering risk assessments, it is important to remember the difference between absolute risk and RR. In the example, the absolute risk for developing CHD is 6.7% in the sedentary group, but that level of risk is 2.68 times that in the active group. In some cases the RR can be extremely high between two groups, but the absolute risk in both groups is rather low. For example, the RR of myocardial infarction after heavy physical exertion in a group of men and women who did not exercise on a regular basis was 107; however, in a 50-year-old, nonsmoking, nondiabetic male, the risk for myocardial infarction during any given hour is approximately

Follow-Up

		Develop disease	Do not develop disease	Total	Incidence rate of disease
Risk factor status	Present	a	b	a + b	a / a + b
	Absent	c	d	c + d	c / c + d

Figure 14.2 Two-by-two table for a prospective cohort study.

Disease Status

		Develop disease	Do not develop disease	Incidence rate of disease
Risk factor	Present (sedentary)	$a = 400$	$b = 5,600$	6.7%
status	Absent (active)	$c = 100$	$d = 3,900$	2.5%
	Total	500	9,500	

Figure 14.3 Hypothetical data—cohort study on physical activity and CHD.

one per million. Thus, even if that risk is elevated 100 times as a result of physical exertion (1 per 10,000), the absolute risk is still very low.

The **odds ratio** (OR) can also be calculated for a prospective study, but as will be evident from the subsequent discussion, this measure is typically associated with the case-control design. The OR is calculated by dividing the odds of exposure to the risk factor in the diseased group by the odds of exposure to the risk factor in the nondiseased group or

$$OR = (a / c) / (b / d)$$
$$= (a \cdot d) / (b \cdot c)$$

(14.3)

Odds and risk are conceptually different. In the standard two-by-two table (see figure 14.1), the risk of disease in the exposed group, as discussed previously, is $[a / (a + b)]$, whereas the odds of disease in the exposed group is simply a / b. Based strictly on the mathematics, if a is small compared with b, which it often is, the odds and the risk are quite similar. This is illustrated by calculating the OR for the hypothetical data in figure 14.3. Here the OR is $(400 \times 3,900) / (100 \times 5,600)$ $= 1,560,000 / 560,000 = 2.79$, which is quite similar to the RR of 2.68. In a case-control study, in which a measure of disease incidence cannot be obtained, the OR provides a reasonable estimate of RR.

Case-Control Study

Recall that in a case-control study, participants are selected based on disease status (present or absent). A two-by-two table illustrating the organization of data from a case-control study is shown in figure 14.4. The analysis in a case-control study is a comparison of the proportion of cases exposed to a suspected risk factor $[a / (a + c)]$ with the proportion of controls exposed to the same risk factor $[b / (b + d)]$. If the exposure to the risk factor is positively related to the disease, then the proportion

Risk factor status		Cases (disease)	Controls (no disease)
	Present	a	b
	Absent	c	d
Proportion exposed		a / a + c	b / b + d

Figure 14.4   Two-by-two table for a case-control study.

of cases exposed to the risk factor should be greater than the proportion of controls' exposure to the risk factor. In the case-control study the only measure of the strength of the association between the risk factor and disease is the OR *(ad / bc)*. Conceptually, the OR is an estimate of the risk of disease given the presence of a particular risk factor compared with the risk of disease if the risk factor is not present. In most instances, the OR from a well-conducted case-control study is a good estimate of the RR that would have been derived from a prospective cohort study provided that the overall risk of disease in the population is low (i.e., less than 5%).

Interpreting Relative Risk and Odds Ratios

If the risk for disease is the same in both the groups exposed and not exposed to the risk factor, the RR and OR will be 1.0. Typically, when calculating RR, the incidence of disease in the exposed group is placed in the numerator and the incidence of disease in the unexposed group is placed in the denominator, as was done in figure 14.3 in which the calculated RR was 2.68. This makes sense because as the impact of the risk factor increases (in this case, the impact of sedentary behavior on CHD incidence), the RR increases. It is also acceptable to reverse the fraction and place the incidence of disease in the sedentary group in the denominator. Then the RR is 2.5 / 6.7 = .37, indicating that the risk in the active group is about one-third that of the sedentary group.

Although the formula for calculating the OR is different from that for calculating RR, the interpretation of the actual value is the same. Usually the OR is expressed with the group exposed to the risk factor in the numerator. However, the interpretation regarding the strength of the observed OR is not changed if the exposed group is placed in the denominator. The 95% confidence intervals are calculated as a measure of the degree of confidence that can be given to the fact that the observed RR or OR is meaningful (Jekel et al., 1996). The confidence interval can be used to determine whether the RR or OR observed differs statistically from 1.0. An RR or OR of 1.0 indicates that there is no difference in the risk of disease between those exposed and those not exposed to the risk factor. If the RR (2.68) for CHD in sedentary people had a 95% confidence interval ranging from 0.72 to 3.15, the RR of 2.68 would not be statistically significant because the 95% confidence interval includes a value of 1.0. However, if the same RR had a 95% confidence interval ranging from 1.5 to 3.5, the RR of 2.68 would be significantly different from 1.0 because the interval does not span 1.0.

Attributable Risk

An important function of epidemiologic research is to estimate the amount of disease burden in a population that is the result of a potentially modifiable risk factor. For example, one might ask: In the U.S. population how much of CHD mortality is the result of sedentary behavior, high blood pressure, or obesity? This type of information is extremely important for making decisions regarding which risk factors to target by intervention efforts to maximize the benefits to public health. Also, the public may be more interested and likely to comply with risk factor modification efforts if the importance of such efforts to their individual health status can be demonstrated.

Several epidemiologic measures are used to assess the impact of exposure to disease risk factors. These include the **attributable risk percent in the exposed group** and **population attributable risk percent.** The following sections describe these measures using the hypothetical data presented in figure 14.3 as an illustration.

Attributable Risk Percent in the Exposed Group (AR%)

The AR% can be calculated to determine the percentage of the total risk for CHD that is due to sedentary behavior among those who are sedentary. Either of two formulas can be used:

$$AR\% = (\text{Risk}_{\text{exposed}} - \text{Risk}_{\text{unexposed}}) / \text{Risk}_{\text{exposed}} \times 100 \qquad (14.4)$$

or

$$AR\% = [(RR - 1) / RR] \times 100 \qquad (14.5)$$

Using the data from figure 14.3 with equation (14.4),

$$AR\% = (6.7 - 2.5) / 6.7 \times 100$$
$$= 4.2 / 6.7 \times 100$$
$$= .6268 \times 100$$
$$= 62.7\%$$

Using equation (14.5),

$$AR\% = [(2.68 - 1) / 2.68] \times 100$$
$$= (1.68 / 2.68) \times 100$$
$$= .6268 \times 100$$
$$= 62.7\%$$

Therefore, among those who are sedentary, 62.7% of the risk for CHD is due to sedentary behavior. The AR% in the group exposed to the risk factor can also be calculated for a case-control study using equation (14.5) and substituting the OR for the RR.

Population Attributable Risk Percent (PAR%)

PAR% addresses the question of the percentage of the risk of disease that is due to a particular risk factor. This is similar to PAR but expresses the result as a percentage rather than an absolute amount. PAR% is calculated using the equation,

$$\text{PAR}\% = [(P_E)\,(RR - 1)] / [1 + (P_E)(RR - 1)] \times 100 \qquad (14.6)$$

where P_E is the proportion of the population exposed to the risk factor, and RR is the relative risk of disease associated with the risk factor. This equation allows for a comparison of the impact of different risk factors on disease risk in a population. For example, the proportion of CHD risk due to sedentary behavior could be compared with that of cigarette smoking. Assume that the RR for CHD associated with sedentary behavior is 2.0, and that 50% of the U.S. population is sedentary (P_E). Also assume that the RR of CHD associated with cigarette smoking is 5.0 and that 20% of the U.S. population smokes. With this information the PAR% for both sedentary behavior and cigarette smoking can be calculated as follows:

PAR% (Sedentary behavior)
$$= [.5\,(2 - 1)] / [1 + (.5)\,(2 - 1)\,] \times 100$$
$$= .5 / 1.5 \times 100$$
$$= .333 \times 100$$
$$= 33.3\%$$

PAR% (Smoking)
$$= [.2\,(5 - 1)] / [1 + (.2)\,(5 - 1)\,] \times 100$$
$$= [.2\,(4)] / [1 + (.2)(4)] \times 100$$
$$= .8 / (1 + .8) \times 100$$
$$= .444 \times 100$$
$$= 44.4\%$$

Therefore, approximately 33% of the CHD in the population is attributable to sedentary behavior, and 44% is attributable to cigarette smoking. Hypothetically, if all those people who were sedentary became active, there would be 33% fewer cases of CHD in the population. Similarly, if all the cigarette smokers quit, there would be 44% fewer CHD cases. Calculation of PAR% thus allows for the comparison of the impact of risk factors that vary in RR and prevalence in the population. Because a risk factor may have a very high RR but a low prevalence in the population, modification of that risk factor would have limited impact from a public health perspective.

DETERMINING CAUSE IN EPIDEMIOLOGIC STUDIES

The goal of many epidemiologic studies is to provide evidence for evaluating the association among variables with the aim of determining cause and effect. In the context of epidemiology, a causal association can be defined as an association between categories of events or characteristics in which an alteration in the frequency or quality of one category is followed by a change in the other category (Jekel et al., 1996). To determine whether an observed association between a risk factor and disease is causal, one must first demonstrate a statistical association between the risk factor and disease outcome. If a statistical association is demonstrated, several criteria, if satisfied, will increase the probability that the association is causal.

Criteria for Causation

The criteria used to assess cause and effect can be traced to the work of John Stuart Mill, first published in 1843 and referred to as Mill's canons (Stebbing, 1875). If the following criteria are true, an observed association is more likely to be causal.

1. *Temporal sequence:* Exposure to the risk factor must precede the development of the disease.

2. *Strength of association:* A large and clinically meaningful difference in disease risk exists between those exposed and those not exposed to the risk factor.

3. *Consistency:* The observed association is always observed when the risk factor is present.

4. *Dose-response relationship:* The risk of disease associated with the risk factor is greater with stronger exposure to the risk factor.

5. *Biological plausibility:* The observed association is consistent with existing knowledge regarding possible biological mechanisms.

The determination of a causal relationship in the absence of direct experimental evidence, as is often the case with the association of physical activity and health outcomes, is neither easy nor entirely objective. People often have different interpretations of the available evidence.

Confounding

When a strong and statistically significant association between a risk factor and disease is demonstrated, there is always the possibility that the observed association is noncausal. An issue referred to in epidemiology as confounding can result in noncausal associations. **Confounding** occurs when a noncausal association between exposure and outcome is observed as the result of the influence of a third variable or group of variables called confounders.

The possibility of confounding is higher in observational studies, which are common in physical activity epidemiology, than in randomized controlled trials. This is because the likelihood that the comparison groups (e.g., exposed/not exposed) differ with regard to a potential confounding variable is minimized in the randomized trial as a result of the random assignment of participants to groups. For example, many observational studies have shown increased physical activity at baseline to be associated with lower risk of all-cause mortality. Does increased physical activity cause the lower mortality risk, or are people with higher levels of physical activity generally in better health and therefore at a lower risk of mortality? Stated differently, does overall heath status confound the observed association between increased physical activity and decreased mortality?

Most well-conducted observational studies use statistical techniques to adjust for indicators of health status such as blood pressure, lipid levels, obesity, and cigarette smoking. Some confounding may still play a role in the observed association if, for example, the measurement of confounders were imprecise or potential confounders were not considered. Nevertheless, many studies report that the association of physical activity and all-cause mortality still persists after adjustment for the measured confounders. This association may be the result of confounding because most studies do not measure all potential confounders. For example, other diseases that may affect both physical activity and mortality, such as diabetes or depression, are often not taken into consideration. This illustrates the difficulty for any single observational study to account for all potential confounders and therefore provide a definitive answer relative to causality. In the study of physical activity and chronic disease risk, the accumulation of results over numerous investigations, conducted in different samples and measuring different confounding variables, adds weight to the evidence for a causal association between physical activity and chronic disease risk.

Effect Modification

In addition to confounding, the issue of **effect modification,** also called interaction, should be considered when evaluating the possibility of causal associations. Effect modification in epidemiology occurs when two or more risk factors modify the effect of each other with regard to the occurrence of an outcome. A minimum of three factors is required for effect modification to occur. For dichotomous variables, effect modification means that the effect of the exposure on the outcome depends on the presence of another variable. For example, the effect of physical activity categories (e.g., sedentary/active) on the presence or absence of CHD may differ depending on some third factor such as gender. If this were the case, gender would be termed an effect modifier.

Effect modification can also be evaluated with continuous variables by determining the effect of exposure on outcome depending on the level of some third variable. For example, the effect of physical activity assessed by questionnaire (kcal/week) on the risk for CHD may vary depending on the level of body mass index (BMI) (body weight [kg] / height [m²]). If so, then BMI is considered an effect modifier in the association between activity and CHD.

Determining the degree to which certain factors act as effect modifiers can provide important information relative to the development of preventive strategies. For example, demonstrating that people with a high BMI are at higher risk of CHD as a result of low physical activity compared to those with normal or low BMI would suggest that strategies specifically designed to increase physical activity in people with high BMI should be developed and evaluated.

The potential for confounding and effect modification increases the difficulty of interpreting the results of epidemiologic studies focused on physical activity and health. Age can be a confounder because of the direct association between inactivity and an increased risk for death and most chronic diseases. Therefore, to examine the association between physical activity and health outcomes, the impact of age needs to be removed. Conversely, age may also change the magnitude of risk associated with other variables and thus is also considered an effect modifier. For example, the risk of CHD increases with increased age, increased blood pressure, and decreased levels of physical activity. Age is also associated with decreased physical activity and increased blood pressure. Therefore, to determine whether an association exists between physical activity and CHD risk, the effect of both age and blood pressure must be controlled. In epidemiologic studies this type of control is typically accomplished by statistical analysis. When reading and interpreting information relative to the association of physical activity and health outcomes, investigators must be aware of the potential consequences of confounding and effect modification and determine whether these issues have been satisfactorily addressed.

SUMMARY

This chapter presented the rudimentary principles and skills of logical inquiry that epidemiologists use to study physical activity. ORs, RR, and attributable risk help determine whether physical activity can reduce the rate of disease or death among people at risk. The research design used in an epidemiologic study—whether a cross-sectional, case-control, observational, or randomized controlled trial—determines the strength of the inference that lower disease risk is explainable solely by physical activity or fitness and not by other confounding or effect-modifying factors. Ultimately, the strength of the evidence for concluding that cause and effect exists is judged by Mill's canons: temporal sequence, strength of association, consistency, dose-response relationship, and biological plausibility. This chapter provided the foundation for understanding how epidemiologists determine whether people's attributes or behavior influences the risk of disease and death.

CHAPTER 15

Measurement Issues in the Clinical Setting

Marilyn A. Looney

The goal for practitioners, whether they are in academic or clinical settings, is to make good decisions. These decisions are usually made after weighing information gathered from a variety of sources. In a clinical setting, the types of information gathered can vary from a description of events preceding an athlete's onset of ankle pain, to a degree of S-T segment depression during an ECG exercise stress test, to a degree of change in joint range of motion after therapeutic exercise, to a patient's proclivity toward conservative treatment plans. The trustworthiness of this information affects the quality of the decisions made by allied health professionals (e.g., certified athletic trainers, physical therapists, and physicians). The use of inaccurate information may lead to inappropriate decisions such as (a) prematurely approving an athlete's return to competition, (b) foregoing further testing such as an angiogram, or (c) recommending a high-risk treatment plan. Not only are good data required to make good decisions, but so are accurate interpretations of these data. A misinterpretation of good data may also lead to bad decisions.

A major focus of recent literature has been to describe (a) the decision-making process in the clinical setting (Bernstein, 1997; Eddy, 1996), and (b) the quality of evidence for diagnostic tests and treatments (Basmajian & Banerjee, 1996; Bossuyt et al., 2003; Kumbhare & Basmajian, 2000). The emphasis is on using evidence from well-designed research studies to make decisions instead of blindly following traditional practice (Ashley, Myers, & Froelicher, 2002; Sackett, Straus, Richardson, Rosenberg, & Haynes, 2000). Take, for example, a comment by Kumbhare and Basmajian (2000): "Most of [sports medicine's] relatively short existence has been dominated by dogmatic approaches aggressively promoted by charismatic trainers and practitioners. . . . A few have relied on whatever research evidence exists in a few narrow niches, but the majority are still flying by the seat of their pants. . . . Our purpose is to sound a clarion call for evidence-based practices and for a wind storm of controlled research" (p. xi).

Allied health professionals must make many decisions as they help patients move from the onset of symptoms to the resolution or acceptable management of the problem. Throughout this chapter, *patient* refers to any person who seeks the services of a certified allied health professional—for example, an athlete evaluated by a certified athletic trainer. Questions allied health professionals ask themselves drive the decision-making process. These questions include the following: What signs or symptoms trigger the administration of a screening test or a referral? What is the best diagnostic test to use? If the screening test is positive, which verification test should be administered? What are the risks associated with the test? Do the risks outweigh the knowledge to be gained? In other words, is the knowledge gained from a high-risk test required to develop an effective treatment plan? Does the patient's medical history increase the probability for a positive diagnosis? What are the patient's major objectives? Is the objective pain reduction at any cost or a return to competition despite increased risk for another injury? Is the treatment plan achieving the desired result?

One of the first decisions an allied health professional faces is selecting an appropriate screening or diagnostic test to administer. Before a decision can be made, evidence must be gathered from well-designed published research studies on the effectiveness of various diagnostic tests. If allied health professionals are to make evidence-based decisions, they must stay current with the medical literature because tradition is no longer an accepted justification for test selection and treatment practices (Sackett et al., 2000). This was the impetus behind two books published for sports medicine (Kumbhare & Basmajian, 2000) and rehabilitation (Basmajian & Banerjee, 1996) professionals. In both books, each chapter is a review of the published research for a specific health problem. The authors provide an evaluation of the research quality and indicate how much importance should be given to the "evidence" in the decision-making process. These books are good resources but will become dated, thus providing an incomplete record of the existing evidence. Thus, allied health professionals must be able to judge for themselves the quality of the new "evidence" provided by the most current research (Bossuyt et al., 2003). This chapter provides information that will help allied health professionals interpret and evaluate evidence regarding the efficacy or effectiveness of diagnostic tests. Because the goal is to develop better consumers of research, formulas and calculations will be minimized.

EVALUATING THE DIAGNOSIC ACCURACY OF TESTS

As outlined in previous chapters (see chapter 2 by David Rowe and Matt Mahar and chapter 3 by Ted Baumgartner), test scores and their interpretations in the clinical setting must be evaluated for evidence of validity and test reliability. Additional resources for psychometric evaluation methodology include Baumgartner, Jackson, Mahar, and Rowe (2003); Greenfield, Kuhn, and Wojtys (1998); McDowell and Newell (1996); and Safrit and Wood (1989).

Clinicians, however, are also interested in diagnostic test performance. Nominal or categorical data are usually the outcome of interest (i.e., the client has or does not have the condition) when evaluating the diagnostic effectiveness of tests. A kappa or weighted kappa coefficient (see chapter 3 by Ted Baumgartner) can be computed to document how well the decisions based on the diagnostic test agree with decisions from a reference test after adjusting for chance agreement. Kappa, however, is not the optimal statistic to evaluate the efficacy of a diagnostic test because the number of true positive results (has the condition) and true negatives (does not have the condition) are combined. How well the test functions separately for each decision cannot be determined from kappa. Common statistics that reflect the accuracy of one of the decisions (i.e., has the condition or does not have the condition) are **sensitivity** (SN), **specificity** (SP), **positive predictive value** and **negative predicted value, likelihood ratios,** and area under the **receiver operating characteristic curve** (AUC) (Altman, 2000a; Altman & Bland, 1994a, 1994b, 1994c; Bossuyt et al., 2003).

Evidence of widespread use of SN and SP was found by conducting a narrow search of the medical literature published in English from 2002 to 2003. More than 27,000 records were found by limiting a MEDLINE search to the MESH phrase "sensitivity and specificity." Because of the widespread use of SN and SP, their meaning and limitations must be understood to appropriately evaluate the literature on diagnostic test effectiveness. The interpretation of these statistics, as well as others, is limited by the research design employed and how representative the sample is of the target population.

Allied health professionals who make evidenced-based decisions must be able to evaluate the quality of published evidence for diagnostic tests. According to an international committee of scientists and journal editors (Bossuyt et al., 2003), "The quality of reporting of studies of diagnostic accuracy is less than optimal" (p. 7). This conclusion was based, in part, on the critiques of the diagnostic test literature conducted by Lijmer and colleagues (1999) and Reid, Lachs, and Feinstein (1995) regarding study design and the absence of relevant information. As a result, the international committee developed the Standards for Reporting of Diagnostic Accuracy (STARD) statement, which includes a checklist of 25 items targeting the relevant information that should be present in all studies reporting the effectiveness of a diagnostic test. When all relevant information is present, allied health professionals should be able to determine how much credibility to give to the published results. The accuracy of diagnostic tests may be overestimated in studies with methodological weaknesses (Lijmer et al., 1999).

Overview of Research Design Issues

The ideal design for determining the **diagnostic accuracy** of a test is to draw a random sample from a specified population; for example, patients seeking assistance for foot injuries at a sports medicine clinic. All patients are administered both the diagnostic test and the reference test that is considered to be the gold standard (i.e., there is an accurate determination of whether the disorder or condition is present or absent). After testing is completed (following standardized procedures), appropriate personnel interpret test results while blinded to the patient's results on the initial test (Jaeschke, Guyatt, & Sackett, 1994a). All personnel who classify patients based on subjective interpretation of test results should have demonstrated very high **intertester** and **intratester reliability** (Begg, 1987; Jaeschke, Guyatt, & Sackett, 1994b; Kumbhare & Basmajian, 2000). Not only should sample sizes be large enough so that unbiased estimates of the statistics can be determined, but they should also be large enough so that the estimates are fairly precise (i.e., narrow 95% confidence intervals exist) (Kent & Larson, 1992). The sample size depends on the prevalence of the disorder or condition in the population. The smaller the percentage of members in the population with the condition, the larger the sample size required to ensure that at least 10 patients with the condition participate in the study (Kraemer, 1992).

Another concern is the observational unit used for data analysis. If there is more than one data point per patient, it is not appropriate to treat each data point as a separate "patient." This introduces dependency among the data points when independence is assumed (Kraemer, 1992). Finally, the omission of patients because they have uninterpretable (ambiguous) test results can bias the calculated statistics. Guidelines that specify the action to take if an uninterpretable test result occurs should be in place prior to the study. Some possible actions include (a) readministering the test, (b) treating ambiguous test results as positive results, (c) treating ambiguous test results as negative results (Kraemer, 1992), or (d) following up at a later time to see whether disease status has developed (Begg, 1987). Regardless of the subsequent action taken, the number of problem cases requires documentation (Kent & Larson, 1992). If there are more than a few problem cases, then the statistical analysis should be conducted with and without these cases in the data set so the impact of dropping these cases can be seen on the evidence for test effectiveness. In practice, the sample is rarely a true representation of the population, and the gold standard rarely identifies patients' true conditions without error. The ramifications for interpreting SN and SP, calculated under less than ideal circumstances, will be discussed after SN and SP have been explained more fully.

A study that illustrates several of the research design issues inherent in examining the diagnostic accuracy of tests was conducted at the United States Military Academy where many cadets experience acute ankle and foot injuries each year (Springer, Arciero, Tenuta, & Taylor, 2000). When clinicians see patients, they must first evaluate the injury using a standardized protocol to determine whether radiographs are warranted to confirm or refute the suspicion that a fracture exists. Although they are worried about ordering unnecessary radiographs, clinicians do not want to miss significant fractures. Thus, the purpose of the study was to evaluate how well the application of modified Ottawa Ankle Rules identified the presence of fractures as verified by ankle and foot radiographs. Over a 7-month period, 156 cadets who had sustained an acute ankle or midfoot injury within the previous 10 days were examined, first by one of two physical therapists and then independently by one of five orthopedic surgeons. Each applied the modified Ottawa Rules to determine whether they would recommend a radiograph. Ankle and foot radiographs (three views each) were obtained for each cadet. Springer and colleagues appear to have assumed that the two physical therapists would have made the same decision using the ankle rules if they had seen the same patient, and also that the surgeons' decisions using the ankle rules were interchangeable. The identity of the physical therapist and surgeon who evaluated a patient did not concern the authors. As a result, the physical therapists' data were combined and the surgeons' data were combined to determine the intertester agreement between physical therapists and surgeons (Fleiss, Levin, & Paik, 2003, pp. 610-617). Unfortunately, intertester and intratester reliability coefficients were not reported for the two physical therapists, nor for the five surgeons or the two radiologists who read the radiographs when evaluating the same set of patients. The radiologists were blinded to the ankle rule predictions, so their interpretations would not be biased by prior knowledge of results. After applying exclusion criteria, 153 patients remained in the study. Out of this number, four had injured both ankles, so the data analysis was based on 157 ankles instead of 153 patients. It was not possible to determine the impact of including dependent data on the SN and SP estimates because of the way the data were reported. The incidence of ankle fractures was approximately 4% for cadets at the United States Military Academy. Based on this estimate and the recommendation that at least 10 patients have the condition (Kraemer, 1992), at least 97 more cadets (10 / .04 − 153) should have been evaluated. The results reported for the ankle evaluations will be used to illustrate the computation of SN and SP.

Computation and Interpretation of Statistical Indexes

There are many ways to document the accuracy of diagnostic test data. In this section, however, only those statistical indexes that appear most often in the allied health literature will be explained with computational examples. These statistics include sensitivity, specificity, predictive values, and likelihood ratios.

Sensitivity and Specificity

Allied health professionals use test results to make decisions about patients' conditions. Each decision represents one of four possible outcomes: (a) correctly decides condition exists (**true positive**), (b) incorrectly decides condition exists (**false positive**), (c) incorrectly decides condition does not exist (**false negative**), and (d) correctly decides condition does not exist (**true negative**). When data are collected following procedures that were outlined earlier for the ankle injury study, they can be arranged in a two-by-two **contingency table** (see figure 15.1). Data residing in cells a and d represent correct decisions (true positives and true negatives, respectively), whereas those in cells b and c represent errors (false positives and false negatives, respectively). Values in cells a and d along with appropriate column marginal values are used to compute SN and SP (see figure 15.1).

SN is the proportion of patients with a condition who had a positive test result, whereas SP is the proportion of patients without the condition who had a negative test result. These proportions are actually probabilities of the test result occurring conditioned on (or given) the true status of the condition (present or not). An ideal

		True condition	
		+	−
Test results	+	a	b
	−	c	d
		a + c	b + d

Sensitivity (SN) = a / (a + c)
Specificity (SP) = d / (b + d)
Prevalence (P) = (a + c) / (a + b + c + d)

Figure 15.1 Diagnostic test results by patients' true condition and test efficacy statistics.

test will have both high SN and high SP. Allied health professionals who use a test with very high SN and get a negative test result will have confidence that the patient does not have the condition in question. This is sometimes referred to as the test's ability to "rule out" the condition. When the concern is how well the test result can "rule in" the condition, false-positive errors should be rare. Confidence placed in a positive test result to "rule in" the condition requires a test with very high SP (Dawson & Trapp, 2001). In the Ottawa Ankle Rules example, clinicians are worried about failing to order radiographs for cadets who actually have fractures. They want to have confidence that a negative result using the modified Ottawa Rules coincides with no clinically significant fractures. In other words, they want high SN.

Out of 157 ankles evaluated with radiographs, there were six fractures, which represents a prevalence rate of 4% (see figure 15.2). This prevalence rate is comparable to rates reported in similar clinical settings (Leddy, Kesari, & Smolinski, 2002). The computations of SN and SP are illustrated in figure 15.2 using data reported by Springer and colleagues (2000). The data represent the physical therapists' decisions after examining the cadets' ankles using the modified Ottawa Ankle Rules. Based on these data, SN is perfect (1.00); that is, a negative result did not fail to identify any clinically significant fractures. On the other hand, there were several false-positive errors in which cadets without fractures had positive test results (1 − SP = .60). This is not surprising given the low incidence rate (.04) of ankle fractures in this clinical population. The lower the incidence or prevalence of a condition in the population, the higher the false-positive rate. This principle is used to guide test-screening policy. It is poor policy to mandate that all members of the general population be tested when the prevalence rate is low because there will be many false-positive results (Brenner & Gefeller, 1997; Dawson & Trapp, 2001; Fleiss et al., 2003). If the incidence rate in the example had been higher (e.g., .15), there would have been 80 false-positive tests instead of 90. Although the ideal test would have both high SN and SP, the application of the modified Ottawa Ankle Rules did prevent the most serious decision error (i.e., failing to order a radiograph when a fracture existed).

		Ankle fracture	
		+	−
Test results	+	6	90
	−	0	61
		6	151

Sensitivity (SN) = 6 / 6 = 1.00

Specificity (SP) = 61 / 151 = .40

Prevalence (P) = 6 / 157 = .04

Figure 15.2   Calculations of sensitivity (SN) and specificity (SP) for physical therapists using modified Ottawa Ankle Rules as adapted from Springer et al. (2000).

See Altman (2000a) for an equation to compute 95% confidence interval for a single proportion.

If another sample of cadets were studied following the same protocol, how much would SN and SP vary from the reported values of this study? In other words, how well do SN and SP estimate the population values? Confidence intervals answer this question with narrow 95% confidence intervals, indicating more precise estimates of the population values. The method used to construct confidence intervals should be reported so the reader can judge whether the correct method was employed. According to Sackett and colleagues (2000), the traditional method for computing the standard error for a single proportion (Fleiss et al., 2003) is not recommended for small sample sizes or when proportions (e.g., SN or SP) are close to 0 or 1. If the traditional method is used under these circumstances, unrealistic confidence intervals will result (Altman, 2000a, 2000b; Newcombe, 1998). Alternative methods for confidence interval construction exist and are described by Newcombe (2004) and Newcombe and Altman (2000).

Receiver Operating Characteristic (ROC) Curve

The use of a two-by-two table to determine SN and SP implies that the diagnostic test results are dichotomous data: positive or negative. This is not true for many diagnostic tests in which numerical scores fall along a continuum. Thus, a cutoff score must be determined that best discriminates between patients who have the condition and those who do not. The **receiver operating characteristic (ROC) curve** is a graphical method to determine the cutoff score (Altman & Bland, 1994c; Dawson & Trapp, 2001). It is illustrated in figure 15.3 by plotting the true-positive rate (SN) against the false-positive rate (1 − SP) for each of 15 scores reported by McCrea (2001). Injured football players identified as having a concussion injury, along with case-matched uninjured teammates, were tested on the sideline immediately after injury with the Standardized Assessment of Concussion (SAC). The SAC is designed to evaluate neurocognitive functioning and neurologic status.

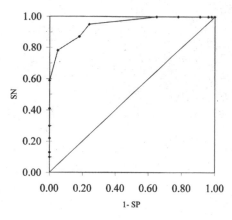

Figure 15.3 Example of an ROC curve for sensitivity (SN) and specificity (SP) values reported by McCrea (2001).

SN and SP were calculated for each SAC difference score (preseason baseline minus postinjury). The more discriminating the SAC, the more the curve should fill the upper left corner. A less discriminating test will have equivalent true-positive and false-positive rates that form a straight line (Altman & Bland, 1994c). In this preliminary investigation, a deficit of one point from baseline had the best combination of SN and SP for classifying players with or without concussion. This is depicted graphically by locating the highest point in the upper left quadrant of figure 15.3. The performance of two or more diagnostic tests can be compared by placing their ROC curves on the same graph. The test whose curve is completely under the curves of the other tests is the less effective diagnostic test (Altman & Bland, 1994c). The difference between areas under the curves (AUC) can be tested for statistical significance (Dawson & Trapp, 2001). See Altman (2000a) for an example of computing a confidence interval for AUC. Lee (1999) proposed an alternative to comparing areas under the curves. Lee's method examines the distance between diseased and diseased-free distributions.

Predictive Values

SN and SP provide information on how well test results can rule in or rule out the condition in question. This does not, however, match the daily decision-making context of the clinician who is interested in answering the following question: What is the probability that a patient with a positive test result has the condition? (See figure 15.4.) The answer to this question is known as the positive predictive value (PPV), or posttest probability. An analogous value can be calculated for a negative predictive value (NPV)—the probability that the patient does not have the condition given a negative test result.

Predictive values should not be computed from the two-by-two table used to determine SN and SP unless the suspected probability that the patient has the

Test Efficacy

Given that the patient has or does not have the disorder or condition, what is the probability that the test result matches the patient's condition?

		True condition	
		+	−
Test result	+	a	
	−		d
		a + c	b + d
		Sensitivity	Specificity

Clinical Decision

Given the patient's pretest probability for the condition, test result, sensitivity, and specificity, what is the probability that the patient does or does not have the disorder or condition?

		True condition			
		+	−		
Test result	+	a*		a* + b*	PPV
	−		d*	c* + d*	NPV
		a + c	b + d		

PPV = positive predictive value; NPV = negative predictive value

Figure 15.4   Conceptual focus of test-oriented and patient-oriented decisions.

*Pretest probability, sensitivity, specificity, and arbitrarily chosen sample size are used to adjust all values within the two-by-two table.

condition prior to testing matches the incidence or prevalence rate for the sample (Dawson & Trapp, 2001). Because this is rarely the case, the predictive values should be determined based on (a) the clinician's best estimate that a patient has the condition before the test is given (known as prior or pretest probability) and (b) reported test SN and SP for a population that matches the patient's characteristics (Altman & Bland, 1994b). These best estimates may be based on the clinician's experience and information from databases and primary studies (Sackett et al.,

2000). By using pretest probability, SN, and SP, cell values for a two-by-two table are computed for a given sample size. The sample size is arbitrarily chosen to make computations easier (Dawson & Trapp, 2001; Jaeschke et al., 1994b).

In the context of the ankle injury example, a cadet is evaluated by a physical therapist (PT) at the U.S. Military Academy. The PT knows that the incidence of ankle fractures is about 4% based on the study conducted by Springer and colleagues (2000). The cadet's symptoms plus the PT's experience result in the PT raising the prior probability to 10% for this cadet. If the PT gets a negative test result using the modified Ottawa Ankle Rules, what is the probability that a fracture does not exist? See figure 15.5 for the computation of the NPV. Is the probability for not having a fracture large enough to forego ordering a radiograph? Because the NPV is 1, the answer is yes. When the 95% confidence intervals for SN and SP, respectively, are moderately wide, allied health professionals may need to compute NPV (PPV) for the upper and lower limits of the SN (SP) confidence interval. This may help them decide how much emphasis to place on the test result in making their next decision:

Predictive values (PV):		True condition		
			+	−	
$a + c = PP (N) = .10 (100) = 10$		+	10	54	64
$a^* = SN (a + c) = 1 (10) = 10$					
$d^* = SP (b + d) = .40(90) = 36$		−	0	36	36
			10	90	100
Positive PV = $a^* / (a^* + b^*) = 10 / 64 = 0.16$					
Negative PV = $d^* / (c^* + d^*) = 36 / 36 = 1.00$					

Likelihood ratios (LR):

Pretest odds = PP / (1 − PP) = .10 / (1 − .10) = 0.11
Positive LR = SN / (1 − SP) = 1 / (1 − .40) = 1.67
Negative LR = (1 − SN) / SP = (1 − 1) / .40 = 0.00

Posttest odds for positive test = Pretest odds (positive LR) = 0.11 (1.67) = 0.18
Posttest probability for positive test = (Posttest odds +) / (1 + Posttest odds +) = 0.18 / (1 + 0.18) = 0.16

Posttest odds for negative test = Pretest odds (negative LR) = 0.11 (0) = 0
Postest probability for negative test = (Posttest odds −) / (1 + Posttest odds −) = 0 / (1 + 0) = 0

Figure 15.5   Adjustment of posttest probabilities using patient's pretest probability given: Pretest probability (PP) = .10; Sensitivity (SN) = 1.00; Specificity (SP) = .40; $N = 100$.

See Altman (2000a) for equations to compute 95% confidence intervals for predictive value and likelihood ratios.

Does the subsequent decision (e.g., order a radiograph or not) differ using the lower limit for SN versus the upper limit in calculating the NPV? According to Kraemer (1992), a legitimate test will have the PPV greater than the prior probability and the NPV greater than 1 minus the prior probability.

Likelihood Ratios

Allied health professionals want to know the probability that a patient will have the suspected condition given the test result. An equivalent way of expressing the same idea from a different perspective uses likelihood ratios (LR). A **positive likelihood ratio** (LR+) is calculated as SN / (1–SP) or the ratio of the probability of getting a positive test result for a patient having the condition to the probability of getting a positive test result for a patient who does not have the condition (Altman & Bland, 1994b) . The greater LR+ is than 1, the greater the posttest probability that the condition is present given a positive test. As LR+ becomes smaller than 1, the condition is less likely to exist with a positive test (Jaeschke et al., 1994b). In essence, LR+ is an indicator of how much the positive test result modifies the prior probability of having the condition. A useful test should result in meaningful differences between prior and posttest probabilities (Altman & Bland, 1994b). Given LR+ and the prior probability, a nomogram is often used to quickly determine the posttest probability of having a condition given a positive test (for more information on nomograms, see Jaeschke et al., 1994b, or Sackett et al., 2000). If a nomogram is not available, posttest probability can be computed using the equations presented in figure 15.5.

LR+ converted to posttest probability is actually the PPV, whereas the **negative likelihood ratio** (LR–) converts to posttest probability of having the condition with a negative test result (1 – NPV) (Dawson & Trapp, 2001). In the example presented in figure 15.5, the pretest odds and LR+ of 0.11 and 1.67, respectively, determine the posttest odds of 0.18. Because this odds is less than 1, we expect the probability of having an ankle fracture given a positive test to be small, which is the case (PPV = .16).

Jaeschke and colleagues (1994b) prefer the LR+ to SN and SP to judge a test that has various levels of interpretation (e.g., very positive, moderately positive, neutral, moderately negative, and extremely negative). Although SN and SP can also be calculated for each level, we cannot use a nomogram to quickly estimate posttest probability. According to Jaeschke and colleagues (1994b) and Sackett and colleagues (2000), LR+ is being used more frequently in the medical literature to further the practice of evidence-based medicine. If PTs had recorded their results as definitely positive, neutral, or definitely negative in the ankle rules study (Springer et al., 2000), a table could have been created that would help PTs quickly determine the posttest probability for fracture given a positive test result and pretest probability for the patient (see figure 15.6 for a hypothetical example) (Sackett et al., 2000). Although LR+ may indicate that the test is useful, it does not necessarily follow that a positive test is a good indicator of whether the condition is present (Altman & Bland, 1994b). The smaller the pretest probability is, the greater the false-positive rate will be (Fleiss et al., 2003).

Test result	Target/Condition		LR +	Posttest probabilities by pretest probability		
	Present	Absent		.05	.10	.20
Definitely positive	.80	.20	4.00	.17	.31	.50
Neutral	.15	.20	0.75	.04	.08	.16
Definitely negative	.05	.60	0.08	.00	.01	.02

Figure 15.6 Hypothetical posttest probabilities for ankle fracture adjusted for test results and pretest probabilities.

MAJOR SOURCES OF BIAS AFFECTING SIZE OF DIAGNOSTIC STATISTICS

Allied health professionals often choose the diagnostic tests they will use based on statistics reported in the literature. Unfortunately, these diagnostic statistics may be biased because of issues related to subject selection or retention, subjectivity of those interpreting test results, and imperfect reference standards. Because these and other sources of bias must be considered when interpreting published reports of test effectiveness, sources of bias will be discussed further in this section.

Spectrum Bias

Spectrum bias deals with the fact that test SN and SP are specific to the population sampled. Just as test reliability and validity evidence are population specific, so are SN and SP. Hlatky and colleagues (1984) investigated factors that might influence SN and SP for exercise stress-test results based on S-T segment depression to indicate the presence of a coronary disease as verified by angiogram results. Because over 3,000 patients received both evaluations, the relationship of age, gender, and severity of disease with S-T segment depression was determined. One of the factors SN was related to was the severity of coronary disease (i.e., number of diseased vessels). SN was .48, .68, and .85 for one-, two-, and three-vessel involvement, respectively.

Studies designed to investigate the efficacy of diagnostic tests need to (a) recruit a very large sample so SN and SP or LR can be determined for subgroups of patients within the sample (Jaeschke et al., 1994b; Kraemer, 1992; Philbrick, Horwitz, Feinstein, Langou, & Chandler, 1982), and (b) recruit a broad spectrum of patients from the population that will "challenge" the test (Randsohoff & Feinstein, 1978). According to Randsohoff and Feinstein, SN evidence should be based on a broad spectrum of patients with the condition that includes three dimensions: variation in severity of the condition, variation in the degree of symptoms, and variation in the severity of a coexisting condition that might cause a false-negative result. In the context of the ankle study example, this means that among cadets who had fractures there were cadets who exhibited variation in the severity of the fracture, variation in pain (mild to severe), and variation in the degree of ankle sprains. With the part of the population that does not have the disease or the condition, researchers should challenge the test to give false-positive results. Within the segment of this population, there needs to be a variation in the severity of a problem that is located in the same area as the target condition or disease, variation in the degree of symptoms in the same location, and variation in the same kind of target problem but different location (Randsohoff & Feinstein, 1978). In the context of the ankle example, this means that cadets who do not have ankle fractures exhibit variation in the severity of ankle sprains, variation in the degree of ankle pain, and variation in the severity of foot fractures.

If a broad spectrum does not exist in both groups to sufficiently "challenge" the test, then falsely inflated SN and SP may result (Brenner & Gefeller, 1997; Randsohoff & Feinstein, 1978). In a case-control study, the data from a group of patients already known to have the condition and a separate group of patients without the condition will result in overestimates of diagnostic accuracy (Lijmer et al., 1999). Published reports of test efficacy studies should include a detailed description of the patients' characteristics with and without the condition. With this information, one can judge whether the test has been sufficiently challenged to give false-positive and false-negative results and whether the patient belongs to the population tested. Not only are SN and SP subject to spectrum bias, but so are LR and predictive values (Brenner & Gefeller, 1997).

Verification Bias

Verification, or **workup, bias** is a concern when the reference standard used to confirm the diagnosis is a high-risk or invasive test. When this is the case, it is common for only those patients who received a positive test result on the diagnostic test to receive the reference standard test. This is often the context for studies looking at the effectiveness of exercise stress tests and may explain the wide range of sensitivities (.35 to .88) and specificities (.41 to 1.00) that Philbrick and colleagues (1982) found in the literature. SN and SP will be biased if the study inclusion criteria filter the patients so only those with the most severe condition are included (Kent & Larson, 1992). Positive tests are overrepresented in the verification sample, which falsely inflates SN and deflates SP (Begg, 1987; Philbrick et al., 1982). The best way to avoid verification bias is to design a prospective study in which all patients receive verification of diagnosis status (Begg, 1987) or treatment outcome. When this is not possible, Begg and Greenes (1983) demonstrated, under certain assumptions, that SN and SP can be mathematically adjusted for verification bias.

Rasch analyses (see chapter 4 by Weimo Zhu) have shown some promise in examining the utility of diagnostic tests (Cipriani, Fox, Khuder, & Boudreau, 2005). These analyses can handle missing reference test data for some patients. Future research is warranted on the effectiveness of Rasch analyses to determine the diagnostic accuracy of tests.

Pseudo-Retrospective Sampling

Kraemer (1992) warned against using a sampling plan that can be susceptible to producing invalid results. In **pseudo-retrospective sampling**, a sample is drawn from a population with high risk of having a condition, and a second sample is drawn from a low-risk population. Patients are often matched across groups according to age, gender, and socioeconomic status. Each patient is tested and diagnosed, and the results are combined into one table. Problematically, the high-risk sample will have more positive diagnoses, and the low-risk group will have the most negative diagnoses. This combination of high-risk and low-risk groups does not produce a representative sample from a single population with positive and negative diagnoses. With this type of sample, the test has not been sufficiently challenged to give false-positive and false-negative results. Thus, SN and SP should not be the basis for selecting a test or computing posttest probabilities for a disorder or condition (Kraemer, 1992).

The concussion study (McCrea, 2001), from which results were used for the ROC curve example in figure 15.3, may have used a form of pseudo-retrospective sampling. All of the high-risk football players were diagnosed with concussion, whereas all low-risk football players were diagnosed as free of concussion. Only those diagnosed with a concussion, along with matched controls, were given the SAC on the sideline. Football players who presented with possible concussion but were not diagnosed appear to have been filtered out of the patient pool. If this happened, the SAC was not sufficiently challenged for its ability to minimize false-positive and false-negative tests. SN and SP values were likely inflated. Future research needs to include all athletes who are checked for signs of concussion. For example, in the McCrea study, all football players would receive preseason baseline testing with SAC, and then every athlete who is suspected of having a concussion would also be given the SAC independent of the certified athletic trainers' initial determination at a predetermined postinjury time. Detailed records would need to be kept for each athlete, including the number of previous concussions, the mechanism of the injury, a list of immediate symptoms and those experienced 48 hr later, and any follow-up medical treatment or complaints related to the event. Given that the incidence rates of concussion of 4 to 6% have been reported for high school and college football players (Barr, 2001; Collins et al., 1999; McCrea, Kelly, Kluge, Ackley, & Randolph, 1997; McCrea et al., 1998), the duration of the study would have to run long enough so that a sufficient number of athletes experienced a concussion with different levels of severity and different types of symptoms. With this information, the efficacy of the SAC to signal the presence of various types of concussion could be determined.

Imperfect Reference Standards

The ideal situation is to judge a new diagnostic test against a reference test that determines the presence of the condition without error. However, the ideal does not occur often in practice. Because reference tests are not always dichotomous, a cutoff score must be used to state a diagnosis. If the cutoff is not standardized, SN and SP will vary across studies. For example, Hlatky and colleagues (1984) used 75% or more blockage of a major coronary artery as the cutoff for a positive diagnosis of coronary disease. If other researchers used 70%, SN and SP would change for the diagnostic test.

Sometimes the best reference test available is not perfect, or the best test is either too risky or too expensive to administer to every patient (Valenstein, 1990). The use of imperfect reference tests introduces bias into SN and SP estimates (Valenstein, 1990); therefore, several approaches have been recommended to deal with the problem. In **discrepant analysis,** another imperfect reference test is administered to patients whose diagnostic and first reference tests results differed (i.e., false-positive and false-negative results). Several researchers have shown, however, that discrepant analysis results in inflated estimates of SN and SP and recommend against its use (Hadgu, 1996; Miller, 1998). Although these values are higher than the original estimates of SN and SP, they may or may not be more biased than the originals (Hawkins, Garrett, & Stephenson, 2001).

Hawkins and colleagues (2001) proposed a method for estimating SN, SP, and their standard errors when patients are sampled from all four cells of the two-by-two table (agreements and disagreements with the imperfect standard). The sample is administered a third test, which is a perfect standard. Hawkins and colleagues also illustrated another method where SN, SP, and standard errors can be estimated when a composite reference standard is formed from the results of two imperfect reference tests. The procedure requires the characteristics of the second reference test to match the characteristics of the first imperfect reference test (e.g., high SN and moderate SP). If this occurs, patients are classified as positive for the condition if results are positive on both imperfect tests, and negative for the condition if results are negative on both or one test.

Latent class analysis represents another approach to the imperfect standard problem. Patients are not retested, but researchers rely on iterative estimation techniques (Garrett, Eaton, & Zeger, 2002; Hui & Walter, 1980; Joseph, Gyorkos, & Coupal, 1995; Qu & Hadgu, 1998). However, Guggenmoos-Holzmann and van Houwelingen (2000) have questioned the validity of latent class analysis. Because imperfect reference tests are common, Guggenmoos-Holzmann and van Houwelingen (2000) and Valenstein (1990) have suggested that the focus of diagnostic studies should change. Instead of examining how well diagnostic test results discriminate between those with and without the condition, the focus should be on how well test results discriminate among treatment or patient outcomes.

Instrument Variability and Bias

An instrument can be either a test that produces an objective score or a person who must make a subjective judgment based on the information at hand. The consistency or reproducibility of the instruments' scores needs to be documented because scores cannot be valid if they are not reproducible. The variability of the scores needs to be documented using the most appropriate statistic (e.g., coefficient of variation, intraclass correlation coefficient, kappa, and so forth; see chapter 3 by Ted Baumgartner).

Many diagnostic tests and reference tests require subjective judgments by the tester as to whether the condition is present. Given that more than one clinician usually administers the test to different patients, it is imperative that intertester reliability (tester agreement) evidence be reported. If intertester reliability is poor (e.g., low kappa coefficient), then SN and SP should not be trusted. In several studies, the reliability of the instruments' scores has been ignored (Glas et al., 2002; Leddy et al., 2002). Reid and colleagues (1995) found that only 21% of the studies they evaluated reported tester agreement, and only 25% reported coefficient of variation for machine instrument reproducibility.

One type of instrument bias occurs when the tester is not blinded to the test result of the first test administered. When this bias exists, the result may be an overestimate of a test's diagnostic accuracy (Lijmer et al., 1999). Instrument bias was a common flaw in 53% of the studies critiqued by Reid and colleagues (1995) ($N = 34$, published between 1990 and 1993), and 69% of the studies ($N = 218$) critiqued by Lijmer and colleagues (1999).

SUMMARY

Informed decision making, whether it is to choose the best diagnostic test or to choose an appropriate course of action based on patients' adjusted posttest probabilities to have the condition, requires trustworthy evidence on test efficacy. The quality of the evidence from research studies depends on the following: (a) reliability and validity of test results, (b) intertester and intratester reliability if subjective test interpretation is required, (c) cutoff scores used for a diagnostic or reference test, (d) blinding of test interpreters from initial test results or the patient's disorder or condition status (or both), (e) an absence of verification bias, (f) sufficient sample size to analyze population subgroups, (g) a detailed description of population characteristics, (h) a sampling plan, (i) a diagnostic test sufficiently challenged for false-positive and false-negative tests, and (j) the addressing of imperfect reference standards.

Although traditional indexes of test efficacy are still prevalent in the literature, there has been a call to move test efficacy studies in a different direction. Instead of examining how well diagnostic tests identify the true status of patients, allied health professionals should shift their focus to patient outcomes (Guggenmoos-Holzmann & van Houwelingen, 2000; Valenstein, 1990). Another focus is to employ prospective studies and use logistic regression models in which the probability of having a disorder is conditioned on the test result adjusted for population characteristics such as age, gender, and other relevant factors (Begg, 1987). As long as a plethora of test efficacy studies continue to report SN and SP, allied health professionals who practice evidence-based decision making will need to filter out the extremely biased estimates based on flawed research designs (Bossuyt et al., 2003).

CHAPTER 16

Preemployment Physical Testing

Andrew S. Jackson

Employers have always used some method to select an employee from potential job applicants. Much of the early preemployment testing focused on cognitive abilities, but with the rise in women seeking employment in physically demanding and male-dominated jobs, the popularity of employment testing for physically demanding jobs increased.[1] Three major forces have motivated employers to use preemployment physical tests: worker productivity, worker safety, and legal requirements. Of the three, the threat of litigation is likely the most powerful societal force motivating employers to use preemployment tests. Failure to hire a member of a protected group can result in discrimination litigation. Females are the primary protected group for physically demanding jobs.

This chapter is a discussion of the process and issues related to developing a valid, legally defensible preemployment test for physically demanding jobs. Although much of the information comes from the professional literature, this chapter is also heavily laden with my professional experience validating legally defensible tests.

LEGAL ISSUES

Employment is one of the most litigated areas in America. Failure to hire a person for reasons other than his or her ability to do the job has the potential for litigation. Arvey and Faley (1988) maintained that the landmark case of *Myart v. Motorola* in 1963 was the beginning of the court system becoming involved in the employment process. Leon Myart, an African-American man, was refused a job in a Motorola plant even though he had previous job-related experience. He was not hired because his score on a 5-min intelligence test was not high enough. Myart filed the complaint with the Illinois Fair Employment Practices Commission and charged racial discrimination. The Illinois Commission ruled that Myart be offered a job and ruled that the test could no longer be used for selection decisions. This landmark case motivated employers to develop preemployment tests that did not discriminate against protected groups (Arvey & Faley, 1988).

Although much of the initial employment litigation was due to racial discrimination, it soon spread to physically demanding jobs. In the 1960s, height and weight standards were a condition of employment for many public safety jobs (i.e., firefighter, police officer, and correctional officer). These standards clearly made it more difficult for women and ethnic minorities such as Asians to meet the standard. Arvey and Faley (1988) reported that in 1973 nearly all the nation's large police departments had a minimum height requirement. In June 1977 the United States Supreme Court in the case *Dothard v. Rawlinson* ruled that height and weight could not be used to select people for the public safety job of correctional-counselor trainee. A female applicant was refused employment because she did not meet the minimum height and weight requirements of 62 in. (157 cm) and 120 lb (54 kg). The standard excluded 33.3% of the females but only 1.3% of the males. The defendant (the employer) argued that the height and weight requirements were job related because they have a relationship to strength. The United States Supreme Court ruled that if strength is a real job requirement, then a direct measure of strength should have been adopted.

A lesson many employers learned is that the failure to use valid preemployment tests can be very expensive. Losing a discrimination case can cost employers millions of dollars. This was painfully evident in the 1982 New York City firefighter case *Berkman v. City of New York*. A leading industrial-organizational psychologist developed the **physical ability test.** None of the women tested passed the test, whereas 46% of the men did. The court stated:

> Nothing in the concepts of dynamic strength, gross body equilibrium, stamina, and the like, has such a grounding in observable behavior of the way firefighters operate that one could say with confidence that a person who possesses a high degree of these abilities as opposed to others will perform well on the job. (Arvey & Faley, 1988, p. 279)

One of the expert witnesses in the case was Dr. William McArdle, a leading exercise physiologist (McArdle, Katch, & Katch, 2002). McArdle argued successfully that the test was not valid because it was not consistent with basic exercise physiology principles. More specifically, his testimony challenged the use of construct measures of physical fitness and exercise performance to assess a candidate's ability to perform the highly specific aerobic, anaerobic, and muscular power tasks associated with firefighting. He maintained that the construct measures were not validated.[2] This case demonstrated that the validity of a physical preemployment test was going to be judged not only in the courts by psychometric criteria, but also by the discipline of exercise physiology.

Federal Employment Laws

Title VII of the Civil Rights Act of 1964; the **Age Discrimination in Employment Act** (ADEA) of 1967; and the **Americans With Disabilities Act** (ADA) of 1990 are the federal laws used for employment litigation. The centerpiece of employment discrimination law is Title VII of the Civil Rights Act of 1964, as amended by Congress on several occasions. Title VII prohibits employment discrimination on the basis of "race, color, religion, sex, and national origin" by employers, labor organizations, and employment agencies. The term *sex* refers to gender and does not include sexual orientation (Rothstein, Craver, Schroeder, & Shoben, 1999). Title VII includes both genders and all majority and minority racial and ethnic groups, as well as religious groups. The act does not apply to military personnel (Rothstein et al., 1999). In 1967 ADEA provided the legal basis for defining job discrimination because of age. The substantive provisions of the act are identical to those of Title VII with the substitution of the word *age* as the prohibited basis for discrimination (Rothstein et al., 1999).

The most recent law used to define discrimination is the ADA of 1990. The ADA is a comprehensive federal law that prohibits discrimination in a wide variety of segments of life. The law has five titles. Title I covers the employment of Americans with physical and mental disabilities (Rothstein et al., 1999). A disabled person is defined as someone with a substantial impairment that significantly limits or restricts a major life activity such as hearing, seeing, speaking, walking, breathing, performing manual tasks, caring for oneself, learning, or working. Under the ADA, preemployment medical examinations and medical inquiry are illegal. Although preemployment medical examinations may not be given, a post-offer medical examination is permitted, and a conditional job offer may be withdrawn if the examination documents that the person cannot perform the essential functions of the job. Although the ADA does not allow a general medical examination, a preemployment physical test can be legally used for employee selection. Under the ADA such a test is not considered a medical examination. If the preemployment test screens out people because of disability defined under the ADA, the employer has the burden of proving that the test is job related and consistent with business necessity (Rothstein et al., 1999). A recurrent theme expressed in this chapter is that the selection device must validly measure the applicant's ability to do the important elements of the job; in other words, the test must be job related.

Discrimination Litigation

An understanding of the litigation process is helpful in designing a preemployment test. The **disparate impact** theory is used to establish discrimination under Title VII, the ADEA, and the ADA. This legal process has a three-part burden of proof:

1. The plaintiff (employee) must establish a disparate impact on a protected group.

2. If disparate impact on a protected group is established, the defendant (employer) must then justify the exclusionary effect with a business necessity. The defendant must show that the selection method is job related.

3. If business necessity is established, the burden of proof shifts back to the plaintiff to demonstrate that the employer failed to use a selection device that is equally effective with a lesser disparate impact.

Disparate Impact

To find an employment selection method discriminatory, the plaintiff (i.e., the job applicant or affected employee) must establish that the method has disparate impact on a protected group. The plaintiff must show that the employment selection method adversely affects the employment opportunities based on race, color, religion, sex, national origin, age, or a qualified disability. The Supreme Court case *Griggs v. Duke Power Co.* was the first case to use a disparate impact theory of discrimination. The power company used standardized aptitude tests for assignment of jobs. The plaintiff class showed that the aptitude tests adversely affected racial groups. Although 58% of Whites passed the test, only 6% of Blacks passed. The courts ruled that under Title VII, employment tests with disparate impact could not be used unless they were job related (Rothstein et al., 1999).

Although disparate impact or adverse impact must be established to prove discrimination, the federal laws do not define explicitly what constitutes adverse impact. The **Equal Employment Opportunity Commission (EEOC)** published the first set of guidelines on employment testing in 1966 and followed with a revision in 1970. This led, in 1978, to the publication of the **Uniform Guidelines on Employee Selection Procedures** (EEOC, 1978). The EEOC, Civil Service Commission, and Departments of Labor and Justice jointly agreed on these federal standards and rules. The EEOC guidelines use the **four-fifths (4/5s) rule** to define adverse impact. Under the 4/5s rule, a selection device has adverse impact when the pass rate for a protected group is less than four-fifths, or 80%, of the pass rate of the group with the highest pass rate. This is illustrated with the *Griggs v. Duke Power Co.* data. The pass rate of the protected group (Black applicants) was 6% compared to 58% for the White group, which resulted in a ratio of 10.3% (6/58 = 0.103), failing the EEOC 4/5s standard. If the pass rate of the Black group was 48%, as an example, the selection device would not have adverse impact. The pass rate of the protected group would be 83% (48/58 = 0.83) of the White group. Rothstein and colleagues (1999) contended that trial courts need not adhere to the 4/5s rule, but legal history shows that the 4/5s rule is viewed with favor.

When physical tests are used for employment decisions, gender is the primary source of adverse impact (Hogan, 1991; Hogan & Quigley, 1986). This potential for adverse impact can be traced to the well-documented male and female differences in strength, maximal oxygen uptake ($\dot{V}O_2max$), and body composition (Baumgartner, Jackson, Mahar, & Rowe, 2003). Although much of the reason for adverse impact in physical testing can be traced to physiological differences between men and women, another factor is the physical demands of the job. The more physically demanding the job is, the less likely it will be to meet the EEOC 4/5s rule. Table 16.1 illustrates this relationship with lifting data obtained on 608 females and males tested at the University of Houston. The lift task was the common floor-to-knuckle-height lift (Jackson & Sekula, 1999). The goal was to continue to lift heavier weights until the lift was failed. Table 16.1 gives

Table 16.1   Male and Female Pass Rates for Floor-to-Knuckle-Height Lifts and the Application of the 4/5s Rule

Lift weight kg (lb)	Female pass rate (%)	Male pass rate (%)	4/5s rule (%)
15 (33)	100.0	100.0	100.0
20 (44)	100.0	100.0	100.0
25 (55)	93.2	100.0	93.2
30 (66)	72.3	100.0	72.3
35 (77)	38.4	99.2	38.7
40 (88)	20.9	96.9	21.6
45 (99)	10.5	87.4	12.0

the percentages of males and females who lifted loads between 15 and 45 kg (33 and 99 lb) and illustrates that adverse impact based on the EEOC 4/5s rule would not occur for lift loads of 25 kg (55 lb) or lower, but would occur for lift loads of 30 kg (66 lb) or higher. The gender difference became progressively larger as the weight load increased.

Business Necessity

Once a plaintiff (applicant or employee) has established that a selection device had adverse impact, the defendant (employer) must prove that the test or selection method was job related (Rothstein et al., 1999). The fact that a test is found to have adverse impact does not mean that the test or selection method is discriminatory. Once an employer becomes aware of the adverse impact, it becomes their responsibility to establish the validity of the requirement. Any adverse effect because of race, gender, national origin, age, or disability is irrelevant as long as the selection device is valid (Rothstein et al., 1999). The United States Supreme Court case *Griggs v. Duke Power Co.* was the first case not only to use a disparate impact theory of discrimination, but also to introduce the concepts of business necessity and job validation as defenses. "Title VII forbids the use of employment tests that are exclusionary in effect unless the employer demonstrates that any given requirement has a 'manifest relationship' to the job" (Rothstein et al., 1999, p. 163). The power company lost the case because it (the defendant) could not show that an aptitude test found to have disparate impact against Black employees was job related.

The EEOC guidelines provide methods of defining the job-relatedness of a test or selection method. The guidelines are used to evaluate whether the selection procedure is job related, or valid, by using the standards published by the American Psychological Association (APA) (APA, 1985, 1987) to judge test validity. Although the EEOC guidelines have had a significant influence in Title VII litigation, the United States Supreme Court ruled that Title VII does not necessarily require employers to introduce formal validation studies (*Albemarle Paper Co. v. Moody*). The legal obligation of employers who use an employment method with adverse impact is to prove its job-relatedness. The position of the court is that discriminatory tests cannot be used unless they are shown to be valid by professionally acceptable methods. In the case *Albemarle Paper Co. v. Moody*, the court used the test validation methods outlined in the EEOC guidelines to evaluate the discriminatory tests used by the paper company to promote workers to more skilled positions and ruled that the validation studies were inadequate in several respects under the EEOC guidelines (Rothstein et al., 1999).

The EEOC guidelines require that both the test and the **cutscore** (i.e., the minimum score an applicant must achieve to be considered successful on a test) must be validated. *Harless v. Duck* was a class action that challenged the physical ability test used by the Toledo (Ohio) Police Department. Applicants were required to complete three of the following four tests: 15 push-ups, 25 sit-ups, a 6 ft standing long jump, and a 25 sec obstacle course. The Sixth Circuit Court endorsed the need for fitness but concluded that there is no justification in the record for the type of exercises chosen or the passing marks for each exercise (Rothstein et al., 1999).

Alternative Selection

If the plaintiff (job applicant) establishes that a test or selection method has disparate impact, but the defendant (employer) proves that the method is valid, the plaintiff can still prove discrimination by demonstrating that a less discriminatory alternative is available. The availability of a less discriminatory alternative was established in *Albemarle Paper Co. v. Moody*. Less discriminatory alternatives pertain not only to the testing device, but also to setting the cutscore. The cutscore is a common source of litigation (Hogan & Quigley, 1986). The data in table 16.1 show that adverse impact against females increases with the physical demands of the task. This is also true in setting a cutscore. The higher a cutscore is set on a physical test, the greater the chances for adverse impact.

VALIDATION STUDY

Men mainly fill physically demanding jobs. Although this may indicate that fewer females apply, it may also be the result of employer hiring practices. Employers with hiring practices that are based on gender are at risk of litigation. Shown in the previous section, if a plaintiff establishes that the method used to select employees has an adverse impact on a protected group, the burden of proof falls on the employer to show that the selection device (e.g., test) is job related (Rothstein et al., 1999).

This is often accomplished with a **validation study.** Although there is no single way of completing a validation study, the *Uniform Guidelines* (EEOC, 1978) give direction and provide psychometric criteria (APA, 1985, 1987) to evaluate validation studies. Although psychometric criteria are academically sound when applied to physical performance tests, it must be remembered that physical tests are different from cognitive tests. This section provides the major components of a physical test validation study and introduces the concept of **physiological validation,** which expands on the psychometric methods.

Job Analysis

The first important component of a validation study is the **job analysis.** The job analysis identifies the important work behaviors demanded by the job. It describes what workers must do on the job and is broadly defined as the collection and analysis of any type of job-related information by any method for any purpose (Gael, 1988). It reduces to words what people do. The *Uniform Guidelines* clearly require that a validity study include a job analysis. Many different methods are available for conducting a job analysis including observing work, recording work activities, interviewing workers or supervisions, using questionnaires to collect data, or using various combinations of these methods (Gael, 1988).

The methods commonly used to validate tests for physically demanding tasks generally come from psychophysical, biomechanical, and physiological sources of data. The methods vary among investigators. A brief summary of the sources of data we have used to conduct job analyses for several validation studies of physically demanding jobs follows (Jackson, 1986; Jackson & Osburn, 1983; Jackson, Osburn, Laughery, & Sekula, 1998; Jackson, Osburn, Laughery, & Young, 1993; Jackson, Osburn, Laughery, Young, & Zhang, 1994).

Task Questionnaire

A **task questionnaire** is a list of statements that define the tasks performed by workers. The process we have used is to interview supervisors and employees and observe the work being done. This information is put into work task statements that workers understand. Following are examples of task statements for oil production workers (Jackson et al., 1998):

- Climbing stairs while carrying loads weighing 50 to 75 lb (23 to 34 kg)
- Breaking/opening/closing 4 to 6 in. (10 to 15 cm) valves in corrosive, dirty, or high-pressure service (over 1000 psi)
- Holding and manipulating heavy fittings while in a crouched or awkward position
- Lifting portable air compressor from the back of a truck
- Lifting bags of cement

- Lifting loads of 50 to 75 lb (23 to 34 kg) from the ground or floor to waist height
- Lifting loads of 75 to 100 lb (34 to 45 kg) from the ground or floor to waist height

The list of statements is given to workers who do that job, and the workers rate the frequency the task is done and its difficulty or physical demand. These are rated with Likert-type rating scales. The data from the worker ratings are used to identify physically demanding tasks performed frequently on the job.

The criticality of a task is often rated for public safety tasks that may not be routinely performed, but are clearly important. For example, a police officer rarely uses a weapon, but if the situation calls for this type of action, an officer must be able to perform it effectively. Similarly, a common test item used to select firefighters is to drag a dummy a specified distance. This test simulates dragging a person out of a burning structure, which is a critical task that all firefighters must be able to do on the job.

Biomechanical and Physiological Analysis

The task questionnaire data identify tasks workers perceive as physically demanding and performed frequently. The next step is to define the important tasks accurately. A biomechanical and physiological analysis provides information to cross-validate the questionnaire and job expert interview data. Biomechanical and physiological data define the true demand of tasks and provide valuable information for setting cutscores on a test.

Materials-handling tasks involve lifting and transporting objects such as freight (Jackson, Osburn, et al., 1993) or humans in a hospital setting (Jackson, Zhang, Laughery, Osburn, & Young, 1993). Ergonomic research (Waters, Putz-Anderson, Garg, & Fine, 1993) shows that many factors affect the difficulty of materials-handling tasks. A biomechanical analysis defines the task in greater detail. This analysis may define the weight of objects lifted and the way a worker must perform the task. The analysis can also involve using an electronic load cell to measure the forces a worker must generate to move a heavy object or crack a valve. Some jobs have a substantial aerobic fitness component. A physiological job analysis helps define the demands of the job. This analysis may range from measuring workers' heart rates (Jackson, Osburn, et al., 1993; Sothmann, Saupe, Jasenor, & Blaney, 1992) to performing an elaborate metabolic analysis of firefighters simulating work (Sothmann et al., 1990). Rayson (2000) provided an excellent overview of physiological methods used to conduct a job analysis.

Psychometric Validation

Validity of a preemployment test concerns the accuracy with which a test or selection device measures the important work behaviors identified by the job analysis. Because validity depends on reliability, whenever feasible, statistical estimates of reliability should also be presented. The *Uniform Guidelines* (EEOC, 1978) list three acceptable validation methods: content, criterion-related, and construct validity studies (see chapter 2 by Matt Mahar and David Rowe).

Content Validity

Content validity is established by replicating the major portions of the job. The goal is to develop a test that is a representative sample of job behavior(s). This validation method involves gathering evidence and establishing a logical relationship between important duties or job behaviors and the preemployment test. There is no agreed-on index of content validity; rather, professional judgment is used. According to the *Uniform Guidelines*, a content validity study needs to present data showing that the content of the selection procedure represents important aspects of performance on the job for which the candidates will be evaluated (EEOC, 1978). The extent that a preemployment test can be judged to be content valid is the extent that the test is a representative sample for the content of the job. The job analysis provides these critical data.

Criterion-Related Evidence for Validity

The *Uniform Guidelines* (EEOC, 1978) state that examination of criterion-related evidence for validity should consist of empirical data demonstrating that the preemployment selection device is predictive of, or significantly correlated with, important elements of job performance. The sample subjects should be representative of candidates normally available in the relevant labor market for the job and should, insofar as feasible, include the race, sex, and ethnic groups normally available in the relevant job market. Statistical significance at the .05 level is accepted as a relationship between criterion and predictor (EEOC, 1978). The performance on a test or other selection device is compared with job effectiveness and may be divided into concurrent and predictive methods. Concurrent methods use current employees and relate tests to current job performance. In the predictive approach, test data are obtained on people prior to hire; performance data are obtained later (EEOC, 1978).

The types of job performance criteria listed in the *Uniform Guidelines* that may be suitable are supervisory ratings, production rate, error rate, tardiness and absenteeism, and success in training. Supervisory ratings are often used, but because of the possible bias of these subjective evaluations, they need to demonstrate that they do not unfairly deny opportunities to members of protected groups. According to the guidelines, this is not an inclusive list of criteria. Other examples of criteria used include injury rates (Doolittle, Spurlin, Kaiyala, & Sovern, 1988), accidents (Reilly, Zedeck, & Tenopyr, 1979), field performance (Reilly et al., 1979), and job-related work tasks (Arnold, Rauschenberger, Soubel, & Guion, 1982; Jackson, Osburn, & Laughery, 1984, 1991; Jackson, Osburn, Laughery, & Vaubel, 1991).

Construct Validity

Construct validity is more theoretical than content validity because it is necessary to establish that a construct is required for job success. One needs to show that the selection device measures the construct used for the job. The data from a construct validation study should show that the preemployment test measures the degree to which candidates have identifiable characteristics that are important for successful job performance (EEOC, 1978). The goal is to establish a linkage between the important constructs and multiple indicators of job performance (Arvey, Nutting, & Landon, 1992). It involves first developing evidence and confirming inferences regarding these constructs and indicants of job performance, and next confirming and disconfirming hypotheses concerning the relationship between physical ability and job constructs.

Test Selection

Preemployment tests use either work-sample or physical ability tests. A **work-sample test** is designed to duplicate or simulate a critical work task or a series of important work tasks. Physical ability tests are common fitness tests.

Work-Sample Tests

A good work-sample test is content valid if it duplicates the actual work task. To illustrate, assume that a job requirement is to lift a 75 lb (34 kg) metal valve from the floor to knuckle height. An example of a content-valid work-sample test would be one that requires the subject to lift a 75 lb valve in the way required on the job. Such a test can be objectively scored as pass or fail. The closer the test duplicates the work behavior, the easier it is to defend it as being content valid. Work-sample tests are commonly used to screen applicants for police officer and firefighter jobs. Arvey and colleagues (1992) reported that most police and firefighter physical ability tests consist of some combination of work-sample tests. Figure 16.1 gives examples of firefighter and police officer work-sample tests.

Although work-sample tests often have excellent content validity, they do have limitations. In some instances, work-sample tests are expensive to create and difficult to transport and set up at different test sites (i.e., a large wooden structure constructed to duplicate the physical environment of the cargo space of aircraft used to transport freight; Jackson, Osburn, et al., 1993). Also, a work-sample test often does not measure an applicant's maximum capacity. Ayoub (1982b) maintained that this introduces two additional limitations. First, applicants seeking employment are likely to be motivated to pass the work-sample test. A highly motivated applicant who lacks the physical capacity to perform the test increases his or her risk of injury (Ayoub, 1982b; Dehlin, Hendenrud, & Horal, 1976; Herrin, Jaraiedi, & Anderson, 1986; Magora, 1970; Snook, Campanelli, & Hart, 1978). Second, a work-sample test is often scored by pass or fail (e.g., could or could not lift a 75 lb [34 kg] valve). Strong applicants will easily pass the test, but weaker applicants may pass but be working near or at their maximum capacity. If a linear relationship between job performance and test performance can be assumed, applicants with the highest test scores can be expected to be the more productive workers. Testing for maximum capacity not only identifies the most potentially productive

Firefighter Work-Sample Test
(Davis, Dotson, O'Connor, and Confessore, 1992)

1. Stair Climb: Carry a 58 lb. hose bundle up five flights of stairs.
2. Hoseline Drag: Drag a 1.75 in. charged hose 100 ft.
3. Rescue Dummy Drag: Lift and drag a 175 lb. dummy 100 ft.
4. Smoke Extractor Carry: Lift and transport a 47.5 lb. fan a distance of 150 ft.
5. Kieser Force Machine: Pound repeatedly on an object with a sledge-hammer until it moves a specified distance.

Police Officer Work-Sample Test
(Weiner, 1994)

1. Obstacle Course: Run 99 yd while changing direction and stepping over objects.
2. Body Drag: Drag a 165 lb. dummy 32 ft.
3. Fence Climbs: Run a short distance, and climb over two fences; one is chain link and the other is wood. Both fences are 6 ft. high.
4. 500-Yard-Run: Run 500 yd. as fast as possible.

Figure 16.1   Examples of firefighter and police officer work-sample tests.

workers, it also provides the opportunity to define a level of reserve needed to reduce the risk of musculoskeletal injury (Cady, Bishoff, O'Connell, Thomas, & Allan, 1979; Cady, Thomas, & Karwasky, 1985; Chaffin, 1974; Chaffin & Park, 1973; Herrin et al., 1986; Keyserling et al., 1980; Keyserling, Herrin, & Chaffin, 1980; Liles, Deivanayagam, Ayoub, & Mahajan, 1984; Snook et al., 1978; Snook & Ciriello, 1991).

Physical Ability Tests

Physical ability tests are commonly fitness tests. The tests most used are strength and aerobic endurance; flexibility and balance tests are used less often. The interested reader is directed to other sources (Baumgartner et al., 2003; Jackson, 2000; National Institute for Occupational Safety and Health, 1977) for a more detailed discussion of preemployment physical ability tests.

Physical ability tests measure an applicant's maximum physical capacity. They are not content valid because they do not duplicate the work task. Evidence needs to be provided to meet the legal criterion of being job related. A common strategy is to establish the relationship between physical ability and work-sample tests. When physical ability and work-sample tests are highly correlated, the physical ability test can replace the work-sample test. Arnold and associates (1982) were among the first to demonstrate that physical ability tests were highly correlated with work-sample tests. They developed work-sample tests for steelworkers and administered them to 81 women and 168 men who were in their first six months of employment. The zero-order correlations between arm strength and work-sample test performance were consistently high (>.82).

Researchers from the University of Houston and Rice University[3] completed a series of preemployment studies in which one step was to examine the relationship between isometric strength and work-sample test performance. The complete research methods appear in technical reports, but a representative sample of this research has been published (Jackson, Borg, Zhang, Laughery, & Chen, 1997; Jackson, Osburn, & Laughery, 1991; Jackson, Osburn, Laughery, & Vaubel, 1991, 1992; Jackson & Sekula, 1999). An outcome of this research was the development of isometric strength test equipment and tests for use in a preemployment setting. The equipment[4] and strength tests are presented in Baumgartner and associates (2003).

Table 16.2 gives the product–moment correlations between isometric strength tests and work-sample test performance. The strength tests were highly correlated with several different types of work-sample tests. The highest correlations were between isometric strength tests and static strength work-sample tests that involved the capacity to generate force in various body positions. These work-sample tests measured the subject's capacity to generate push, pull, and lift force and generate valve-cracking torque while in the positions required by the work environment. The isometric strength tests were also correlated with absolute endurance work-sample tests. These dynamic work-sample tests involved valve turning, shoveling, and repetitive materials-handling tasks. The dynamic tasks were performed either to exhaustion (e.g., valve turning), or at a "comfortable rate" set by the person being tested (e.g., shoveling 600 lb [272 kg] of material over a 3.5 ft [1 m] wall). The correlations between strength and the absolute endurance work tests ranged from .67 to .83. Strength is also related to lifting capacity (Jackson et al., 1997; Jackson & Sekula, 1999) and firefighter work-sample test performance (Jeanneret & Associates, 1999). The correlations between the elapsed time to complete the five-item firefighter work-sample test shown in figure 16.1 and four isometric strength tests ranged from –.61 for leg and torso strength to –.90 for arm strength.

Several investigators (Barnard & Duncan, 1975; Lemon & Hermiston, 1977a, 1977b; Manning & Griggs, 1983; O'Connell, Thomas, Caddy, & Karwasky, 1986; Sothmann et al., 1992) published data showing that fire suppression work tasks have a substantial aerobic endurance component. Sothmann and colleagues (1990) defined the minimum aerobic capacity required to meet the demands of firefighting. The authors used a work-sample test involving seven job-related firefighter tasks and found that people with a $\dot{V}O_2$max below 33.5 ml/kg/min could not do the work effectively. The 1.5 mi (2.4 km) distance run tends to be the aerobic test used to evaluate aerobic capacity in field settings. The correlation between 1.5 mi (2.4 km) run time and elapsed time to complete the five-item firefighter work-sample test given in figure 16.1 was .59 (Jeanneret & Associates, 1999). This supported the laboratory data showing that firefighting ability is a function of aerobic power.

Table 16.2   Correlations Between the Sum of Isometric Strength and Simulated Work Sample Tests

Reference	Sample test	Type of test	r_{xy}
Jackson & Osburn, 1983	One-arm push force	Isokinetic—peak torque	.91
Jackson, 1986	Push force	Static—max force	.86
Jackson et al., 1993	Push force	Static—max force	.76
Jackson, 1986	Pull force	Static—max force	.78
Jackson et al., 1993	Pull force	Static—max force	.67
Laugherty & Jackson, 1984	Lifting force	Static—max force	.93
Jackson et al., 1998	Valve cracking	Static—max force	.91
Jackson et al., 1992	Valve turning	Dynamic—endurance	.83
Jackson et al., 1993	Box transport	Dynamic—endurance	.76
Jackson et al., 1993	Moving document bags	Dynamic—endurance	.70
Jackson et al., 1991	Shoveling coal	Dynamic—endurance	.71
Jackson et al., 1993	50 lb bag carry	Dynamic—endurance	.63
Jackson & Osburn, 1983	70 lb block carry	Dynamic—endurance	.87

Percent body fat and fat-free mass are common body composition variables. Percent body fat tends to be negatively correlated with work-sample tests that required moving the body. Because fat-free mass is the body's force-producing component, it tends to be correlated with work tasks related to strength. The gender differences in body composition are well documented (Jackson et al., 2002; McArdle et al., 2002; Wilmore & Costill, 1994). Body composition variables are not typically used in the public sector to make preemployment decisions because of the potential of litigation for gender discrimination, but they have been used to evaluate the physiological capacity to perform military jobs (Hodgdon, 1992; Vogel & Friedl, 1992).

The major limitation of using basic ability tests is that, unlike well-constructed work-sample tests, they are not content valid. The linkage between the test and job must be established. The principal advantages of common physical ability tests such as the 1.5 mi (2.4 km) distance run and isometric strength are that (a) the applicant's maximum capacity is measured, (b) the tests are easy to administer and transportable to a variety of test sites, and (c) the same tests can be used for a variety of work tasks (see table 16.2).

Physiological Validation

The *Uniform Guidelines* validation methods (EEOC, 1978) were derived from the standards advanced by the APA (1985, 1987) for educational and psychological tests. Although psychometric methods provide sound scientific criteria to evaluate psychological and educational tests, physical preemployment tests are neither educational nor psychological tests; they are physiological tests. The term *physiological validation* was introduced by Hodgdon and Jackson (2000) to differentiate between psychometric and physiological validity issues.

The exercise science literature provides excellent examples of the use of psychometric methods to validate physiological tests. To illustrate, concurrent validation strategies have been used extensively to validate field measures of body composition (Brozek & Keys, 1951; Durnin & Wormsley, 1974; Jackson & Pollock, 1978; Jackson, Pollock, & Ward, 1980) and aerobic capacity (Bruce, Kusumi, & Hosmer, 1973; Foster, Jackson, & Pollock, 1984; Jackson et al., 1990; Pollock et al., 1976). The goal of these validation studies was to find tests that were correlated with established physiological criteria—laboratory measured percent body fat and $\dot{V}O_2$max. Physiological test validation involves applying relevant exercise science research and theory to support the test validation results. At least three important differences delineate psychometric and physiological test validation of employment tests. These include defining the test metric in scientific terms, defining the work task, and matching the worker to the demands of the task.

Typically, physiological tests are scored on a ratio measurement scale, whereas psychological tests use an ordinal or interval scale. Physiological tests are scaled in a scientific unit of measurement, such as oxygen uptake, caloric expenditure, force exerted, and pounds lifted. In contrast, the unit of measurement of psychological tests is typically a person's response to some type of scale (e.g., Likert scale) or number of correct answers. The unit of measurement on psychological tests is of little importance and is often transformed from the original metric into some form of standard score with a known mean and standard deviation (e.g., mean = 500, SD = 100). The physiological test metric has scientific meaning such as pounds of force exerted for an isometric strength test or pounds moved per minute for a materials transfer work-sample test.

A second difference between physiological and psychological preemployment test validation is that the physical demands of work tasks can often be clearly defined. This is largely because of the capacity to define the physical demands of the work task in accepted scientific metric. Extensive physiological research has defined the energy cost of a host of occupational, recreational, and fitness tasks in the metric of oxygen expenditure (L/min) or caloric expenditure (kcal/min) (Durnin & Passmore, 1967; Passmore & Durnin, 1955). These energy cost tables are published in basic exercise physiology texts (Åstrand & Rodahl, 1986; Brooks & Fahey, 1984; McArdle et al., 2002; Wilmore & Costill, 1994). It is also possible to measure the force demands placed on workers required to perform many industrial tasks. The torque or force required to "crack" valves or to push or pull objects can be measured with torque wrenches and electronic load cells (Jackson et al., 1992; Jackson, Osburn, Laughery, & Sekula, 1998). It is often possible to define the demands of materials-handling tasks by weight, type of lift, and distance transported (Waters et al., 1993). A materials-handling task of a freight moving company was defined by the power-output metric of pounds-moved-per-minute (Jackson, Osburn, et al., 1993). These data define the physical stress demanded by a work task in a common scientific unit of measurement as compared to a psychological test where a person's score is interpreted by its place within a distribution.

A final difference between physiological and psychological test validation is that not only can the physiological capacity of the worker be measured, but their physiological capacity can also be linked to the task. Once the physiological demands of a work task are defined, the next step of a physiologically based validation strategy is to determine whether a worker has the capacity to meet those demands. The ergonomic goal is to match the worker to the task. This has been the method used to define the minimum energy cost demanded of firefighting (Sothmann et al., 1990) and to define strength levels demanded of an industrial task (Keyserling et al., 1980). An applicant with a $\dot{V}O_2max$ of 25 ml/kg/min would not have the physiological capacity to fight fires. A person without sufficient strength will not be able to lift a 75 lb (34 kg) load in a manner required by the task. The goal of setting a cutscore is to define the minimum physiological capacity needed to be an effective worker. In contrast, psychological scores compare the individual with others from the population. This important issue is addressed in more detail in the section related to setting a cutscore.

Test Fairness

An important issue of preemployment test development is test fairness. **Test unfairness** occurs when members of a protected group obtain lower scores on a preemployment test compared to members of another group, and the difference in scores is not reflected in differences in the measure of job performance (EEOC, 1978). The common method used in physical test validation is to determine whether a common regression equation can be used to explain the relationship between the predictor and criterion tests of two groups. This is called the Cleary test of fairness. In physical test validation, the two groups are likely male and female applicants. The strategy is to first determine whether the two groups share a common regression slope, and then decide whether the groups' regression intercepts are within chance variation.

Multiple regression is the statistical model used to determine test fairness. This analysis involves dummy coding the group variable (e.g., female = 0, male = 1) and forming a group by predictor-test interaction term (Pedhazur, 1997). The statistical strategy used is to form a full model consisting of the three variables: (a) predictor test; (b) dummy coded group variable; and (c) interaction term, the product of the group and predictor test. The next step is to generate two restricted models: (a) a model with two independent variables, the group variable and the predictor test; and (b) a model with just the predictor variable. The statistical test used to evaluate group differences in slopes and intercepts is to analyze changes in R^2 between the full and restricted models. Pedhazur outlined these statistical methods and tests of significance. This is illustrated next with physiological data.

Group Difference in Regression Slopes

A task analysis of freight-moving tasks showed that moving packages rapidly from a container to a conveyor belt was a physically demanding task (Jackson, Osburn, et al., 1993). A work-sample test was developed to duplicate the demands of the repetitive transport task. The task involved moving packages that ranged in weight from 15 to 80 lb (6.8 to 36 kg). The distribution of package weights was representative of the weight distribution encountered by workers. Exercise heart rate was measured to ensure that the work rate of the simulation test was representative of the actual work rate that was measured with the task analysis. The subjects were instructed to work at a brisk rate consistent with their fitness and not to move packages that exceeded their capacity.

Figure 16.2 shows the bivariate relationship between the predictor test (sum of isometric strength) and criterion test (materials transport expressed by power output [lb/min]). The data are contrasted by gender. The R^2 change between the full model and restricted model of the strength test and dummy-coded gender variable was .04, which was statistically significant ($F[1,199] = 18.96$; $p < .01$). This step-down change in R^2 demonstrated that the male and female regression lines were not parallel. The graph shows that the slope for the female subjects (.534) is over twice as steep as that for the male subjects (.208).

Closer examination of the task and data showed that the reason for the steeper female slope was that weaker females were not able to lift and transport the heaviest freight packages. Data in our lab showed that only about 41% of females can lift 80 lb (36 kg) loads compared to more than 99% of males. Although the data in figure 16.2 suggest that the strength test would be biased toward females, a physiological interpretation provides a clearer picture. The data show that the minimal level of strength needed to meet the demands of the task is about 400 lb (181 kg). If the employment goal were to hire employees below the 400 lb (181 kg) level, the ergonomic strategy would be to redesign the job by reducing the maximum weight of packages moved in this manner.

Intercept Differences

The second part of the Cleary test is to evaluate differences in regression intercepts. Figure 16.3 shows an example of intercept differences with published male and female body composition data (Jackson & Pollock, 1978; Jackson et al., 1980). The

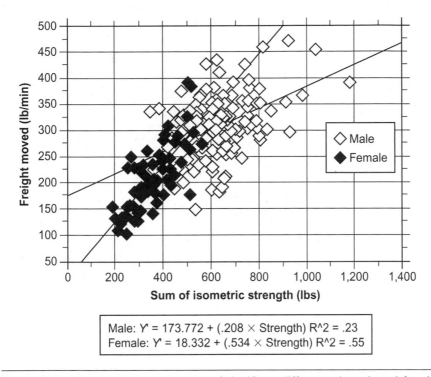

Male: Y' = 173.772 + (.208 × Strength) R^2 = .23
Female: Y' = 18.332 + (.534 × Strength) R^2 = .55

Figure 16.2 Test for fairness: examples of significant differences in male and female regression slopes.

independent variable is the sum of seven skinfold measurements, and the dependent variable is hydrostatically measured percent body fat. The figure shows that the slopes of the male and female regression lines were parallel. The differences in slopes (R^2 change < .001) were within random variation ($F[1,675] = 1.25$; $p >$ .05). Adding the dummy-coded gender variable to the sum of skinfolds accounted for over 12% of percent fat variance ($F[1,675] = 398.75$; $p < .01$). As these data show, the significant intercept difference indicates that for a given sum of skinfold fat, the percent body fat of females was systematically higher than that of males with the same sum of seven skinfolds. The regression lines differed by an average percent body fat of about 6%.

Although skinfold fat should not be used for preemployment testing, the data in figure 16.3 provide an excellent example of common slopes with different intercepts. The example also shows the need to interpret the results with physiological evidence. The significant difference in intercepts can be traced to the well-known gender differences in body composition. The body has two types of fat, subcutaneous and essential. Skinfold fat measures subcutaneous fat, whereas hydrostatically determined percent body fat measures both fat sources. It is well established that the essential fat of women is higher than that of men (McArdle et al., 2002).

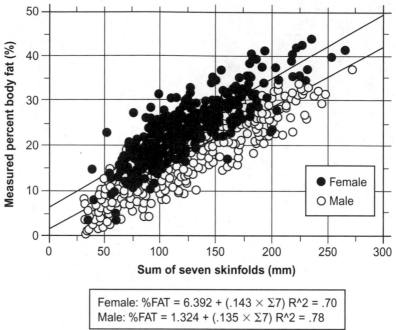

Figure 16.3 Test for fairness: example of parallel regression slopes, but significant differences in male and female regression lines.

The physiological explanation for the gender difference in intercepts was gender differences in essential fat.

Common Slope and Intercept

Figure 16.4 illustrates the homogeneity of male and female regression lines for the predictor and criterion tests. The figure gives the scatter plot of the male and female relationship between isometric strength and peak push force. A task analysis showed that push force was a physically demanding task required of workers who moved freight containers (Jackson, Osburn, et al., 1993). The figure shows that the male and female regression lines were similar. Statistical analysis revealed that the slopes ($F[1,203] = 1.50; p > .05$) and intercepts ($F[1,203] = 2.00; p > .05$) of the male and female regression lines were not statistically significant. The group and group-by-strength variables accounted for less than 0.1% of the push force variable.

The regression analysis demonstrated that differences in the male and female regression lines shown in figure 16.4 were random and that a single regression line could be used to estimate push force from isometric strength. The figure illustrates that the push force and isometric strength of the males exceeded those of the females. The nonsignificant difference in slopes and intercepts confirmed that the

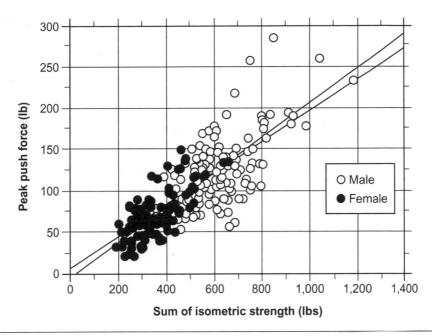

Figure 16.4   Test for fairness: example of homogeneity of male and female regression slopes and intercepts.

gender difference in work-test performance was due to strength and not gender. This supports the use of the strength tests to select applicants with the minimum physiological capacity to do the job.

Setting Cutscores

The final step in constructing a preemployment test is to set the cutscore. This is a crucial step because the cutscore is the standard used to select or reject an applicant and a source of potential discrimination litigation. Establishing a cutscore can be straightforward for a well-defined, content-valid work-sample test scored pass or fail, but this often is not the case. Figure 16.1 gives examples of common firefighter and police work-sample tests that were developed using a content validity methodology. All these tests are scored by elapsed time, and the applicant's final score is the sum of all items. Although an applicant can be failed for not being able to perform the task (e.g., climb a 6 ft [1.8 m] high fence), most applicants complete all items. This creates a performance distribution. The task then becomes one of defining a valid cutscore within the distribution. Setting the cutscore too low increases the chances of selecting workers who cannot do the work, whereas setting the cutscore too high increases the likelihood of adverse impact against females (see table 16.1). If the cutscore produces adverse impact, the test developers need to be in a position to defend the selection. Validity in this context refers to defining the minimum level of physical capacity required by the job.

It is my experience that several factors need to be considered when setting a defensible cutscore:

- *Adverse impact.* The higher a cutscore of a physical test is, the more likely adverse impact against females will be. Consideration must be given to the protected group that the standard screens out.

- *Risk of injury.* Subjecting workers to a task without the fitness needed to meet the physical demands of the task increases the risk of injury. Unfit workers not only lower productivity, but also increase the risk of musculoskeletal injuries, particularly low-back injuries. The total compensable cost for low-back disorders in the United States is in the billions of dollars, and about 50% of work-related back injuries are linked to materials-handing tasks (Ayoub, Dempsey, & Karwowski, 1997; Snook, Campanelli, & Hart, 1978; Waters et al., 1993). Ergonomic research demonstrated that the risk of back injury was not just a function of the demands of the materials-handling task, but also the physiological capacity of the worker. The risk of musculoskeletal injury increases as the demands of the task approach the worker's maximum physiological capacity. A detailed discussion of this topic is beyond the scope of this chapter. The interested reader is directed to other sources (Cady et al., 1979, 1985; Chaffin, 1974; Chaffin & Park, 1973; Herrin et al., 1986; Keyserling et al., 1980; Keyserling, Herrin, & Chaffin, 1980; Liles et al., 1984; Snook et al., 1978; Snook & Ciriello, 1991).

- *Physiological interpretation of the validation results.* An important element of a physical test validation study is to establish the congruence among the statistical results, published research, and physiological theory. It is critical to provide a sound physiological explanation of the validation study results and selected cutscore. Failure to interpret the results by accepted academic standards leaves the decision more difficult to defend.

- *Environmental conditions.* Often, the location of the validation study will be different from the work environment (e.g., firefighter tests are not administered in burning buildings). A more rigorous cutscore may be warranted when the demands of the work environment exceed those of the test environment.

- *Workforce numbers.* The number of workers available at the work site can affect the rigor of a cutscore. A more lenient standard might be considered when several workers are available to do the work. Although a lenient selection standard would increase the probability that a worker cannot meet the most physical demands of the job, it may not be a serious problem if others are available to help. In contrast, a more rigorous standard might be considered if a worker does not have help.

- *Criticality of the job.* It is imperative that all workers be physically capable of performing critical tasks. The firefighter dummy drag test is defended on the basis of criticality.

General Cutscore Methods

Although there are various ways to establish a cutscore, two general methods have been used. One emphasizes distribution norms of relevant groups, and the other attempts to match the worker to the demands of the job. An example of the normative approach is the method used to set the cutscore for the POST work-sample test for police officers (Weiner, 1994). All applicants must pass the POST to become a California police officer. Figure 16.1 shows that the POST test is a four-item test scored by elapsed time on all items. Each item is converted to a standard score that is summed for a final score. An applicant must complete all items and achieve a total score of 384 points. The mean (±SD) of the normative sample of 527 males was 573 (±39) and of the 103 females was 483 (±60). With a cutscore of 384, over 99% of male and about 95% of female applicants can be expected to pass the POST. The pass rate of the protected group, females, is about 95% that of the pass rate of males, which meets the EEOC 4/5s rule. Although the cutscore meets the 4/5s criterion, no rationale is given for selecting the cutscore of 384. It is likely that the major consideration was to meet the EEOC 4/5s standard.

The method of matching the worker to the demands of the job can be traced to the disciplines of work physiology and ergonomics. A well-established principle of work physiology is to define the physiological demands of common work tasks. This concept was expressed in the early work of Passmore and Durnin (1955). They defined common work tasks in the metric of energy expenditure. This approach is also documented in the **National Institute for Occupational Safety and Health (NIOSH)** preemployment recommendations for materials-handling tasks (NIOSH, 1977). Arm, torso, and leg isometric strength tests were used to measure the force a worker could generate in common lift positions. Prospective ergonomic research established that the risk of musculoskeletal injury, especially to the low back, increased as workers were required to lift loads that exceeded their isometric lift strength capacity (Chaffin, 1974; Chaffin & Park, 1973; Herrin et al, 1986; Keyserling, Herrin, & Chaffin, 1980; Liles et al., 1984). The goal was to select workers who had the strength required by the lift task.

Although both normative and physiological methods have merit, the results of a recent court action (*Lanning et al. v. SEPTA*)[5] for setting the cutscore of a police officer physical test suggests that the legal system may favor the physiological approach. Applicants were seeking employment in the Southeastern Pennsylvania Transportation Authority (SEPTA). The task analysis revealed that officers of this subway system often had to run substantial distances while chasing a suspect. The test developers recommended the basic ability test of 1.5 mi (2.4 km) run and set the cutscore to represent an aerobic capacity of 42.5 ml/kg/min. This standard did have an adverse impact on women and was at issue in the case. The expert witness for the plaintiffs argued that transportation authority police officers needed this level of aerobic fitness because they were often required to run after suspects and needed a reserve to subdue the fleeing suspect. Evidence was presented showing that officers who met the aerobic fitness standard did have arrest records three times higher than officers who did not meet the standard. This suggested that the test and standard were job related. The defendant's expert recommendation was to use normative data to set the cutscore and testified that "he typically sets a cut-point for a physical abilities test one standard deviation below the average performance of incumbents." The judge found this position "without merit" and did not credit his testimony. Although the arguments are more complex than briefly described here, the position taken by the court was that a work-related rationale needed to be considered when establishing an employment standard on a physical test. Paramount in this ruling were the physiological data supporting the requirement that a standard must be consistent with the minimum physical demands of the job.

Matching the Worker to the Job

The approach of matching the worker to the job necessitates that the physiological demands of the job be clearly defined. This information comes from the job analysis. Although this may be easy for materials-handling jobs in which the weights and objects transported can be determined, it is difficult for other jobs. For example, what is the physiological demand associated with the police officer task of subduing a combative suspect?

Once the physiological demands are defined, the test items are selected to measure the physiological ability required by the task. The relationship between the physical ability and work-sample tests establishes this relationship and shows that the test is valid. The final step is to define the minimum physiological level required by the task. As table 16.2 shows, strength is correlated with several work tasks, but the minimum level required to do different tasks will vary. Lifting a 50 lb (23 kg) load, for example, requires less strength than lifting a 75 lb (34 kg) load. The aerobic capacity for firefighters would need to be higher in major cities where they may have to climb stairs in tall buildings (e.g., the World Trade Center in New York on September 11, 2001) compared to residential environments consisting mainly of one- and two-story dwellings. Both the validity of the test and minimum physiological capacity need to be defined.

Provided next are strategies that can be used to define the task's minimum physiological demand. This is a two-step process: identifying candidate cutscores and setting the cutscore.

Identifying Candidate Cutscores

In our research we administered both physical ability strength tests and work-sample tests that simulated the demands of the job. Regression models were used to link the work-sample tasks scored on a continuous or pass–fail scale. The regression models were used to identify candidate cutscores. These methods are illustrated next.

Continuous Scale

Simple linear and multiple regression models (Pedhazur, 1997) were used to define the relationship between isometric strength and the work-sample strength test. The approach is illustrated with data used to define the isometric strength cutscore for workers required to move containers full of freight (Jackson, Osburn, et al., 1993).

The job analysis indicated that one physically demanding job of freight workers was pushing or pulling containers loaded with freight. The weight of the containers was available because these are critical data used to load aircraft. An electronic load cell was used to define the peak force[6] required to move the freight containers. This provided a distribution that defined the peak force a worker would need to generate on the job. The subject's peak push force was measured with a work-sample push test that measured the subject's push capacity while in the position required on the job. An electronic load cell measured this peak force. Figure 16.4 shows the scatter grams with the male and female regression lines. As shown earlier, the difference between the slopes and intercepts of the male and female regression lines was within chance variation, which supported the use of a single regression line to define this relationship. The regression equation is

$$\text{Push force (lb)} = 2.031 + (0.198 \times \text{Strength})$$
$$(R = .78, SEE = 20.0 \text{ lb}) \tag{16.1}$$

Equation (16.1) provides a valid model for defining the strength required to generate levels of push force. Assuming that the job-related push force was defined to be 100 lb, equation (16.1) shows that a strength score of 495 estimates a push force of 100 lb.

Pass or Fail

Often the criterion of job performance is scaled as a dichotomous variable. Many manual lifting tasks are scored pass or fail; the applicant could or could not lift a given weight load (Jackson et al., 1992, 1998). Logistic regression analysis (Hosmer & Lemeshow, 1989; Pedhazur, 1997) provides a model to identify candidate cutscores for a dichotomous variable. A logistic regression model estimates the probability of group membership (e.g., criterion variable of pass or fail) given a score on the predictor variable. Logistic regression analysis, like regression models with continuous variables, documents the validity of the independent variable and provides an empirical model for defining the likelihood of being able to perform a task. The application of simple logistic regression analysis[7] is illustrated next with a lifting task.

A task analysis of an oil production plant showed that lifting heavy valves from the floor to knuckle height was an important, physically demanding work task

(Jackson et al., 1998). The production employees worked alone, which prevented them from seeking help to do the work. A work-sample test was developed to simulate the task of lifting several loads varying in weight. The physical dimensions of the lift duplicated the work task. The test was scored pass or fail depending on the subject's ability to complete the lift. The predictor test was the sum of four isometric strength tests: arm, shoulder, torso, and leg strength (Baumgartner et al., 2003). Three weight loads—60, 90, and 120 lb (27, 41, and 54 kg)—are used to illustrate the approach. These weights represent industrial lifts ranging from moderately heavy to very heavy. Logistic regression analysis showed that the regression weight for strength was significantly different from zero, indicating that lift success was a function of strength. The logistic equations for these three lifts are as follows:

$$\text{Logit (60 lb)} = (.020 \times \text{Strength}) - 3.926 \qquad (16.2)$$

$$\text{Logit (90 lb)} = (.017 \times \text{Strength}) - 5.689 \qquad (16.3)$$

$$\text{Logit (120 lb)} = (.023 \times \text{Strength}) - 10.334 \qquad (16.4)$$

Once the logistic equation was defined, the probability of success (P) can be estimated with the following equation:

$$P = \frac{e^{ax+b}}{1 + e^{ax+b}} \qquad (16.5)$$

where the term e is the base of the natural logarithm, a value of 2.718, and $ax + b$ is the logit for a given strength value.

Figure 16.5 graphically shows the probability of success for completing the three lift weights for selected levels of strength. The logistic probability curves show that, as would be physiologically expected, the strength needed to lift the load increased with the weight of the lift. There was a 50% probability, for example, that someone with 200 lb of strength could lift a 60 lb load, but only a 10% probability of that person lifting 90 lb. The likelihood of someone with 200 lb of strength lifting 120 lb is zero. The physiological levels required to be 50% confident of lifting the 90 and 120 lb loads are about 350 and 450 lb of strength. These data can be used to identify candidate cutscores consistent with issues related to the work environment.

Defining a Final Cutscore

The regression models provide valid information that can be used to select candidate cutscores that define a person's physiological capacity to do the work. This section provides an empirical strategy for defining a final cutscore. The variables required are a valid criterion that is a dichotomous variable and a predictor test that is continuously scaled. If the criterion test were a continuous variable (e.g., the push capacity test; see figure 16.4), it would need to be transformed to a dichotomous scale at the level demanded by the task (e.g., 100 lb of push force required to move a container). This biostatistical approach involves using receiver operating characteristic (ROC) curve analysis (Rosner, 2000) to select a cutscore based on **test sensitivity** and **test specificity.** The interested reader is directed to literature for research examples of the use of test sensitivity and test specificity (Vecchio, 1986; also see chapter 15 by Marilyn Looney) and ROC analysis (Wellens et al., 1996).

This approach is illustrated using the 120 lb lift test as the criterion and the sum of four strength tests as the predictor test (see figure 16.5). The 120 lb lift is a work-sample test that validly represents a demanding work task (Jackson, 1998). The sample included 136 subjects. Of these, 71 (52.2%) lifted the 120 lb load, and 65 (47.8%) failed the lift. The sum of isometric strength was the continuously scaled physical ability test. The mean ($\pm$ SD) strength of the group that lifted the load was 644.2 ($\pm$ 108.1), which was significantly higher ($F[1, 134] = 250.69$; $p < .0001$) than the fail group mean of 301.1 ($\pm$ 140.7).

An ROC analysis is a plot of true positives and false positives (1-Specificity). Figure 16.6 provides the computer-generated ROC curve for these data (SAS, 2002). The ROC curve was developed with the lift test being the dichotomous criterion variable and every possible dichotomous combination of the strength test. With two dichotomous variables, there are four possible outcomes. With a goal

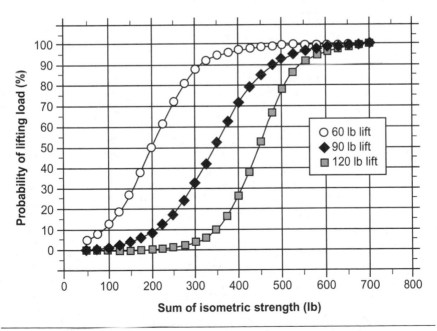

Figure 16.5 Logistic curves of the probability of being able to lift the weight load as a function of lifter strength.

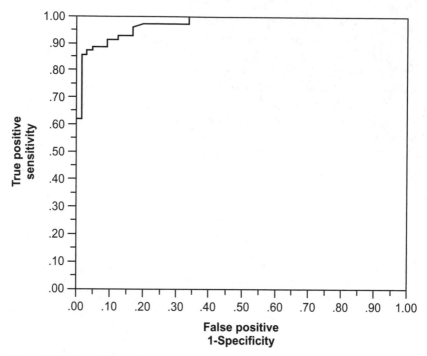

Using YN/120 = 'YES' to be the positive level
Area under curve = 0.97400

Figure 16.6   ROC analysis (SAS, 2002) for the 120 lb lift test and the sum of four isometric strength tests. The plot shows the change in the rate of false positives for all possible levels of test sensitivity. Standard output for the JMP program is the test sensitivity and false-positive percentages for every pair of scores. These data can be used generate a cutscore for any test sensitivity.

of identifying people who have sufficient strength to lift the 120 lb load, the four possible outcomes are as follows:

1. *True positive (TP):* The strength test indicated that the person could lift the load, and the person completed the lift.

2. *True negative (TN):* The strength test indicated that the person could not lift the load, and the person failed the lift.

3. *False positive (FP):* The strength test indicated that the person could lift the load, but the person failed the lift.

4. *False negative (FN):* The strength test indicated that the person could not lift the load, but the person completed the lift.

Test sensitivity and specificity are calculated with the following equations:

$$\text{Sensitivity} = [TP / (TP + FN)] \times 100 \qquad (16.6)$$

$$\text{Specificity} = [TN / (TN + FP)] \, 3 \, 100 \qquad (16.7)$$

The ROC provides a graphic model for selecting the most sensitive cutscore with the lowest FP rate. Table 16.3 gives the data used to calculate test sensitivity and specificity for selected cutscores. Figure 16.6 shows that the FP rate sharply declined when test sensitivity decreased from 100% to about 85%, strength cutscores of 350 and 475. After this point, there was a minimal decrease in the FP rate with a decrease in test sensitivity. A test sensitivity of 62% is the cutpoint where the FP rate is zero. This is a strength cutscore of 625 lb. In contrast, the strength cutpoint 350 with a sensitivity of 100% is associated with an FP rate of 35.4%. Once the ROC trend is known, the next step is to develop the strength cutscore for the desired TP and FP combination.

Table 16.3 gives the cutscores with a range of test sensitivity and specificity values to 100%. The sensitivity and specificity values change in different directions with the change in cutscore. To illustrate, the sensitivity for a 350 lb cutscore is 100%, indicating that all subjects who completed the 120 lb lift had a strength score of ≥350 lb. Test specificity, however, was 64.6%, showing that the 350 lb candidate cutscore produced a high rate (35.4%) of FPs. FPs are candidates with strength levels ≥350 lb who could not lift the 120 lb load. In contrast, test sensitivity for the 625 lb cutscore is 62.0% with a test specificity of 100%, resulting in no FPs. All candidates with a strength ≥625 lb lifted the 120 lb load. The problem of setting the cutscore at 625 is that 38% of the candidates who lifted the load would have failed to reach the strength cutscore of 625 lb (i.e., 1 − .62). They could do the work, but would not have been selected. They are FNs. The goal is to select a

Table 16.3   Test Sensitivity, Test Specificity, and False-Positive Rate (1-Specificity) for Isometric Strength Candidate Cutscores

Cutscore (lbs)	TP (n)	FN (n)	TN (n)	FP (n)	Sensitivity (%)	Sensitivity (%)	FP Rate (1-Specificity)
350	71	0	42	23	100.0	64.6	35.4
400	66	5	54	11	93.0	83.1	16.9
450	63	8	62	3	88.7	95.4	4.6
475	61	10	63	2	85.9	96.9	3.1
480	59	12	64	1	83.1	98.5	1.5
500	58	13	64	1	81.7	98.5	1.5
625	44	27	65	0	62.0	100.0	0.0

cutscore that minimizes the errors. The data in table 16.3 and figure 16.6 support setting the cutscore in the 450 to 500 lb range.

Standard output of the ROC analysis is the percentage under the curve, which for this example (figure 16.6) was 97.4%. The percentage represents test accuracy. This statistic indicates that there is a 97.4% probability of correctly distinguishing between two people who can and cannot lift the 120 lb load based on their strength level (Rosner, 2000). The percentage under the curve is a useful tool for comparing the accuracy of different tests. Figure 16.7 illustrates this. A second ROC analysis is provided for the same 120 lb lift test with just one strength test—the shoulder lift. The area under the curve for the single strength test is 92.5%, about 5% less than the sum of the four tests (figure 16.6). Comparing the FP rate for cutscores at common sensitivity values shows the difference in test accuracy. To illustrate, selecting a shoulder strength cutscore with a sensitivity of 85% had an FP rate of

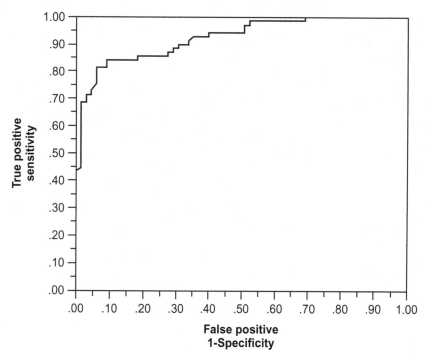

Using YN/120 = 'YES' to be the positive level
Area under curve = 0.92546

Figure 16.7 ROC analysis (SAS, 2002) for the 120 lb lift test and a single isometric strength test (shoulder lift). The area under the curve of 92.5% indicates that a single test is less accurate than the sum of four strength tests, which has 97.4% of the area under the curve. A comparison of figures 16.6 and 16.7 shows that for common levels of test sensitivity, the shoulder lift test has a higher rate of false positives than the sum of four strength tests.

18.5%, over 12 times higher than the FP rate for the sum of four tests, which was just 1.54%. This showed that the sum of four strength tests is a more accurate (i.e., valid) selection device than just the single shoulder lift test. This is consistent with physiological theory. The lift task involves not only the shoulders, but also the legs, low back, and arms.

SUMMARY

The United States Supreme Court ruling that height and weight could not be used to select people for public safety jobs and the NIOSH publication on preemployment strength testing for materials-handling tasks (NIOSH, 1977) led to the increased use of testing for physically demanding jobs. Both events occurred in the late 1970s. Numerous validation studies have been completed, but just a few (Arnold et al., 1982; Reilly et al., 1979) are published in peer-reviewed journals. Typically, a company or governmental agency contracts with professionals to develop the test. The validation report is often considered "Privileged and Confidential." The report is the property of the funding agency and distributed in a manner of their choice. It is often difficult or impossible to obtain a copy of the validation study. This has hindered the scientific evolution and credibility of preemployment research.

The need for materials-handling preemployment tests will likely decrease. The objective of ergonomics is to create a safe work environment and reduce the risk of musculoskeletal injury. A primary ergonomic approach advanced by NIOSH in the 1970s was preemployment testing, matching the worker to the demands of the job. The more recent approach is job redesign (Ayoub, 1982a). The goal is to identify physically demanding work tasks (NIOSH, 1981; Waters et al., 1993) and engineer the stress and demand out of the work. Physically demanding work tasks will likely decline, and physical testing will remain for those jobs in which redesign is not a choice.

Numerous physical ability tests have been developed for public safety jobs, and they have been a major source of litigation (Hogan & Quigley, 1986). A central issue in this litigation is the use of a common cutscore for all applicants. Although this is practical for materials-handling tasks in which the demand can be clearly defined, it is difficult for police officers. The results of two cases[8] at the state level suggest that police departments may use a different approach. Both cases involved the use of physical fitness tests. The critical difference was to define the cutscore based on gender and age norms. The source of the litigation in both cases were from males who failed the tests but scored higher than some females who passed. This signaled a change in test philosophy. The emphasis shifted from defining the minimum level of fitness required by the job to one of health and fitness. This was expressed in the Scott case involving a state trooper who challenged disciplinary actions imposed on him for failing to pass the Vermont State Police physical fitness test:

The goal of the Department's fitness program is to improve the overall health and physical fitness of the Department. It is also 'the Department's philosophy that physical fitness is vital to satisfactory job performance.' The fitness standards are based on those developed by the Cooper Institute for Aerobics Research, and measure overall physical fitness rather than the ability to perform specific tasks. (2001 WL 727008, p. 1)

Alspaugh and Kujawa were two males who failed the preemployment physical fitness test required to attend the police academy of the Michigan State Police. The Michigan Court of Appeals endorsed the use of age- and gender-normed tests for selecting police officers as a means of obtaining a larger applicant pool of eligible candidates. The court stated:

Defendant's practice of gender norming the performance skills test is to segregate the most physically fit candidates within each respective group by controlling for the innate physiological difference between the genders to expand the entire pool of qualified applicants. Thus, defendant's practice of gender norming the performance skill test is an act of inclusion rather than exclusion.

. . . we hold that defendant's gender norming procedure is designed to control for the immutable physiological differences between males and females and to, thus, determine and attain the same levels of general physical fitness within each of the relative gender classes such that the most physically fit individuals within each group are identified and become eligible for eventual certification as police officers. (2001 WL 7373421, p. 35)

The use of gender and age standards for entry into police academies is not universal. The more common practice is a single cutscore that defines the minimum physical demands of the job. A possible reason for the change is that the minimum demands of police work are difficult to define and defend. In addition there is a growing body of evidence showing the relationship between fitness and health (United States Department of Health and Human Services, 1996). The defensibility of gender and age norming for police work will eventually be decided in the federal courts.

Notes

1. What I consider mandatory reading is the physical ability testing chapter published by Hogan (1991).

2. Personal communication with Dr. William McArdle, August, 2002.

3. These investigators included Dr. Andrew S. Jackson, Department of Health and Human Performance, University of Houston; Dr. Hobart Osburn,

Department of Psychology, University of Houston; and Dr. Kenneth R. Laughery, Department of Psychology, Rice University.

4. Information on this equipment can be obtained from the Lafayette Instrument Company at www.licmef.com. Go to the link on strength testing.

5. The complete "Findings of Fact and Conclusions of Law" on *Lanning et al. v. SEPTA* can be found at www.paed.uscourts.gov/opinions/00D0916P.HTM.

6. The greatest amount of push force required to move a container was the force required to overcome inertia. It takes less force to move a container once it is moving. For this reason, peak force became the important variable.

7. Multiple independent variables can be used with logistic regression. This was the statistical procedure used to establish cardiovascular disease risk factors with the Framingham population.

8. Scott Appeal to the Supreme Court of Vermont 2001 WL 727008 and *Alspaugh and Likawa v. Michigan Law Enforcement Officers*, 2001 WL 737342. These came from Westlaw (WL) citations.

Glossary

24 hr physical activity recall—An indirect method of acquiring detailed information, typically via telephone interview, regarding the type, purpose, duration, and intensity of a respondent's physical activity during the past 24 hours.

absolute intensity—A standard or actual rate of energy expenditure (e.g., L O_2 uptake $\cdot$ min^{-1}, METs, kcal $\cdot$ min^{-1}) assumed for a specific activity.

absolute reliability—An index of reliability reflecting an estimate of measurement error. (*See* **standard error of measurement.**)

accelerometers—Electronic devices worn on the body that measure in a single or multiple plane the rate and magnitude of displacement in the body's center of mass or a body limb. The absolute value and frequency of the acceleration forces are summed over a defined observation period (e.g., per minute) to provide a direct measure of the volume of physical activity performed.

age-based scale—A score scale that bases test performance on a typical performance for a test taker at a given age.

Age Discrimination in Employment Act (ADEA)—The federal law that is the legal basis for defining job discrimination because of age.

alpha coefficient—A common estimate of the internal consistency reliability of written test scores.

alternative assessment—An assessment that integrates higher-order cognitive, affective, and psychomotor processes in assessment tasks.

Americans With Disabilities Act (ADA)—The 1990 comprehensive federal law that prohibits discrimination in many aspects of life, including both physical and mental disabilities.

anchor-test design—One of the common equating data collection designs, in which a common set of items is administered with each of two or more tests or forms to equate the scores of the tests or forms.

artificial neural networks (ANNs)—A set of mathematical models for information processing based on how neurons and synapses work in the human brain.

assessment-driven instruction—Instructional strategies focused on helping students master performance assessments.

attributable risk—An estimate of the amount of risk attributable to the risk factor.

attributable risk percent in the exposed group—An estimated percent of risk in a group with that particular risk factor. For example, one could determine the percentage of the total risk for coronary heart disease that is due to sedentary behavior among those who are sedentary.

authentic assessments—Performance assessments administered in real-life settings.

autoregressive model—A statistical model that predicts a variable from itself measured at a different occasion.

Bayes estimation—A method of estimating a person's level on a given trait based on previously collected population data. Bayes estimation is typically used at the beginning of a test or survey when too few items have been administered to make an accurate assessment based only on the person's responses.

benchmarks—Specific measurable signposts that describe progress toward a content standard.

between-subjects designs—Study designs in which the sampling units (e.g., the research participants) are exposed to a single condition each; in other words, they are measured only once along the dependent variable.

blocking—Dividing sampling units (for example, research participants) into groups with similar scores on a matching variable (e.g., same gender).

calorimetry—Methods used to calculate the rate and quantity of energy expenditure when the body is at rest and during physical activity.

case-control study—A study in which subjects are selected based on the presence or absence of a disease. A comparison of cases (disease) and controls (no disease) is made relative to the frequency of past exposure to potential risk factors for the disease.

classical test theory—A measurement theory that derives evidence for validity and reliability by attempting to link observed scores and true scores by generalizing from a sample to a universe. It is also known as true score theory because it is based on the belief that a person's observed score (X) on a test is the sum of a true score (T), plus an independent measurement error (E): $X = T + E$.

classroom assessment—Assessment developed and conducted by the classroom teacher for monitoring student progress and grading.

coefficient of variation—A reliability estimation technique commonly used with scores from research apparatus; obtained by dividing the standard deviation of scores by the group mean.

cohort—A generational or demographic group large enough to permit the statistical control of a number of confounding variables, thus permitting inference to a population.

common-item, nonequivalent-groups design—An application of the anchor-test design to two nonequivalent groups of test takers.

computer-based testing (CBT)—Any method of administering tests electronically using a computer as opposed to printed or oral tests; also called *computerized testing* and *computer-administered testing*.

computerized adaptive testing (CAT)—A sequential computer-based test in which successive items are selected from a pool of items based on the test taker's performance on previous items. Based in item response theory (IRT), this type of testing is intended to select items that are of appropriate difficulty (or relevance) for the test taker.

concurrent evidence of validity—Evidence that test scores correlate highly with a criterion measure, when scores on both tests are obtained at approximately the same time.

confidence limits for R—The degree of confidence that the population value of R is spanned by a lower and upper value.

confirmatory factor analysis—An analysis that defines a factor structure or model on an a priori (theoretical or empirical) basis.

confirmatory stage—An intermediate stage of validation in which evidence is collected to confirm or disconfirm the definition of the nature of a construct and instruments are evaluated to determine whether they measure the construct.

confounding—A situation that occurs when a noncausal association between exposure and outcome is observed as the result of a third variable or group of variables called confounders.

consequential validity—Analysis of the positive and negative, intended and unintended consequences of assessments.

construct— The theoretical representation of a characteristic of people (e.g., sport anxiety, strength, intelligence, body image); a concept that can be defined but not directly measured.

construct-irrelevant variation—An indication that some of the variability in the task or test is irrelevant to the purpose of the test. In chapter 7, this is referred to as a *nuisance dimension.*

construct underrepresentation—Defining a construct too narrowly, resulting in a lack of inclusion of all important components.

construct validation—A continuous process of collecting data to understand the nature of a construct and to evaluate the use and interpretation of measures of that construct.

construct validity—Marshaling evidence to support the appropriateness, meaningfulness, and usefulness of specific inferences made from test scores.

content balancing—The process of including items in proportion to a test plan that contains multiple content areas.

content relevance—The process of defining and evaluating the universe of content and behaviors assessed by a test.

content representativeness—The extent to which assessment tasks are representative of a defined universe of content and behaviors.

content standards—Broad descriptions of what students should know and be able to do at each grade level.

content validity—In employment settings, the degree to which the selection device (e.g., test) replicates major job behaviors.

contingency table—A two-way table representing the interaction between mutually exclusive responses or outcomes for two variables.

convergent evidence of validity—Evidence that different measures of the same construct are moderately to highly correlated.

criterion-referenced test standard—A standard that is used to classify people as either proficient or nonproficient, passing or failing, and so on.

criterion-related evidence of validity—The relationship between scores on a test and scores on a criterion measure of the construct of interest.

criterion score—The test score recorded for a person based on at least two scores.

Cross-Industry Standard Process for Data Mining (CRISP-DM)—A commonly used process model for data mining applications that includes six phases: business understanding, data understanding, data preparation, modeling, evaluation, and development.

cross-sectional surveys—Sometimes called *prevalence studies;* measure both risk factors and the presence or absence of disease at the same point in time. This approach is expedient and relatively inexpensive, but it cannot determine the temporal sequence of a potential cause-and-effect relationship.

cross-validation—A method to test the accuracy of a regression equation on a different sample from the one on which it was developed.

crude rates—Rates based on a total population, without consideration of any population characteristic such as the distribution of age, gender, or ethnicity.

cutscore—The score on the test that determines whether a job applicant passes the test.

data—A set of scores that may be collected using many different methods.

database mining—A type of data mining with a focus on the information stored in a database; database mining consists of three components: data sources, data warehousing, and data analysis.

data mining—A process that uses a variety of data analytical tools to discover patterns and relationships in data to enable more accurate and valid descriptions and predictions.

decision tree—A method for classification and prediction that is based on a set of mathematical application methods for classification and prediction that are based on rules representing cause-and-effect relationships.

definitional stage—An early stage of validation in which evidence is collected to determine the nature (definition) of a construct and instruments are designed or selected to measure the construct.

developer-defined scaling—The process of transforming raw scores onto a scale defined by test developers.

diagnostic accuracy—The degree to which test results can correctly determine whether the disorder or condition is absent or present.

dichotomously scored items—Items or tasks scored in two categories.

differential item functioning—A situation that occurs when items function differentially for examinees of the same ability or trait level but from different demographic groups.

direct calorimetry—The most precise method for directly measuring energy expenditure by assessing heat production at rest or during exercise.

discrepant analysis—An analysis that requires that a second imperfect reference test (diagnosis errors can occur) be given to patients; is used only after the diagnostic and first reference test results differ (i.e., false-positive and false-negative results).

discriminant analysis—A multivariate procedure used to quantify the extent to which each of a number of variables contributes to identify differences among groups.

discriminant evidence of validity—Evidence that measures of different constructs do not correlate as highly as measures of the same construct.

discrimination parameter—A statistical index describing the discrimination, or sensitivity to distinguish test takers, of a test item; defined differently under the frameworks of various testing theories.

disparate impact—The legal term that defines job discrimination based on race, color, religion, sex, national origin, age, or a qualifying disability.

dose-response—A relationship in which increasing levels, or "doses," of physical activity result in corresponding changes in the expected levels of a defined physical performance or health parameter.

double-blind trial—A randomized trail in which neither participants nor observers who collect the data are aware of group assignment.

doubly labeled water—A form of indirect calorimetry designed to estimate total energy expenditure using stable water isotopes (^2H_2O and $H_2{}^{18}O$) that are dosed according to body weight. The difference in isotope turnover rates is used to provide a measure of $\dot{V}CO_2$, which is then used to estimate oxygen uptake and energy expenditure.

duration—The dimension of physical activity referring to the amount of time (e.g., min, hr, d) an activity is performed.

ecologic study—A specific type of cross-sectional survey used to compare the frequency of some risk factor of interest (e.g., sedentary behavior) with an outcome measure such as obesity in the same geographic region.

effect modification—A situation that occurs when two or more risk factors modify the effects of each other with regard to the occurrence of an outcome. For example, the effect of physical activity categories (sedentary/active) on the presence or absence of coronary heart disease may differ by some third variable or effect modifier such as gender.

effect size—A measure of the magnitude of the experimental effect; for example, the standardized mean difference between a treatment and control group.

endogenous variable—A variable whose value is derived, either directly or indirectly, from other variables in a model.

energy expenditure—The total exchange of energy required to perform a specific type of biological work; often used to express the volume of physical activity performed during a defined time frame.

epidemiology—The study of the nature, cause, control, and determinants of the frequency and distribution of disease, disability, and death in human populations.

Equal Employment Opportunity Commission (EEOC)—The federal agency with the legal power to examine job discrimination.

equating—The process of statistical adjustment to place scores of two or more alternate forms of a test onto a common scale. The forms are constructed to the same or similar content and statistical specifications and are administered under identical procedures.

equipercentile equating—A traditional equating method in which score distributions are set to be equal so that the same percentile ranks from different tests are considered to indicate the same level of performance.

equivalent-groups design—A common equating data collection design in which two or more tests or forms are administered to two equivalent groups of test takers.

error score—In reliability theory, the score indicating the amount of error in the obtained score of a person.

error variance—Unexplained variance arising from unspecified sources. Error variance contains a composite of random measurement error and specific measurement error associated with a particular measuring instrument. (*See* **measurement error.**)

ethics—A set of moral ideas and principles that guide our conduct; in testing settings, principles that guide the decision-making process.

examinee-centered method—A standard-setting method by which standards are set based on the difference of the data distribution among groups with known ability or status (e.g., instructed vs. noninstructed groups).

exercise—Planned, structured, and repetitive bodily movement performed to improve or maintain one or more components of physical fitness. Exercise is a specific subcategory of physical activity.

exogenous variable—A variable whose value is not derived from any other variable in a model (i.e., a wholly independent variable).

experimental study—A study in which change in the independent variable is manipulated by the investigator.

exploratory factor analysis—An analysis that identifies a factor structure or model for a set of variables without an a priori specification.

exposure control—A form of control used to minimize the use of individual items, typically to limit the likelihood that an examinee will be administered an item he or she or a peer has seen in a previous administration. The frequency of item administration is monitored, and items with lower frequency are selected for administration before items that have been "overexposed" (used too often).

factor analysis—A statistical technique used to evaluate the internal structure, or dimensionality, of a multidimensional construct.

false negative—A test result that states incorrectly that a condition does not exist. In employment settings, a false negative occurs when a selection device indicates that an applicant cannot perform the work task but the applicant can.

false positive—A test result that states incorrectly that a condition exists. In employment settings a false positive occurs when selection device indicates that an applicant can perform the work task but the applicant cannot.

familywise (FW) error rate—The probability of committing at least one type I error in a study that involves more than one statistical test.

focal group—The group against which the studied item is suspected of being biased.

four-fifths (4/5s) rule—The criterion used by the EEOC to determine whether a selection device (e.g., test) has adverse impact. The selection rate of a protected group cannot be less than 80% of that of the nonprotected group.

frequency—The dimension of physical activity referring to how often an activity is performed.

generalizabililty—In the context of test validation, the extent to which inferences from test scores can be generalized across different assessment tasks, contexts, and raters.

genetic algorithms (GAs)—A set of mathematical algorithms based on an analogy to the biological process, which uses selection, crossover, and mutation operations to evolve successive generations of solutions.

global activity questionnaires—short surveys (e.g., one to four items) used to obtain a general index of physical activity level.

goodness-of-fit—The degree to which the actual or observed covariances matrix is predicted by the estimated model.

gross energy expenditure—The total amount of energy expended for a specific activity including the resting energy expenditure. Gross energy expenditure is typically used for between-persons comparisons.

group invariance—A feature of item response theory that states that parameters of items (e.g., difficulty, discrimination, or guessing) will not be affected by different samples employed in parameter estimations.

growth curve model—A model that analyzes change in a construct or outcome over time. The model provides an estimation of the characteristics of growth with respect to group-level change (fixed effects) and individual variation in change (random effects).

growth mixture model—A model that estimates population heterogeneity in change over time by delineating individual differences in development in subpopulations as described by latent trajectory classes, in which each class may have a different random effect growth model.

hierarchical linear models (HLMs)—Linear regression models involving more than one level of data in which lower-level data are nested within higher-level data (see Bryk & Raudenbush, 1987, 1992).

high-stakes assessment—Large-scale district or statewide assessments used to inform educational policy, reform curriculum, and monitor accountability.

incidence rates—The number of new cases of a disease occurring in a specific population in a specified time period.

incident cases—New cases of a disease, disorder, or condition that occur in a population over a specified time period.

indirect calorimetry—Methods used to estimate energy expenditure by measuring oxygen uptake ($\dot{V}O_2$) and carbon dioxide production ($\dot{V}CO_2$) based on expired ventilatory gas analysis or radioisotope concentrations from serial urine samples.

intensity—The dimension of physical activity referring to the level of effort or physiological demand required to perform the activity. Intensity can be expressed as an absolute or relative term.

interclass correlation coefficient—A classification of correlation coefficients that uses the Pearson product–moment coefficient.

internal consistency reliability—The consistency of test scores within days or across trials in a multitrial test.

intertester reliability—The consistency of scoring among different testers. (*See* **rater reliability.**)

intraclass correlation coefficient—A classification of correlation coefficients that uses analysis of variance in its computation.

intratester reliability—The consistency of scoring over repeated measures by a single tester. (*See* **rater reliability.**)

IRT equating—Equating methods based on item response theory.

item—A question or statement on a test, assessment, or survey.

item bank—A collection of items or questions organized and cataloged to take into account the content of each item, as well as other measurement characteristics (e.g., difficulty and discrimination).

item bias—A term used to indicate judgmental detection of differential performance on an item (or task) for examinees of the same ability.

item-centered method—A commonly used standard-setting method in which an expert panel is employed to evaluate each test item and associate the item with a criterion behavior.

item characteristic curve (ICC)—A mathematical function relating the probability of a certain item response, usually a correct response, to the level of the attribute measured by the item; known also as *item response curve* or *function.*

item difficulty—A statistical index describing the trait level difficulty, or "hardness," of a test item. It is defined differently under the frameworks of various test theories.

item enemies—Questions that should not appear in the same assessment. Item enemies are a concern for all examinations in which one item might cue the response for another item, and of particular concern for computer-generated forms because a human might not review the final group of items administered.

item information function—An index used in IRT to describe how precisely abilities can be estimated by an item at a certain ability level.

item response theory (IRT)—A relatively new test theory that employs mathematical models to describe the relationship between performance on a test item and an estimate of the test taker's true ability on the trait or proficiency being measured.

job analysis—A systematic method of defining the important work behaviors demanded by a job. It is a crucial step of a validation study.

kappa—A reliability coefficient used with criterion-referenced test standards. Kappa statistically adjusts the proportion of agreement for chance classifications. (*See* **proportion of agreement coefficient.**)

kilocalories (kcal)—1,000 calories, or 4.184 kilojoules; used to express the energy expended during physical activity.

kinesmetrics—A measurement theory applicable to the movement sciences.

known-difference evidence—Evidence that an instrument measures a given construct by demonstrating that groups that differ on the construct obtain different scores on the instrument.

Kuder-Richardson formula 20—A formula which uses the proportion of subjects passing and proportion failing an item along with the variance of the total test scores to calculate an estimate of the internal consistency reliability of knowledge test scores.

Kuder-Richardson formula 21—A formula which uses mean score on a test along with the variance of the total test scores to calculate an estimate of the internal consistency reliability of knowledge test scores.

kurtosis—The degree to which the peakedness (or flatness) of a data distribution deviates from that of a normal distribution.

latent class analysis—A version of factor analysis using categorical data.

latent curve models (LCM)—Statistical models in which latent change variables (factors) are modeled to explain the change in observed (measured) variable(s).

latent trait—An underlying characteristic of a person that is not directly observable but manifests itself by influencing the person's responses to an instrument designed to measure the trait.

latent variable—A variable that is not directly observed but is inferred or estimated from observed variables.

likelihood ratio test—A ratio formed from two likelihoods. This can be used for detection of DIF and is asymptotically distributed as a chi-square and, therefore, can be tested for significance.

linear equating—A traditional equating method in which the means and standard deviations of the two tests for a particular group of test takers are set equal.

link analysis—A mathematical application derived from graph theory that follows relationships among records to develop problem-solving models.

linking—The statistical process that places two or more tests onto a common scale so that scores can be compared to each other; includes equating and calibration.

local independence—One of the important assumptions of item response theory stating that an examinee's responses to the testing items are statistically independent.

logit—The unit of a log-odds scale in item response theory; the contraction of "log-odds unit."

longitudinal data analyses—Analyses of data measured for a relatively long period of time. In most cases, longitudinal data are measured repeatedly from the same participants.

Lord's chi-square—A chi-square statistic for detection of DIF that compares the item parameter estimates obtained for an item in the reference and in the focal group.

Mantel-Haenszel statistic—A nonparametric statistic used for detection of DIF.

market basket analysis—A market research method that uses the information about customers' purchases to give an insight into who they are and why they make certain purchases, as well as into which products tend to be purchased together and which are most amenable to promotion.

mastery cutoff score—The minimum test score needed to be classified as a master, passing, or proficient.

matching—Grouping participants with similar scores on a concomitant variable so that both the treatment and control groups contain subjects with similar characteristics on the matching variable (e.g., within the same age range). Often the matching is done with other participants who have the same score on the matching variable.

maximum likelihood estimation—An estimation method that iteratively improves parameter estimates to minimize a specified fit function.

mean square—An index of score variability obtained by applying analysis of variance to the data that are used to estimate test score reliability; a sum of squares divided by its degrees of freedom.

measurement error—The degree to which the latent variable(s) of interest are not perfectly described by the observed variable(s); under classical test theory, the difference between the observed score and the hypothetical true score.

measurement model—A model that contains a series of measurement equations describing how latent variables are measured or operationalized by corresponding observed variables, and the links between the latent variables.

metabolic equivalent (MET)—A unit used to estimate the metabolic cost (oxygen consumption) of physical activity. One MET equals the resting metabolic rate of approximately 3.5 ml $O_2 \cdot kg^{-1} \cdot min^{-1}$, or 1 kcal $\cdot kg^{-1} \cdot hr^{-1}$.

mode—The dimension of physical activity that identifies the specific type of activity being performed (e.g., walking, bicycling, jumping, weightlifting, bowling).

model identification—The degree to which there is a sufficient number of equations to "solve for" each of the coefficients (unknowns) to be estimated. Models can be underidentified (i.e., they cannot be solved), just-identified (i.e., the number of equations equals the number of estimated coefficients with no degrees of freedom), or overidentified (i.e., there are more equations than estimated coefficients and the degrees of freedom are greater than zero).

model respecification—Modification of an existing model with estimated parameters to correct for inappropriate parameters encountered in the estimation process or to create a competing model for comparison.

multilevel model—An analysis of hierarchical data (e.g., educational systems with a hierarchy of pupils within classes within schools) that allows simultaneous modeling of between-groups and within-groups variability in outcome.

multiple comparison tests—Statistical tests used to examine which pair of group means are significantly different following an ANOVA (F-test) procedure.

multitrait–multimethod (MTMM) matrix—A matrix of intercorrelations used to evaluate convergent and discriminant evidence of validity.

multivariate tests—Statistical tests involving multiple dependent variables.

National Institute for Occupational Safety and Health (NIOSH)—The federal agency with the responsibility for studying and promoting safety in the workplace.

negative likelihood ratio—The ratio of the probability of a patient who does not have the condition getting a negative test result to the probability of a patient who has the condition getting a negative test result . This converts to the probability of having the condition with a negative test result.

negative predictive value—The probability that a patient with a negative test result does not have the condition.

nested model—A model that has the same constructs or variables but differs in terms of the number or types of relationships represented. The most common form of nested model occurs when a single relationship is deleted (i.e., fixed to zero) from another model. Thus, the model with fewer estimated relationships is "nested' within the more general model.

net energy expenditure—Energy expenditure associated exclusively with the activity itself; computed as gross energy expenditure minus a person's resting energy expenditure. This term is used to compare the energy costs of specific activities.

nomological network—The underlying theory and relationships that govern a construct.

noncentrality parameter—A measure representing the factor by which the F ratio departs from the central F distribution when a difference between treatment means actually exists.

norm-referenced test standard—A standard used to rank order people from best to worst.

objectivity—See **rater reliability.**

observational study—A study in which change in the independent variable occurs as a result of natural history (i.e., it is self-initiated by the people being studied).

observed score—See **obtained score.**

observed variable—A variable that is directly measured in a study.

obtained score—In reliability theory, the score that is recorded for a person.

odds ratio (OR)—A ratio calculated by dividing the odds of exposure to the risk factor in the diseased group by the odds of exposure to the risk factor in the nondiseased group.

online analytical processing (OLAP)—An integration and application of a number of data mining techniques and data warehouses that enable information users to gain insight into data through fast, consistent, and interactive access to a wide variety of possible views of information.

opportunity to learn (OTL)—In the educational context, the availability of appropriate instruction, curriculum, and resources enabling all students to meet content standards and benchmarks.

oxygen uptake ($\dot{V}O_2$)—The extraction of oxygen via the respiratory system, which is then delivered via the circulatory system and used by working tissues to support biological processes.

path diagram—A graphical portrayal of the complete set of relationships among a model's constructs or variables.

pedometers—Electronic devices typically worn at the waist and designed to measure ambulatory activity as steps per unit time of observation (e.g., per day).

percentile—The score on a test below which a given percentage of test takers' scores fall.

percentile rank—The percentage of scores in a specified distribution that fall below the point at which a given score lies.

performance assessment—An assessment in which high congruency exists between the performance (behavior) desired by the tester and the assessment tasks.

performance criteria—The specific behaviors assessed in an alternative assessment scoring guide.

performance cutscore—The numerical value associated with each level of the performance scale in an alternative assessment scoring guide.

performance mastery cutoff (PMC)—The performance cutscore representing testee mastery on an alternative assessment scoring guide.

performance scale—Values (commonly numeric) assigned to performance criteria in an alternative assessment scoring guide.

performance standard—The qualitative description associated with each performance cutscore in an alternative assessment scoring guide.

physical ability test—A common fitness tests (e.g., isometric strength test) used to determine a person's capacity to perform physically demanding work tasks.

physical activity—Bodily movement that is produced by the contraction of skeletal muscle and that substantially increases energy expenditure.

physical activity logs—A direct method of measuring physical activity in which respondents identify the type and duration of physical activities performed over a specific period of time. They are considered to be a simplified form of the physical activity record.

physical activity records—A direct method of measuring physical activity in which respondents keep an ongoing account, in diary format, of all activities performed over a defined observation period. Typically, respondents record information about the type, purpose, duration, intensity, and body position for each activity.

physical fitness—A set of attributes (e.g., muscle strength and endurance, cardiorespiratory fitness, flexibility) that relate to one's ability to perform physical activity.

physiological validation—A concept recently introduced in the literature to differentiate between psychometric and physiological validity issues. Physiological validation uses a test metric scored on a ratio scale that has physiological meaning, defines the physical demands of the work task in scientific terms, and links the physiological capacity of applicants to the demands of the work task.

polytomous items—Items or tasks scored in more than two categories.

population attributable risk percent—The percentage of risk of disease that is due to a particular risk factor.

population-based scaling—Transforming raw scores onto a new scale based on their relative positions in the sample or population used to define the scale.

positive likelihood ratio—The ratio of the probability of a patient who has the condition getting a positive test result to the probability of a patient who does not have the condition getting a positive test result.

positive predictive value—The probability that a patient with a positive test result has the condition.

predictive evidence of validity—Evidence that test scores can be used to make decisions about future behavior or future outcomes.

prevalence rate—The incidence rate multiplied by the average duration of the disease (i.e., how many people have a particular disease or engage in a particular behavior, such as smoking or physical activity, at a certain point in time).

prevalent cases—People within a population who have a certain disease disorder or condition at a given point in time.

process criteria—The components of student evaluation that assess the work habits and affective components supporting mastery of content standards and benchmarks.

product criteria—The components of student evaluation that assess mastery of benchmarks and content standards via results of performance assessments.

progress criteria—The components of student evaluation that indicate how far students have come in meeting content standards and benchmarks.

proportion of agreement coefficient—A reliability coefficient used with criterion-referenced test standards that indicates the proportion of people classified the same on two occasions.

prospective cohort studies—Incidence or longitudinal follow-up studies that involve the selection of a group of people at random from some defined population or a selection of groups exposed or not exposed to a risk factor of interest. These studies require the researcher to collect baseline information of potential disease risk factors and then follow the cohort over time to track the incidence of disease. A major advantage of these studies is that the risk factor profile is established before the outcome is assessed.

prospective study—A study of future happenings, events, and findings; any study following a condition, concern, or disease into the future over time.

pseudo-retrospective sampling—The sampling of two independent groups—one from a population at high risk of having a condition and the other from a population at low risk.

psychometric—Measurement theory applied to psychological constructs.

quantitative history—An instrument used to indirectly measure the frequency and duration of physical activities performed over the past year or lifetime.

Raju's area measure—A measure of the area between two item response curves.

randomized controlled trial—A trial in which participants are selected for study and randomly assigned to receive either an experimental manipulation or a control condition. Measurements are taken before and after the intervention period in both groups to assess the degree of change in outcome between the intervention and control condition.

Rasch model—One of the most commonly used one-parameter logistic IRT models.

rater reliability—An estimate of the degree to which several raters, scorers, or judges agree on the scores of the people tested.

recall bias—The notion that people who have experienced an adverse event (cancer, heart attack) may think more about why this problem occurred than healthy people may and thus might be more likely to recall exposure to potential risk factors.

recall questionnaires—An indirect method of assessing physical activity; generally use 10 to 20 items designed to extract information about the frequency, duration, and types of activities performed over a defined recall period.

receiver operating characteristic (ROC) curve—A graphic method for selecting a test cutscore that plots sensitivity values on the y-axis and 1-specificity values on the x-axis; used to determine the test cutoff score that yields the best sensitivity and specificity values.

reference group—A group that provides the basis for comparison in a test for DIF.

relative efficiency (RE)—A statistical index comparing the information function of one test with that of another at a particular ability level.

relative intensity—The rate of energy expenditure for a specific activity expressed as a percentage of the person's maximal capacity to do physical work (e.g., % maximal oxygen uptake, % heart rate reserve).

relative reliability—A ratio of true score variability to obtained score variability for a group of participants; estimated by calculating a reliability coefficient.

reliability coefficient—A coefficient used as an estimate of the reliability of a set of test scores.

repeated measures—A measurement design in which the same variable(s) is measured repeatedly over time from the same participants.

residual error variance—Also called disturbance; reflects the unexplained variance in latent endogenous variable(s) due to all unmeasured causes.

retrospective study—A study that looks back in time after the occurrence of injury, disease, or death in an attempt to reconstruct the risk habits.

risk difference—The risk of disease in the group exposed to the risk factor minus the risk of disease in the unexposed group.

risk factor—A characteristic that, if present, increases the probability of disease in a group of people who have that characteristic compared with a group of people who do not have that characteristic.

rubrics—Scoring guides used in alternative assessments.

scale score—A score on a test that is expressed on some defined score of measurement or a transformed raw test score.

scaling—The process of creating a scale score, which may enhance test score interpretation by placing scores from different tests or scales onto a common scale.

score—*See* **obtained score.**

sensitivity—The proportion of patients with the condition who had positive test results.

simultaneous item-bias test—A nonparametric statistic used for detection of DIF.

single-blind trial—A trial in which only the data collection personnel are unaware of group assignment.

single-group design—A common equating data collection design in which two or more tests or forms are administered to the same group of test takers.

skewness—The degree to which a data distribution departs from a symmetric distribution.

Spearman-Brown prophecy formula—A formula used to predict the reliability of test scores when the length of the test is changed.

specificity—The proportion of patients without the condition who had negative test results.

specific rates—Rates calculated separately for population subgroups, typically age, gender, ethnicity (called age-specific rates, gender-specific rates, etc.).

spectrum bias—A source of bias that affects sensitivity and specificity statistics because these statistics apply to the specific population sampled and criteria used to define the presence of a condition.

sphericity—A statistical assumption required in a univariate repeated-measures ANOVA. To obtain a valid statistical probability, the variances of repeatedly measured variables should be approximately equal, and the covariances between repeatedly measured variables should be approximately equal.

stability reliability—The consistency of test scores across days. (*See* **test–retest reliability.**)

standard error (SE)—A statistical precision index defined differently under various testing and statistical theories and applications; provides local precision information in item response theory.

standard error of measurement—An estimate of the amount of measurement error to expect in a test score.

standardized rates—Crude rates that have been adjusted for some population characteristic such as age or gender to allow for valid comparisons of rates among populations in which the distribution of age and gender many be quite different.

standards-based education—An educational model in which mastery of content standards becomes the focus of curriculum, instruction, and assessment.

statistical power—The probability of detecting a treatment effect when one actually exists; defined as the probability of correctly rejecting the null hypothesis (H_0).

stratifying—Creating homogeneous groups of sampling units (e.g., research participants) based on a concomitant variable in which each group has more participants than conditions.

structural equation modeling—A multivariate technique combining aspects of multiple regression and factor analysis to estimate a series of interrelated dependent relationships simultaneously; commonly used to examine links in a nomological network.

structural model—A model that contains a series of structural (i.e., regression) equations describing the relationships among latent variables.

task questionnaire—A list of statements that define the tasks done by workers. It often is used to conduct a job analysis.

test—A data measurement or data collection method.

test bias—A situation that occurs when meanings and implications of test scores are different for a particular subgroup than the meanings and implications for the rest of the test takers.

test equating—*See* **equating.**

test information function—A sum of all item information functions in a test or instrument at a particular ability level.

test item—A statement, question, exercise, or task on a test for which the test taker is to select or construct a response or perform a task.

testlets—A group of items administered as a block. For a reading test, this would refer to a group of items associated with a single reading passage. For some item response theory models, items are grouped into testlets for psychometric concerns.

test reliability—An index of the precision of test scores; under classical test theory, the degree to which obtained scores reflect a hypothetical true score.

test–retest reliability—Administering a test on each of two days to estimate stability reliability.

test sensitivity—The proportion of people who pass a test at the selected cutscore and can do the work. (*See* **sensitivity**.)

test specificity—The proportion of people who fail an employment test at the selected cutscore and cannot do the work. (*See* **specificity**.)

test theory—Statistical and mathematical procedures for estimating the key characteristics of a test or measure such as validity and reliability.

test unfairness—A situation in which members of a protected group obtain lower scores on a preemployment test compared to members of another group and the difference in scores is not reflected in differences in the job performance measure.

test validity—*See* **validity.**

text mining—A type of data mining with a focus on text that is unstructured, ambiguous, language dependent, and heterogeneous; examples of text include incident reports, patents, e-mails, and text-based records.

theory-testing stage—A later stage of validation in which the underlying theories explaining how the construct of interest fits into a broader area of knowledge are investigated.

Title VII—The 1964 federal law that prohibits employment discrimination on the basis of race, color, religion, sex, and national origin by employers, labor organizations, and employment agencies.

traditional equating—An equating method based on classical test theory in which the score correspondence of tests is established by setting the characteristics of the score distributions equal for a specified group of test takers.

trait—A relatively stable characteristic of a person (e.g., analytic capability).

true negative—A test result that states correctly the condition does not exist. In employment settings a true negative indicates that an applicant cannot perform the work task and the applicant cannot.

true positive—A test result that states correctly the condition does exist. In employment settings a true positive indicates that an applicant can perform the work task and the applicant can.

true score—In reliability theory, a hypothetical score indicating the true or actual ability of a person; can be conceptualized as the average score over an infinite number of independent repeated observations for a single person.

T-score scale—A commonly used population-based scale with a mean of 50 and a standard deviation of 10.

Type I error—Rejecting the null hypothesis (H_0) when in fact the null hypothesis is true.

Type II error—Failing to reject the null hypothesis (H_0) when in fact the null hypothesis is false.

unidimensionality—One of the important assumptions of IRT stating that only a single latent trait or ability is being measured by items or tasks in a test.

Uniform Guidelines on Employee Selection Procedures—The document published by the EEOC that provide standards to judge the validity of an employment selection device such as a test.

univariate tests—Statistical tests involving a single dependent variable.

validation study—Research designed to develop a selection device (e.g., test) that is job related in an employment setting.

validity—Evidence for the appropriate interpretation of test scores.

verification or workup bias—A situation that exists when only patients who received a positive test result receive the reference test. The ideal situation is to have patients receive both the diagnostic test and the reference test.

Web mining—A type of data mining with a focus on the World Wide Web; it looks for patterns and their relationship in both structured and unstructured data stored on the Web.

within-subjects designs—Study designs in which the sampling units (e.g., the research participants) receive two or more measurements (i.e., are studied in two or more conditions); also known as repeated measures designs

work-sample test—An employment test that duplicates actual work tasks.

z-score scale—A common standard score transformation with a mean of 0 and a standard deviation of 1; used as the basis for many other standard score transformations (e.g., the *T*-score scale).

References

Chapter 1

American Educational Research Association, American Psychological Association, & National Council on Measurement in Education. (1999). *Standards for educational and psychological testing.* Washington, DC: American Educational Research Association.

King, H.A. (1984). Measurement and evaluation as an area of study: A plea for new perspectives. In C.M. Tipton & J.G. Hay (Eds.), *Specialization in physical education: The Alley legacy* (pp. 53-64). Iowa City: University of Iowa.

Looney, M.A. (1996). "Home" improvement: The task for measurement specialists. In T.M. Wood (Ed.), Exploring the kaleidoscope. *Proceedings of the 8th Measurement and Evaluation Symposium* (pp. 141-151). Corvallis, OR: Oregon State University College of Health and Human Performance.

Lord, F.M., & Novick, M.R. (1968). *Statistical theories of mental test scores.* Reading, MA: Addison-Wesley.

McDonald, R.P. (1999). *Test theory: A unified treatment.* Hillsdale, NJ: Erlbaum.

Safrit, M.J. (1989). An overview of measurement. In M.J. Safrit & T.M. Wood (Eds.), *Measurement concepts in physical education and exercise science* (pp. 3-20). Champaign, IL: Human Kinetics.

Spray, J.A. (1989). New approaches to solving measurement problems. In M.J. Safrit & T.M. Wood (Eds.), *Measurement concepts in physical education and exercise science* (pp. 229-248). Champaign, IL: Human Kinetics.

Suen, H.K. (1990). *Principles of test theories.* Hillsdale, NJ: Erlbaum.

Wood, T.M. (1989). Measurement and change: The sound of one hand clapping. In M.J. Safrit (Ed.), Measurement theory and practice in exercise and sport science. *Proceedings of the 6th Measurement and Evaluation Symposium* (pp. 92-108). Madison, WI: University of Wisconsin.

Chapter 2

American Psychological Association, American Educational Research Association, & National Council on Measurements Used in Education. (1954). Technical recommendations for psychological tests and diagnostic techniques. *Psychological Bulletin, 51*(2, Part 2), 1-38.

American Psychological Association, American Educational Research Association, & National Council on Measurement in Education. (1966). *Standards for educational and psychological tests and manuals.* Washington, DC: American Psychological Association.

American Psychological Association, American Educational Research Association, & National Council on Measurement in Education. (1974). *Standards for educational and psychological tests.* Washington, DC: American Psychological Association.

American Psychological Association, American Educational Research Association, & National Council on Measurement in Education. (1985). *Standards for educational and psychological testing.* Washington, DC: American Psychological Association.

American Psychological Association, American Educational Research Association, & National Council on Measurement in Education. (1999). *Standards for educational and psychological testing.* Washington, DC: American Educational Research Association.

Angoff, W.H. (1988). Validity: An evolving concept. In H. Wainer & H.I. Braun (Eds.), *Test validity* (pp. 19-32). Hillsdale, NJ: Erlbaum.

Benson, J. (1995, July). *Construct validation: Where are we after 40 years and what does the future hold?* Presidential address delivered to the 16th annual meeting of the Stress and Anxiety Research Association, Prague, Czech Republic.

Benson, J. (1998). Developing a strong program of construct validation: A test anxiety example. *Educational Measurement: Issues and Practice, 17*(1), 10-17, 22.

Baumgartner, T.A., Jackson, A.S., Mahar, M.T., & Rowe, D.A. (2003). *Measurement for evaluation in physical education and exercise science* (7th ed.). Boston: McGraw-Hill.

Blair, S.N., Kohl, H.W., III, Paffenbarger, R.S., Jr., Clark, D.G., Cooper, K.H., & Gibbons, L.W. (1989). Physical fitness and all-cause mortality: A prospective study of healthy men and women. *Journal of the American Medical Association, 262,* 2395-2401.

Bollen, K.A. (1989). *Structural equations with latent variables.* New York: Wiley.

Bollen, K.A., & Long, J.S. (1993). *Testing structural equation models.* Thousand Oaks, CA: Sage.

Campbell, D.T., & Fiske, D.W. (1959). Convergent and discriminant validation by the multitrait-multimethod matrix. *Psychological Bulletin, 56,* 81-105.

Cronbach, L.J. (1971). Test validation. In R. L. Thorndike (Ed.), *Educational measurement* (2nd ed., pp. 443-507). Washington, DC: American Council on Education.

Cronbach, L.J. (1988). Five perspectives on validity argument. In H. Wainer & H. I. Braun (Eds.), *Test validity* (pp. 3-17). Hillsdale, NJ: Erlbaum.

Cronbach, L.J. (1989). Construct validation after thirty years. In R. Linn (Ed.), *Intelligence: Measurement, theory, and public policy (Proceedings of a symposium in honor of Lloyd Humphreys)* (pp. 147-171). Urbana, IL: University of Illinois.

Cronbach, L.J., & Meehl, P.E. (1955). Construct validity in psychological tests. *Psychological Bulletin, 52,* 281-302.

Cureton, K.J., Sloniger, M.A., O'Bannon, J.P., Black, D.M., & McCormack, W.P. (1995). A generalized equation for prediction of $\dot{V}O_2$ peak from 1-mile run/walk performance. *Medicine and Science in Sports and Exercise, 27,* 445-451.

Guion, R.M. (1977). Content validity—The source of my discontent. *Applied Psychological Measurement, 1,* 1-10.

Hoyle, R.H. (1995). *Structural equation modeling: Concepts, issues, and applications.* Thousand Oaks, CA: Sage.

Jackson, A.S., & Coleman, A.E. (1976). Validation of distance run tests for elementary school children. *Research Quarterly, 47,* 86-94.

Kelly, T.L. (1927). *Interpretation of educational measurements.* Yonkers-on-Hudson, NY: World Book.

Leger, L.A., Mercier, D., Gadoury, C., & Lambert, J. (1988). The multistage 20 metre shuttle run test for aerobic fitness. *Journal of Sports Science, 6,* 93-101.

Loevinger, J. (1957). Objective tests as instruments of psychological theory. *Psychological Reports, 3,* 635-694.

Messick, S. (1975). The standard problem: Meaning and values in measurement and evaluation. *American Psychologist, 30,* 955-966.

Messick, S. (1980). Test validity and ethics of assessment. *American Psychologist, 35,* 1012-1027.

Messick, S. (1981a). Constructs and their vicissitudes in educational measurement. *Psychological Bulletin, 89,* 575-588.

Messick, S. (1981b). Evidence and ethics in the evaluation of tests. *Educational Researcher, 10*(9), 9-20.

Messick, S. (1988). The once and future issues of validity: Assessing the meaning and consequences of measurement. In H. Wainer & H. I. Braun (Eds.), *Test validity* (pp. 33-45). Hillsdale, NJ: Erlbaum.

Messick, S. (1989). Validity. In R. Linn (Ed.), *Educational measurement* (3rd ed., pp. 13-103). New York: ACE/Macmillan.

Messick, S. (1994). Foundations of validity: Meaning and consequences in psychological assessment. *European Journal of Psychological Assessment, 10,* 1-9.

Messick, S. (1995a). Validity of psychological assessment: Validation of inferences from persons' responses and performances as scientific inquiry into score meaning. *American Psychologist, 50,* 741-749.

Messick, S. (1995b). Standards of validity and validity of standards in performance assessment. *Educational Measurement: Issues and Practice, 14*(4), 5-8.

Messick, S. (2000). Consequence of test interpretation and use: The fusion of validity and values in psychological assessment. In R.D. Goffin & E. Helmes (Eds.), *Problems and solutions in human assessment: Honoring Douglas N. Jackson at seventy* (pp. 3-20). Boston: Kluwer Academic.

Morrow, J.R., Jr., Jackson, A.W., Disch, J.G., & Mood, D.P. (2000). *Measurement and evaluation in human performance* (2nd ed.). Champaign, IL: Human Kinetics.

Myers, J., Prakash, M., Froelicher, V., Do, D., Partington, S., & Atwood, J.E. (2002). Exercise capacity and mortality among men referred for exercise testing. *New England Journal of Medicine, 346,* 793-801.

Popper, K.R. (1959). *The logic of scientific discovery.* New York: Basic Books.

Rejeski, W.J., & Focht, B.C. (2002). Aging and physical disability: On integrating group and individual counseling with the promotion of physical activity. *Exercise and Sport Science Reviews, 30,* 166-170.

Rudisill, M.E., Mahar, M.T., & Meaney, K.S. (1993). The relationship between children's perceived and actual motor competence. *Perceptual and Motor Skills, 76,* 895-906.

Safrit, M.J. (1981). *Evaluation in physical education* (2nd ed.). Englewood Cliffs, NJ: Prentice Hall.

Safrit, M.J., Cohen, A.S., & Costa, M.G. (1989). Item response theory and the measurement of motor behavior. *Research Quarterly for Exercise and Sport, 60,* 325-335.

Safrit, M.J., & Wood, T.M. (Eds.). (1989). *Measurement concepts in physical education and exercise science.* Champaign, IL: Human Kinetics.

Safrit, M.J., & Wood, T.M. (1995). *Measurement in physical education and exercise science* (3rd ed.). Boston: McGraw-Hill.

Safrit, M.J., Zhu, W., Costa, M.G., & Zhang, L. (1992). The difficulty of sit-ups tests: An empirical investigation. *Research Quarterly for Exercise and Sport, 63,* 277-283.

Sallis, J.F., & Saelens, B.E. (2000). Assessment of physical activity by self-report: Status, limitations, and future directions. *Research Quarterly for Exercise and Sport, 71,* S1-S14.

Schumacker, R.E., & Lomax, R.G. (1996). *A beginner's guide to structural equation modeling.* Hillsdale, NJ: Erlbaum.

Shepard, L.A. (1993). Evaluating test validity. *Review of Research in Education, 19,* 405-450.

Slaughter, M.H., Lohman, T.G., Boileau, R.A., Horswill, C.A., Stillman, R.J., Van Loan, M.D., & Bemben, D.A. (1988). Skinfold equations for the estimation of body fatness in children and youth. *Human Biology, 60,* 709-723.

Spray, J. (1989). New approaches to solving measurement problems. In M.J. Safrit & T.M. Wood (Eds.), *Measurement concepts in physical education and exercise science* (pp. 229-248). Champaign, IL: Human Kinetics.

Whittington, D. (1998). How well do researchers report their measures? An evaluation of measurement in published educational research. *Educational and Psychological Measurement, 58,* 21-37.

Wood, T.M. (1989). The changing nature of norm-referenced validity. In M.J. Safrit & T.M. Wood (Eds.), *Measurement concepts in physical education and exercise science* (pp. 23-44). Champaign, IL: Human Kinetics.

Wood, T.M. (1990). Measurement and change: The sound of one hand clapping. In M.J. Safrit (Ed.), *Proceedings of the Sixth Measurement and Evaluation Symposium: Measurement theory and practice in exercise and sport* (pp. 92-108). Madison, WI: University of Wisconsin.

Zhu, W. (1998). Test equating: What, why, how? *Research Quarterly for Exercise and Sport, 69,* 11-23.

Chapter 3

Alsawalmeh, Y.M., & Feldt, L.S. (1992). Tests of the hypothesis that the intraclass reliability coefficient is the same for two measurement procedures. *Applied Psychological Measurement, 16*(2), 195-205.

Atkinson, G. (2003). What is this thing called measurement error? In T. Reilly and M. Marfell-Jones (Eds.), Kinanthropometry VIII. *Proceedings of the 8th International Conference of the International Society for the Advancement of Kinanthropometry (SAK)* (pp. 3-14). London: Taylor and Francis.

Baumgartner, T.A. (2000). Estimating the stability reliability of a score. *Measurement in Physical Education and Exercise Science, 4,* 175-178.

Baumgartner, T.A., & Chung, H. (2001). Confidence limits for intraclass reliability coefficients. *Measurement in Physical Education and Exercise Science, 5,* 179-188.

Baumgartner, T.A., & Jackson, A.S. (1982). *Measurement for evaluation in physical education* (2nd ed.). Dubuque, IA: Brown.

Baumgartner, T.A., Jackson, A.S., Mahar, M.T., & Rowe, D.A. (2003). *Measurement for evaluation in physical education and exercise science* (7th ed.). Madison, WI: McGraw-Hill.

Brennan, R.L. (2001). *Generalizability theory.* New York: Springer-Verlag.

Darracott, S.H. (1995). *Individual differences in variability and pattern of performance as a consideration in the selection of a representative score from multiple trial physical performance data.* Unpublished doctoral dissertation, University of Georgia, Athens.

Feldt, L.S. (1990). The sampling theory for the intraclass reliability coefficient. *Applied Measurement in Education, 3,* 361-367.

Ferguson, G.A., & Takane, Y. (1989). *Statistical analysis in psychology and education* (6th ed.). New York: McGraw-Hill.

Fleiss, J.L. (1981). *Statistical methods for rates and proportions* (2nd ed.). New York: Wiley.

Haggard, E.A. (1958). *Intraclass correlation and the analysis of variance.* New York: Dryden Press.

Hopkins, W.G. (2000). *A new view of statistics.* Internet Society for Sport Science: www.sportsci.org/resource/stats/.

Looney, M.A. (1989). Criterion-referenced measurement: Reliability. In M.J. Safrit & T.M. Wood (Eds.), *Measurement concepts in physical education and exercise science* (pp. 137-152). Champaign, IL: Human Kinetics.

Looney, M.A. (2000). When is the intraclass correlation coefficient misleading? *Measurement in Physical Education and Exercise Science, 4,* 73-78.

McDonald, R.P. (1999). *Test theory: A unified treatment.* Hillsdale, NJ: Erlbaum.

McGraw, K.O., & Wong, S.P. (1996). Forming inference about some intraclass correlation coefficients. *Psychological Methods, 1,* 30-46.

Morrow, J.R. (1989). Generalizability theory. In M.J. Safrit & T.M. Wood (Eds.), *Measurement concepts in physical education and exercise science* (pp. 73-96). Champaign, IL: Human Kinetics.

Morrow, J.R., Fridye, T., & Monaghen, S.D. (1986). Generalizability of the AAHPERD Health Related Physical Fitness test. *Research Quarterly for Exercise and Sport, 57,* 187-195.

Morrow, J.R., & Jackson, A.W. (1993). How "significant" is your reliability? *Research Quarterly for Exercise and Sport, 64,* 352-355.

Morrow, J.R., Jackson, A.W., Disch, J.G., & Mood, D.P. (2000). *Measurement and evaluation in human performance* (2nd ed.). Champaign, IL: Human Kinetics.

Nunnally, J.C., & Bernstein, I.H. (1994). *Psychometric theory* (3rd ed.). New York: McGraw-Hill.

Safrit, M.J. (1973). *Evaluation in physical education.* Englewood Cliffs, NJ: Prentice Hall.

Safrit, M.J., & Wood, T.M. (1995). *Introduction to measurement in physical education and exercise science* (3rd ed.). St. Louis: Times Mirror/Mosby.

Shavelson, R.J., & Webb, N.M. (1991). *Generalizability theory: A primer.* Thousand Oaks, CA: Sage.

Spray, J.A. (1982). Effects of autocorrelated errors on intraclass reliability estimation. *Research Quarterly for Exercise and Sport, 53,* 226-236.

Stegner, A.J., Tobar, D.A., & Kane, M.T. (1999). Generalizability of change scores on the body awareness scale. *Measurement in Physical Education and Exercise Science, 3,* 125-140.

Thompson, B. (Ed.). (2003). *Score reliability: Contemporary thinking on reliability issues.* Thousand Oaks, CA: Sage.

Tobar, D.A., Stegner, A.J., & Kane, M.T. (1999). The use of generalizability theory in examining the dependability of scores on the profile of mood states. *Measurement in Physical Education and Exercise Science, 3,* 141-156.

Traub, R.E. (1994). *Reliability for the social sciences: Theory and applications.* Thousand Oaks, CA: Sage.

Turner, A.A., & Bouffard, M. (1998). Comparison of modified to standard bioelectrical impedance errors using generalizability theory. *Measurement in Physical Education and Exercise Science, 2,* 177-196.

Vacha-Haase, T. (1998). Reliability generalization: Exploring variance in measurement error affecting score reliability across studies. *Educational and Psychological Measurement, 58,* 6-20.

Zhu, W. (1996). Should total score from a rating scale be used directly?. *Research Quarterly for Exercise and Sport, 67,* 363-372.

Zhu, W. (2001). ITRS: A program to compute interrater reliability. *Measurement in Physical Education and Exercise Science, 5,* 57-62.

Chapter 4

American Educational Research Association, American Psychological Association, & National Council on Measurement in Education. (1999). *Standards for educational and psychological testing.* Washington, DC: American Educational Research Association.

Andrich, D. (1978). A rating formulation for ordered response categories. *Psychometrika, 47*(1), 561-573.

Baker, F.B. (1985). *The basics of item response theory.* Portsmouth, NH: Heinemann.

Baker, F.B. & Kim, S.-H. (2004). *Item response theory: Parameter estimation techniques* (2nd ed.). New York: Marcel Dekker.

Berk, A.R. (Ed.). (1980). *Criterion-referenced measurement: The state of the art.* Baltimore: Johns Hopkins University Press.

Bezruczko, N. (2002). A multi-factor Rasch scale for artistic judgment. *Journal of Applied Measurement, 3,* 360-399.

Crocker, L., & Algina, J. (1986). *Introduction to classical and modern test theory.* New York: Harcourt Brace Jovanovich College.

Embretson, S.E., & Reise, S.P. (2000). *Item response theory for psychologists.* Hillsdale, NJ: Erlbaum.

Gulliksen, H. (1950). *Theory of mental tests.* New York: Wiley.

Hambleton, R.K. (1989). Principles and selected applications of item response theory. In R.L. Linn (Ed.), *Educational measurement* (pp. 147-200). New York: Macmillan.

Hambleton, R.K., & Cook, L.L. (1983). The robustness of item response models and effects of length and sample size on the precision of ability estimation. In D. Weiss (Ed.), *New horizons in testing* (pp. 31-49). New York: Academic Press.

Hambleton, R.K., & Murray, L. (1983). Some goodness of fit investigations for item response models. In R. K. Hambleton (Ed.), *Application of item response theory* (pp.71-94). Vancouver, BC: Educational Research Institute of British Columbia.

Hambleton, R.K., & Swaminathan, H. (1985). *Item response theory: Principles and applications*. Boston: Kluwer-Nijhoff.

Hambleton, R.K., Swaminathan, H., & Rogers, J.H. (1991). *Fundamentals of item response theory*. Newbury Park, CA: Sage.

Hattie, J.A. (1984). An empirical study of various indices for determining unidimensionality. *Multivariate Behavioral Research, 19,* 49-78.

Linacre, J.M. (1989). *Many-faceted Rasch measurement.* Chicago: MESA Press.

Looney, M.A. (1997). Objective measurement of figure skating performance. *Journal of Outcome Measurement, 1,* 143-163.

Lord, F.M. (1952). *A theory of test scores* (Psychometric Monograph No. 7). Psychometric Society.

Lord, F.M. (1975). *Evaluation with artificial data of a procedure for estimating ability and item characteristic curve parameters* (ETS RB-75-33). Princeton, NJ: Educational Testing Service.

Lord, F.M. (1977). Practical applications of item characteristic curve theory. *Journal of Educational Measurement, 14,* 117-138.

Lord, F.M. (1980). *Applications of item response theory to practical testing problems.* Hillsdale, NJ: Erlbaum.

Lord, F.M., & Novick, M.R. (1968). *Statistical theories of mental test scores.* Reading, MA: Addison-Wesley.

Ludlow, L.H. (1986). Graphical analysis of item response theory residuals. *Applied Psychological Measurement, 10,* 217-229.

Mantel, N., & Haenszel, W. (1959). Statistical aspects of the analysis of data from retrospective studies of disease. *Journal of the National Cancer Institute, 22,* 719-748.

Markward, N.J. (2004). Establishing mathematical laws of genomic variation. *Journal of Applied Measurement, 5,* 1-14.

McDonald, R.P. (1999). *Test theory: A unified treatment.* Hillsdale, NJ: Erlbaum.

McDonald, R.P. (2000). A basis for multidimensional item response theory. *Applied Psychological Measurement, 24,* 99-114.

Murray, L.N., & Hambleton, R.K. (1983). *Using residual analyses to assess item response model-test data fit* (Laboratory of Psychometric and Evaluative Research Report No. 140). Amherst, MA: School of Education, University of Massachusetts.

Nichols, P.D., Chipman, S.F., & Brennan, R.L. (Eds.) (1995). *Cognitively diagnostic assessment.* Hillsdale, NJ: Erlbaum.

Pastor, D.A. (2003). The use of multilevel item response theory modeling in applied research: An illustration. *Applied Measurement in Education, 16,* 223-243.

Rasch, G. (1960). *Probabilistic models for some intelligence and attainment tests.* Copenhagen: Danish Institute for Educational Research. [Reprinted, Chicago, IL: University of Chicago Press, 1980.]

Roscoe, J.T. (1975). *Fundamental research statistics for the behavioral sciences* (2nd ed.). Chicago: Holt, Rinehart & Winston.

Rosenbaum, P.R. (1984). Testing the conditional independence and monotonicity assumptions of item response theory. *Psychometrika, 49*(3), 425-435.

Safrit, M.J., Cohen, A.S., & Costa, M.G. (1989). Item response theory and the measurement of motor behavior. *Research Quarterly for Exercise and Sport, 60,* 325-335.

Safrit, M.J., Zhu, W., Costa, M.G., & Zhang, L. (1992). The difficulty of sit-up tests: An empirical investigation. *Research Quarterly for Exercise and Sport, 63,* 277-283.

Sijtsma, L., & Molenaar, I.W. (2002). *Introduction to nonparametric item response theory.* Thousand Oaks, CA: Sage.

Spray, J.A. (1987). Recent developments in measurement and possible applications to the measurement of psychomotor behavior. *Research Quarterly for Exercise and Sport, 58,* 203-209.

Spray, J.A. (1990). One-parameter item response theory models for psychomotor tests involving repeated, independent attempts. *Research Quarterly for Exercise and Sport, 61,* 162-168.

Suen, H.K. (1990). *Principles of test theories.* Hillsdale, NJ: Erlbaum.

Swaminathan, H., & Gifford, J.A. (1979). *Estimation of parameters in the three-parameter latent-trait model* (Laboratory of Psychometric and Evaluation Research Report No. 90). Amherst, MA: University of Massachusetts, School of Education.

Tenenbaum, G., & Fogarty, G. (1998). Application of the Rasch analysis to sport and exercise psychology measurement. In J.L. Duda (Ed.), *Advances in sport and exercise psychology measurement* (pp. 409-421). Morgantown, WV: Fitness Information Technology.

Umar, J. (1997). Item banking. In J. P. Keeves (Ed.), *Educational research, methodology, and measurement: An international handbook* (2nd ed., pp. 923-930). New York: Elsevier Science.

van der Linden, W.J., & Hambleton, R.K. (Eds.). (1997). *Handbook of modern item response theory.* New York: Springer.

Wainer, H. (1990). *Computerized adaptive testing: A primer.* Hillsdale, NJ: Erlbaum.

Wainer, H., & Wright, B.D. (1980). Robust estimation of ability in the Rasch model. *Psychometrika, 45*(3), 373-391.

Wingersky, M.S., Barton, M.A., & Lord, F.M. (1982). *LOGIST user's guide.* Princeton, NJ: Educational Testing Service.

Wood, T.M. (1987). Putting item response theory into perspective. *Research Quarterly for Exercise and Sport, 58,* 216-220.

Wright, B.D. (1977). Solving measurement problems with the Rasch model. *Journal of Educational Measurement, 14,* 97-116.

Wright, B.D., & Masters, G.N. (1982). *Rating scale analysis: Rasch measurement.* Chicago: MESA Press.

Wright, B.D., & Stone, M.H. (1979). *Best test design.* Chicago: MESA Press.

Wu, M.L., Adams, R.J., & Wilson, M.R. (1998). *ACER ConQuest: Generalised item response modeling software manual.* Camberwell, Melbourne, Victoria: Australian Council for Educational Research.

Yen, W.M. (1981). Using simulation results to choose a latent trait model. *Applied Psychological Measurement, 5,* 245-262.

Zhu, W. (1996). Should total scores from a rating scale be used directly? *Research Quarterly for Exercise and Sport, 67,* 363-372.

Zhu, W. (2002). A confirmatory study of Rasch-based optimal categorization of a rating scale. *Journal of Applied Measurement, 3*(1), 1-15.

Zhu, W., & Cole, E.L. (1996). Many-faceted Rasch calibration of a gross-motor instrument. *Research Quarterly for Exercise and Sport, 67*(1), 24-34.

Zhu, W., & Kurz, K.A. (1994). Rasch Partial Credit analysis of gross motor competence. *Perceptual and Motor Skill, 79,* 947-961.

Zhu, W., & Safrit, M.J. (1993). The calibration of a sit-ups task using the Rasch Poisson Counts model. *Canadian Journal of Applied Physiology, 18,* 207-219.

Zhu, W., Safrit, M.J., & Cohen, A.S. (1999). *FitSmart test user manual: High school edition.* Champaign, IL: Human Kinetics.

Zhu, W., Timm, G., & Ainsworth, B.A. (2001). Rasch calibration and optimal categorization of an instrument measuring women's exercise perseverance and barriers. *Research Quarterly for Exercise and Sport, 72*(2), 104-116.

Zhu, W., Updyke, W., & Lewandowski, C. (1997). Post-hoc Rasch analysis of optimal categorization of an ordered-response scale. *Journal of Outcome Measurement, 1*(4), 286-304.

Chapter 5

American Psychologocial Association & National Council on Measurement in Education. (1985). *Standards for educational and psychological testing.* Washington, DC: American Psychological Association.

American Educational Research Association & National Council on Measurement in Education. (1999). *Standards for educational and psychological testing.* Washington, DC: American Psychological Association.

American Psychological Association. (1953). *Ethical standards of psychologists.* Washington, DC: Author.

American Psychological Association. (1967). *Casebook on ethical standards of psychologists.* Washington, DC: Author.

American Psychological Association.(1977). *Ethical standards of psychologists.* Washington, DC: Author.

American Psychological Association.(1981). *Ethical principles of psychologists.* Washington, DC: Author.

American Psychological Association, American Educational Research Association, & National Council on Measurement Used in Education. (1954). *Technical recommendations for psychological tests and diagnostic techniques.* Washington, DC: American Psychological Association.

American Psychological Association, American Educational Research Association, & National Council on Measurement Used in Education. (1966). *Standards for educational and psychological tests.* Washington, DC: American Psychological Association.

American Psychological Association, American Educational Research Association, & National Council on Measurement Used in Education. (1974). *Standards for educational and psychological tests.* Washington, DC: American Psychological Association.

Bird, S.J., & Housman, D.E. (1996). Conducting and reporting research, *Professional Ethics, 4,* 127-154.

Campbell, D.T., & Fiske, D.W. (1959). Convergent and discriminant validation by the multitrait-multimethod matrix. *Psychological Bulletin, 56,* 81-105.

Carlson, S.B., Bridgeman, B., Camp, R., & Waanders, J. (1983). *Relationship of admission test scores to writing performance of native and nonnative speakers of English* (RR-85-21). Princeton, NJ: Educational Testing Service.

Carron, A.V., Widmeyer, W.N., & Brawley, L.R. (1985). The development of an instrument to assess cohesion in sport teams: The Group Environment Questionnaire. *Journal of Sport Psychology, 7,* 244-266.

Casperson, C.J., Powell, K.E., & Christenson, G.M. (1985). Physical activity, exercise, and physical fitness: Definitions and distinctions for health-related research. *Public Health Report, 100,* 126-131.

Cates, J.A. (1999). The art of assessment in psychology: Ethics, expertise, and validity. *Journal of Clinical Psychology, 55,* 631-641.

Chelladurai, P., & Saleh, S.D. (1980). Dimensions of leader behavior in sports: Development of a leadership scale. *Journal of Sport Psychology, 2,* 34-45.

Churchman, C.W. (1971). *The design of inquiring systems: Basic concepts of systems and organization.* New York: Basic Books.

Cole, N.S., & Moss, P.A. (1989). Bias in test use. In R.L. Linn (Ed.), *Educational Measurement* (pp.201-220). New York: American Council on Education/Macmillan Series on Higher Education.

Cremin, L.A. (1989). *Popular education and its discontents.* New York: Harper & Row.

Cronbach, L.J. (1980). Validity on parole: How can we go straight? In W.B. Schrader (Ed.), New directions for testing and measurement: Measuring achievement over a decade. *Proceedings of the 1979 ETS Invitational Conference* (pp. 99-108). San Francisco: Jossey-Bass.

Dewey, J. (1939). Theory of valuation. In O. Neuroth (Ed.), *International encyclopedia of unified science.* Chicago: University of Chicago Press.

Etzel, E., Yura, M.T., & Perna, F. (1998). Ethics in assessment and testing in sport and exercise psychology. In J.L. Duda (Ed.), *Advances in sport and exercise psychology measurement* (pp. 423-432). Morgantown, WV: Fitness Information Technology.

George, S.L. (1997). Perspectives on scientific misconduct and fraud in clinical trials. *Chance, 10,* 3-5.

Goodwin, L. (1999). The role of factor analysis in the estimation of construct validity. *Measurement in Physical Education and Exercise Science, 3*(2), 85-100.

Haney, W., & Madaus, G. (1991). The evolution of ethical and technical standards for testing. In R.K. Hambleton & J.N. Zaal (Eds.), *Advances in educational and psychological testing* (pp. 395-425). Boston: Kluwer Academic.

Kaplan, A. (1964). *The conduct of inquiry.* San Francisco: Chandler.

Kroll, W. (1993). Ethical issues in human research. *Quest, 45,* 32-44.

Linn, R.L. (2002). Constructs and values in standards-based assessment. In H. L.Braun, D.N. Jackson, & D.E. Wiley (Eds.), *The role of constructs in psychological and educational measurement* (pp. 231-254). Hillsdale, NJ: Erlbaum.

Linn, R.L., Baker, E.L., & Dunbar, S.B. (1991). Complex, performance-based assessment: Expectations and validation criteria. *Educational Researcher, 20,* 5-21.

Looney, M.A. (2000). When is the intraclass correlation coefficient misleading? *Measurement in Physical Education and Exercise Science, 4,* 73-78.

Lynn, J., & Virnig, B.A. (1995) Assessing the significance of treatment effects: Comments from the perspective of ethics. *Medical Care, 33,* AS292-AS298.

Messick, S. (1975). The standard problem: Meaning and values in measurement and evaluation. *American Psychologist, 30,* 955-966.

Messick, S. (1980). Test validity and the ethics of assessment. *American Psychologist, 35,* 1012-1027.

Messick, S. (1981). Evidence and ethics in the evaluation of tests. *Educational Researcher, 10,* 9-20.

Messick, S. (1989a). Meaning and values in test validation: The science and ethics of assessment. *Educational Researcher, 18,* 5-11.

Messick, S. (1989b). Validity. In R.L. Lynn (Ed.), *Educational measurement* (pp. 13-104). New York: American Council on Education/Macmillan Series on Higher Education.

Messick, S. (1994). The interplay of evidence and consequences in the validation of performance assessments. *Educational Researcher, 23,* 13-23.

Moss, P.A. (1992). Shifting conceptions of validity in education measurement: Implications for performance assessment. *Review of Educational Research, 62,* 229-258.

National Association for Sport and Physical Education. (2004). *Moving into the future: National standards for physical education* (2nd ed.). New York: McGraw-Hill.

Patterson, P. (2000). Reliability, validity and methodological response to assessment of physical activity via self-report (Response to Sallis and Saelens). *Research Quarterly for Exercise and Sport, 71,* 15-20.

Pedhazur, E.J., & Schmelkin, L.P. (1991). *Measurement, design, and analysis: An integrated approach.* Hillsdale, NJ: Erlbaum.

Resnick, L.B. (1990). Tests as standards of achievement in school. In J. Pfleiderer (Ed.), *The uses of standardized tests in American education. Proceedings of the 1989 ETS invitational conference* (pp. 63-80). Princeton, NJ: Educational Testing Service.

Rikli, R.E., & Jones, C.J. (1999a). Development and validation of a functional fitness test for community-residing older adults. *Journal of Aging and Physical Activity, 7,* 129-161.

Rikli, R.E., & Jones, C.J. (1999b). Functional fitness normative scores for community-residing older adults, ages 60-94. *Journal of Aging and Physical Activity, 7,* 162-181.

Roberts, G.C. (1993). Ethics in professional advising and academic counseling of graduate students. *Quest, 45,* 78-87.

Safrit, M.J. (1993). Oh what a tangled web we weave. *Quest, 45,* 52-61.

Shepard, L.A. (1993). Evaluating test validity. In L. Darling-Hammond (Ed.), *Review of research in education* (pp. 405-450). Washington, DC: American Educational Research Association.

Thomas, J.R., & Nelson, J.K. (2001). *Research methods in physical activity* (4th ed.). Champaign, IL: Human Kinetics

Wiggins, G. (1993). Assessment: Authenticity, context, and validity. *Phi Delta Kappan, 75,* 200. (Reprinted by permission from Wiggins, G., 1993, *Assessing student performance: Exploring the purpose and limits of testing.* San Francisco: Jossey-Bass.)

Zelaznik, H.N. (1993). Ethical issues in conducting and reporting research: A reaction to Kroll, Matt, and Safrit. *Quest, 45,* 62-68.

Chapter 6

American Educational Research Association, American Psychological Association, & National Council on Measurement in Education. (1999). *Standards for educational and psychological testing.* Washington, DC: American Educational Research Association.

Angoff, W.H. (1971). Scales, norms, and equivalent scores. In R.L. Thorndike (Ed.), *Educational measurement* (2nd ed., pp. 508-600). Washington, DC: American Council on Education.

Baker, F.B. (1993). EQUATE 2.0: A computer program for the characteristic curve method of IRT equating. *Applied Psychological Measurement, 17,* 20.

Burton, A.W., & Miller, D.E. (1998). *Movement skill assessment.* Champaign, IL: Human Kinetics.

Cizek, G.J. (Ed., 2001). *Setting performance standards: Concepts, methods, and perspectives.* Hillsdale, NJ: Erlbaum.

Cook, L.L., & Eignor, D.R. (1983). Practical considerations regarding the use of item response theory to equate tests. In R.K. Hambleton (Ed.), *Applications of item response theory* (pp. 175-195). Vancouver, BC: Educational Research Institute of British Columbia.

Cook, L.L., & Eignor, D.R. (1991). An NCME instructional module on IRT equating methods. *Educational Measurement: Issues and Practice, 10,* 37-45.

Crocker, L., & Algina, J. (1986). *Introduction to classical and modern test theory.* New York: Holt, Rinehart & Winston.

Davier, A.A. won, Holland, P.W., & Thayer, D.T. (2004). *The Kernel method of test equating.* New York: Springer.

Docherty, D. (Ed.). (1996). *Measurement in pediatric exercise science.* Champaign, IL: Human Kinetics.

Dorans, N.J. (2004). Equating, concordance, and expectation. *Applied Psychological Measurement, 28*(4), 227-246.

Feuer, M.J., Holland, P.W., Green, B.F., Bertenthal, M.W., & Hemphill, F.C. (Eds.) (1999). *Uncommon measures: Equivalence and linkage among educational tests.* Washington, DC: National Academy Press.

Hambleton, R.K., & Jones, R.W. (1993). An NCME instructional module on comparison of classical test theory and item response theory and their applications to test development. *Educational Measurement: Issues and Practice, 12,* 38-47.

Hambleton, R.K., & Swaminathan, H. (1985). *Item response theory: Principles and applications.* Boston: Kluwer-Nijhoff.

Hambleton, R.K., Swaminathan, H., & Rogers, H.J. (1991). *Fundamentals of item response theory.* Thousand Oaks, CA: Sage.

Harris, D.J., & Crouse, J.D. (1993). A study of criteria used in equating. *Applied Measurement in Education, 6,* 195-240.

Heyward, V.H. (1991). *Advanced fitness assessment and exercise prescription* (2nd ed.). Champaign, IL: Human Kinetics.

Kang, S.J., & Zhu, W. (1998). An empirical examination of the equivalence of three upper body strength tests [Abstract]. *Research Quarterly for Exercise and Sport, 69*(Suppl. 1), A-57.

Kolen, M.J. (1981). Comparison of traditional and item response theory methods for equating tests. *Journal of Educational Measurement, 18,* 1-11.

Kolen, M.J. (1984). Effectiveness of analytic smoothing in equipercentile equating. *Journal of Educational Statistics, 9,* 25-44.

Kolen, M.J. (1988). An NCME instructional module on traditional equating methodology. *Educational Measurement: Issues and Practice, 7,* 29-36.

Kolen, M.J. (2004). Linking assessments: Concept and history. *Applied Psychological Measurement, 28*(4), 219-226.

Kolen, M.J., & Brennan, R.L. (2004). *Test equating, scaling, and linking: Methods and practices* (2nd ed.). New York: Springer.

Kriska, A.M., & Caspersen, C.J. (1997). Introduction to a collection of physical activity questionnaires. *Medicine and Science in Sports and Exercise, 29*(Suppl.), 5-9.

Lewis, D.M., Mitzel, H.C., & Green, D.R. (1996, June). Standard setting: A bookmark approach. In D.R. Green (Chair), *IRT-based standard-setting procedures utilizing behavioral anchoring.* Symposium conducted at the Council of Chief State School Officers National Conference on Large-Scale Assessment, Phoenix, AZ.

Livingston, S.A. (1993). Small-sample equating with log-linear smoothing. *Journal of Educational Measurement, 30,* 23-29.

Looney, M.A. (1989). Criterion-referenced measurement: Reliability. In M.J. Safrit & T.M. Wood (Eds.). *Measurement concepts in physical education and exercise science.* (pp. 137-152). Champaign, IL: Human Kinetics.

Lord, F.M. (1980). *Applications of item response theory to practical testing problems.* Hillsdale, NJ: Erlbaum.

Lyman, H.B. (1971). *Test scores and what they mean* (2nd ed.). Englewood Cliffs, NJ: Prentice Hall.

Mislevy, R.J., Sheehan, K.M., & Wingersky, M. (1993). How to equate tests with little or no data. *Journal of Educational Measurement, 30,* 55-78.

Montoye, H. J., Kemper, H.C.G., Saris, W.H.M., & Washburn, R.A. (1996). *Measuring physical activity and energy expenditure.* Champaign, IL: Human Kinetics.

Petersen, N.S., Kolen, M.J., & Hoover, H.D. (1989). Scaling, norming, and equating. In R.L. Linn (Ed.), *Educational measurement* (3rd ed., pp. 221-262). New York: Macmillan.

Ross, J.G., & Gilbert, G.G. (1985). The national children and youth fitness study: A summary of findings. *Journal of Physical Education, Recreation and Dance, 56,* 45-50.

Safrit, M.J. (1989). Criterion-referenced measurement: Validity. In M.J. Safrit & T.M. Wood (Eds.), *Measurement concepts in physical education and exercise science* (pp. 119-135). Champaign, IL: Human Kinetics.

Safrit, M.J., & Wood, T.M. (1995). *Introduction to measurement in physical education and exercise science* (3rd ed.). St. Louis: Mosby.

Safrit, M.J., Zhu, W., Costa, M.G., & Zhang, L. (1992). The difficulty of sit-up tests: An empirical investigation. *Research Quarterly for Exercise and Sport, 63,* 277-283.

Shin, S., & Zhu, W. (1998). A comparison of international children's running performance [Abstract]. *Research Quarterly for Exercise and Sport, 69*(Suppl. 1), A-57.

Stocking, M.L., & Lord, F.M. (1983). Developing a common metric in item response theory. *Applied Psychological Measurement, 7,* 201-210.

Ulrich, D.A. (2000). *Test of gross motor development* (2nd ed.). Austin, TX: Pro-Ed.

Wright, B.D. (1968). Sample-free test calibration and person measurement. *Proceedings of the 1967 Invitational Conference on Testing Problems* (pp. 85-101). Princeton, NJ: Educational Testing Service.

Yen, W.M. (1983). Tau-equivalence and equipercentile equating. *Psychometrika, 48,* 353-369.

Yen, W.M. (2002). Scaling and equating. Paper presented at the New York State Technical Conference. [Online]. Available: www.emsc.nysed.gov/osa/scalequate/se.ppt.

Zeng, L., Kolen, M.J., & Hanson, B.A. (1995). Random groups equating program (RAGE, Version 2.0) [Computer software]. Iowa City: American College Testing.

Zhu, W. (1998a). Test equating of commonly used physical fitness tests [Abstract]. *Research Quarterly for Exercise and Sport, 69* (Suppl. 1), A-56.

Zhu, W. (1998b). Test equating: What, why, how. *Research Quarterly for Exercise and Sport, 69,* 11-23.

Zhu, W. (2001). An empirical investigation of Rasch equating of motor function tasks. *Adapted Physical Activity Quarterly, 18,* 72-89.

Zhu, W. (2002). Equating and linking of physical activity questionnaires. In G. J. Welk (Ed.), *Physical activity assessments for health-related research* (pp. 81-92). Champaign, IL: Human Kinetics.

Zhu, W., & Kang, S.J. (1998). Equivalence of three commonly used sit-up tests [Abstract]. *Research Quarterly for Exercise and Sport, 69*(Suppl. 1), A56-A57.

Zhu, W., Safrit, M.J., & Cohen, A.S. (1999). *FitSmart test user manual: High school edition.* Champaign, IL: Human Kinetics.

Chapter 7

Ackerman, T. (1992). A didactic explanation of item bias, item impact, and item validity from a multidimensional perspective. *Journal of Educational Measurement, 29,* 67-91.

Bejar, I.I., & Wingersky, M.S. (1981). *An application of item response theory to equating the Test of Standard Written English* (College Board Report No. 81-8). Princeton, NJ: Educational Testing Service.

Bolt, D.M. (2000). A SIBTEST approach to testing DIF hypotheses using experimentally designed test items. *Journal of Educational Measurement, 37,* 307-327.

Bolt, D.M., Cohen, A.S., & Wollack, J.A. (2001). A mixture model for multiple choice data. *Journal of Educational and Behavioral Statistics, 26*(4), 381-409.

Bolt, D.M., Cohen, A.S., & Wollack, J.A. (2002). Item parameter estimation under conditions of test speededness: Application of a mixture Rasch model with ordinal constraints. *Journal of Educational Measurement, 39,* 331-348.

Candell, G.L., & Drasgow, F. (1988). An iterative procedure for linking metrics and assessing item bias in item response theory. *Applied Psychological Measurement, 12,* 253-260.

Chang, H.H., Mazzeo, J., & Roussos, L. (1996). Detecting DIF for polytomously scored items: An adaptation of the SIBTEST procedure. *Journal of Educational Measurement, 33,* 333-353.

Cohen, A.S., & Bolt, D.M. (2005). A mixture model analysis of differential item functioning. *Journal of Educational Measurement, 42*(2), *133-148.*

Cohen, A.S., & Kim, S.-H. (1993). A comparison of Lord's chi-square and Raju's area measures in detection of DIF. *Applied Psychological Measurement, 17*(1), 39-52.

Cohen, A.S., Kim, S.-H., & Baker, F.B. (1993). Detection of differential item functioning in the graded response model. *Applied Psychological Measurement, 17*(4), 335-350.

Cook, L.L., Eignor, D.R., & Hutton, L.R. (April, 1979). *Considerations in the application of latent trait theory to objectives-based criterion-referenced tests.* Paper presented at the annual meeting of the American Educational Research Association, San Francisco, CA.

Divgi, D.R. (1980, April). *Evaluation of scales for multilevel test batteries.* Paper presented at the annual meeting of the American Educational Research Association, Boston.

Educational Testing Service. (1987). *ETS sensitivity review process: An overview.* Princeton, NJ: Author.

French, A.W., & Miller, T.R. (1996). Logistic regression and its use in detecting differential item functioning in polytomous items. *Journal of Educational Measurement, 33,* 315-332.

Haebara, T. (1980). Equating logistic ability scales by a weighted least squares method. *Japanese Psychological Research, 22,* 144-149.

Holland, P.W., & Thayer, D.T. (1988). Differential item performance and the Mantel-Haenszel procedure. In H. Wainer & H.I. Braun (Eds.), *Test validity* (pp. 129–145). Hillsdale, NJ: Erlbaum.

Ibarra, R.A., & Cohen, A.S. (1999, February). *Multicontextuality: A hidden dimension in testing and assessment.* Paper presented at the ETS Invitational Conference on Fairness, Access, Multiculturalism, and Equity (FAME), Princeton, NJ.

Kang, T., & Cohen, A.S. (2003, April). *A mixture model analysis of ethnic group DIF.* Paper presented at the annual meeting of the National Council on Measurement in Education, Chicago, IL.

Kim, S.-H., & Cohen, A.S. (1992). Effects of linking methods on detection of DIF. *Journal of Educational Measurement, 29,* 51-66.

Kim, S.-H., Cohen, A.S., & Kim, H.-O. (1994). An investigation of Lord's procedure for the detection of differential item functioning. *Applied Psychological Measurement, 18*(3), 217-228.

Kolen, M.J., & Brennan, R.L. (2004). *Test equating, scaling, and linking: Methods and practices.* New York: Springer.

Li, Y. (2001). *Detecting differences in item response as a function of item characteristics.* Unpublished amsters thesis, Department of Educational Psychology, University of Wisconsin, Madison.

Li, Y., Cohen, A.S., & Ibarra, R.A. (2004). Characteristics of mathematics items associated with gender DIF. *International Journal of Testing, 4*(2), 115-136.

Linn, R.L. (1993). The use of differential item functioning statistics: A discussion of current practice and future implications. In P.W. Holland & H. Wainer (Eds.), *Differential item functioning* (pp. 349-366). Hillsdale, NJ: Erlbaum.

Linn, R.L., Levine, M.V., Hastings, C.N., & Wardrop, J.L. (1981). Item bias in a test of reading comprehension. *Applied Psychological Measurement, 5,* 159-173.

Looney, M.A., Spray, J.A., & Castelli, D. (1996). The task difficulty of free-throw shooting for males and females. *Research Quarterly for Exercise and Sport, 67,* 265-271.

Lord, F.M. (1980). *Applications of Item Response Theory to practical testing problems.* Hillsdale, NJ: Erlbaum.

Loyd, B.H., & Hoover, H.D. (1980). Vertical equating using the Rasch model. *Journal of Educational Measurement, 17,* 169-194.

McLaughlin, M.E., & Drasgow, F. (1988). Lord's chi-square test of item bias with estimated and with known person parameters. *Applied Psychological Measurement, 11,* 161-173.

Messick, S. (1989). Validity. In R.L. Linn (Ed.), *Educational measurement* (3rd ed., pp. 13-103). Washington, DC: American Council on Education.

Pine, S.M. (1977). Application of item characteristic curve theory to the problem of test bias. In D. J. Weiss (Ed.), *Application of computerized adaptive testing: Proceedings of a symposium presented at the 18th annual convention of the Military Testing Association.* (Research Report No. 77-1, pp. 37-43). Minneapolis: University of Minnesota, Department of Psychology, Psychometric Methods Program.

Portenza, M.T., & Dorans, N.J. (1995). DIF assessment for polytomously scored items: A framework for classification and evaluation. *Applied Psychological Measurement, 19,* 23-37.

Raju, N.S. (1988). The area between two item characteristic curves. *Psychometrika, 53,* 495-502.

Raju, N.S. (1990). Determining the significance of estimated signed and unsigned areas between two item response functions. *Applied Psychological Measurement, 14,* 197-207.

Rao, C.R. (1973). *Linear statistical inference and its applications.* New York: Wiley.

Roussos, L., & Stout, W.F. (1996). A multidimensionality-based DIF analysis paradigm. *Applied Psychological Measurement, 20,* 355-371.

Shealy, R., & Stout, W.F. (1993). An item response theory model for test bias. In P.W. Holland & H. Wainer (Eds.), *Differential item functioning.* Hillsdale, NJ: Erlbaum.

Shepard, L.A., Camilli, G., & Williams, D.M. (1984). Accounting for statistical artifacts in item bias research. *Journal of Educational Statistics, 9,* 93-128.

Stocking, M.L., & Lord, F.M. (1983). Developing a common metric in item response theory. *Applied Psychological Measurement, 7,* 201-210.

Stout, W., & Roussos, L. (1996). *SIBTEST* [Computer program]. Statistical Laboratory for Educational and Psychological Measurement, Department of Statistics, University of Illinois at Urbana-Champaign.

Thissen, D. (1991). *MULTILOG* [Computer program]. Chicago, IL: Scientific Software.

Thissen, D., Steinberg, L., & Gerrard, M. (1986). Beyond group mean differences: The concept of item bias. *Psychological Bulletin, 99,* 118–128.

Thissen, D., Steinberg, L., & Wainer, H. (1988). Use of item response theory in the study of group differences in trace lines. In H. Wainer & H.I. Braun (Eds.), *Test validity* (pp. 147-169). Hillsdale, NJ: Erlbaum.

Thissen, D., Steinberg, L., & Wainer, H. (1993). Detection of differential item functioning using the parameters of item response models. In P.W. Holland & H. Wainer (Eds.), *Differential item functioning* (pp. 67-113). Hillsdale, NJ: Erlbaum.

Zhu, W. (2000). Score equivalence is at the heart of international measures of physical activity. *Research Quarterly for Exercise and Sport, 71,* 121-128.

Ziecky, M. (1993). Practical questions in the use of DIF statistics in test development. In P.W. Holland & H. Wainer (Eds.), *Differential item functioning* (pp. 337–347). Hillsdale, NJ: Erlbaum.

Zwick, R., Thayer, D.T., & Mazzeo, J. (1997). Descriptive and inferential procedures for assessing differential item functioning in polytomous items. *Applied Measurement in Education, 10,* 321-334.

Chapter 8

American Academy of Orthopaedic Surgeons. (2004). *Normative data study and outcomes instruments.* www.AAOS.org. Washington, DC: Author.

American Psychological Association. (1997). *Computer adaptive testing: From inquiry to operation.* Washington, DC: Author.

Andrich, D. (1978). A rating formulation for ordered response categories. *Psychometrika, 31,* 84-98.

Angoff, W.F., & Huddleston, E.M. (1958). *The multi-level experiment: A study of a two-level test system for the College Board Scholastic Aptitude Test* (Statistical Report No. SR-58-21). Princeton, NJ: Educational Testing Service.

Bergstrom, B., & Cline, A. (2003, Summer). Clear exam review. *Beyond multiple choice: Innovating in professional testing, 14.*

Bergstrom, B., & Lunz, M.E. (1992). Confidence in pass/fail decisions for computer adaptive and paper and pencil examinations. *Evaluation and the Health Professions, 15,* 453-464.

Bergstrom, B., & Stahl, J. (2003). *Virtual test development.* www.promissor.com .

Bergstrom, B.A. & Lunz, M.E. (1999). CAT for certification and licensure. In F. Drasgow & J.B. Olson-Buchanan (Eds.), *Innovations in computerized assessment* (pp. 67-92). Hillsdale, NJ: Erlbaum.

Bergstrom, B.A., Lunz, M.E., & Gershon, R.C. (1994). An empirical study of computerized adaptive test administration conditions. *Journal of Educational Measurement, 31,* 251-263.

Binet, A. (1908). Le développement de l'intelligence chez les enfants [The development of intelligence in children]. *L'Année Psychologique, 14,* 1-94.

Bock, R.D. (1972). Estimating item parameters and latent ability when responses are scored in two or more nominal categories. *Psychometrika, 37,* 29-51.

Choppin, B. (1985). Principles of item banking. *Evaluation in Education, 9,* 87-90.

Daniel, M. (1983). *Correlations of aptitude tests with high school grades* (Technical Report 1983-4). Chicago: Johnson O'Connor Research Foundation.

Dijkers, M.P. (2003). A computer adaptive testing simulation applied to the FIM instrument motor component. *Archives of Physical Medicine and Rehabilitation, 84,* 384-393.

Gershon, R., Cella, D., Dineen, K., Rosenbloom, S., Peterman, A., & Lai, J.-S. (2003). Item response theory and health related quality of life in cancer. *Expert Review Pharmacoeconomics Outcomes Research, 3*(6), 783-791.

Gershon, R.C. (1995). *CAT Software System* [Computer software]. Chicago: Computer Adaptive Technologies.

Gershon, R.C. (1996). *The effect of individual differences variables on the assessment of ability for computerized adaptive testing.* Unpublished doctoral dissertation, Evanston, IL: Northwestern University.

Gershon, R.C., & Bergstrom, B.A. (1991, April). *Individual differences in computer adaptive testing: Anxiety, computer literacy and satisfaction.* Paper presented at the annual meeting of the National Council of Measurement in Education, Chicago, IL.

Gershon, R.C., & Bergstrom, B.A. (1995). *Does cheating on CAT pay: NOT!* (Report No. TM024692).

Hambleton, R.K., Swaminathan, H., & Rogers, J.H. (1991). *Fundamentals of item response theory.* Thousand Oaks, CA: Sage.

Huff, K.L., & Sireci, S.G. (2001). Validity issues in computer-based testing. *Educational Measurement: Issues and Practice, 20*(3), 16-25.

Krathwohl, D.R., & Huyser, R.J. (1956). The sequential item test (SIT). *American Psychologist, 2,* 419.

Lord, F.M. (1952). A theory of test scores. *Psychometric Monographs, 7,* 1-84.

Lord, F.M. (1980). *Applications of item response theory to practical testing problems.* Hillsdale, NJ: Erlbaum.

Lunz, M.E., Bergstrom, B.A., & Wright B.D. (1992). The effect of review on student ability and test efficiency in computer adaptive tests. *Applied Psychological Measurement, 16,* 33-40.

Masters, G., & Evans, J. (1986). Banking non-dichotomously scored items. *Applied Psychological Measurement, 10,* 355-367.

McBride, J.R., Corpe, V.A., & Wing, H. (1987, August). *Equating the computerized adaptive edition of the differential aptitude tests.* Presented at the annual meeting of the American Psychological Association, New York, NY.

Olsen, J.B., Maynes, D., Slawson, D., & Ho, K. (1986, April). *Comparisons of paper-administered, computer-administered and computerized adaptive tests of achievement.* Paper presented at the annual meeting of the American Educational Research Association, San Francisco, CA.

Parshall, C.G. (2001). Automated test assembly for online administration. In C.G. Parshall, T.A. Davey, J.A. Spray, & J.C. Kalohn (Eds.), *Practical considerations in computer-based testing* (pp. 106-125). New York: Springer Verlag.

Rasch, G. (1960). *Probabilistic models for some intelligence and attainment tests.* Copenhagen: Dansmarks Paedagogiske Institut.

Reckase, M.D. (1989). Adaptive testing: The evolution of a good idea. *Educational Measurement Issues and Practice, 8,* 11-15.

Spray, J.A. (1987). Recent developments in measurement and possible applications to the measurement of psychomotor behavior. *Research Quarterly for Exercise and Sport, 58,* 203-209.

Stahl, J., Bergstrom, B., & Gershon, R. (2000). CAT administration of language placement examinations. *Journal of Applied Measurement, 1,* 292-302.

Stocking, M.L. (1997). Revising item responses in computerized adaptive tests: A comparison of three models. *Applied Psychological Measurement, 21,* 129-142.

Stocking, M.L., & Lewis, C. (2000). Methods of controlling the exposure of items in CAT. In W.J. van der Linden (Ed.), *Computerized adaptive testing* (pp. 163-182). Boston: Kluwer Academic.

Vispoel, W.P. (1999). Creating computerized adaptive tests of musical aptitude: Problems, solutions, and future directions. In F. Drasgow & J.B. Olson-Buchanan (Eds.), *Innovations in computerized assessment* (pp. 151-176). Hillsdale, NJ: Erlbaum.

Vispoel, W.P., Hendrickson, A.B., & Bleiler, T. (2000). Limiting answer review and change on computerized adaptive vocabulary tests: Psychometric and attitudinal results. *Journal of Educational Measurement, 37,* 21-38.

Vispoel, W.P., Rocklin, T.R., Wang, T., & Bleiler, T. (1999). Can examinees use a review option to obtain positively biased ability estimates on a computerized adaptive test? *Journal of Educational Measurement, 36,* 141-157.

Wainer, H. (2000). *Computerized adaptive testing: A primer (*2nd ed.) (with N. Dorans, D. Eignor, R. Flaugher, B. Green, R. Mislevy, L. Steinberg, & D. Thissen). Hillsdale, NJ: Erlbaum.

Wainer, H., Bradlow, E.T., & Du, Z. (2000). Testlet response theory: An analog for the 3PL model useful in testlet-based adaptive testing. In W.J.van der Linden & C.A.W. Glas (Eds.), *Computerized adaptive testing: Theory and practice* (pp. 245-270). Boston: Kluwer Academic.

Wainer, H., & Kiely, G.L. (1987). Item clusters and computerized adaptive testing: A case for testlets. *Journal of Educational Measurement, 24,* 185-201.

Ware, J.E., Jr. (2003). Conceptualization and measurement of health-related quality of life: Comments on an emerging field. *Archives of Physical Medicine and Rehabilitation, 84*(4), Suppl. 2, S43-S51.

Weiss, D.J., & Kingsbury, G. (1984). Application of computerized adaptive testing to educational problems. *Journal of Educational Measurement, 21,* 361-375.

Wright B.D., & Bell, S.R. (1984). Item banks: What, why, how. *Journal of Educational Measurement, 21,* 331-345.

Wright B.D., & Stone, M.H. (1979). *Best test design.* Chicago: MESA Press.

Wright, B.D., & Masters, G.N. (1982). *Rating scale analysis: Rasch measurement.* Chicago: MESA Press.

Zara, A.R. (1992, April). *An investigation of computerized adaptive testing for demographically diverse candidates on the national registered nurse licensure examination.* Paper presented at the annual meeting of the National Council on Measurement in Education, San Francisco, CA.

Zenisky, A., & Sireci, S.G. (2002). Technological innovations in large-scale assessment. *Applied Measurement in Education, 14,* 337-362.

Zhu, W. (1992). Development of a computerized adaptive visual testing model. In G. Tenenbaum, T. Baz-Liebermann, & Z. Artai (Eds.), *Proceedings of the International Conference on Computer Application in Sport and Physical Education* (pp. 260-267). Natanya, Israel: Wingate Institute for Physical Education and Sport and the Zinman College of Physical Education.

Zhu, W., Safrit, M.J., & Cohen, A.S. (1999). *FitSmart test user manual: High school edition.* Champaign, IL: Human Kinetics.

Chapter 9

Aiken, L., & West, S. (1991). *Multiple regression.* Thousand Oaks, CA: Sage.

Algina, J., & Moulder, B.C. (2001). A note on estimating the Jöreskog-Yang model for latent variable interaction using LISREL 8.3. *Structural Equation Modeling, 8,* 40-52.

Allison, P.D. (1987). Estimation of linear models with incomplete data. In C.C. Clogg (Ed.), *Sociological methodology* (pp. 71-103). San Francisco: Jossey-Bass.

Anderson, J.C., & Gerbing, D.W. (1988). Structural equation modeling in practice: A review and recommended two-step approach. *Psychological Bulletin, 10,* 411-423.

Arbuckle, J.L. (1996). Full information estimation in the presence of incomplete data. In G.A. Marcoulides & R.E. Schumacker (Eds.), *Advanced structural equation modeling: Issues and techniques* (pp. 243-277). Hillsdale, NJ: Erlbaum.

Arbuckle, J.L., & Wothke, W. (1999). *AMOS 4.0 user's guide*. Chicago: Smallwaters.

Bandura, A. (1986). *Social foundations of thought and action*. Englewood Cliffs, NJ: Prentice Hall.

Bauer, D.J., & Curran, P.J. (2003). Distributional assumptions of growth mixture models: Implications for overextraction of latent trajectory classes. *Psychosocial Methods, 8,* 338-363.

Bauman, A., Sallis, J.F., & Owen, N. (2002). Environmental and policy measurement in physical activity research. In G.J Welk. (Ed.), *Physical activity assessments for health-related research* (pp. 241-251). Champaign, IL: Human Kinetics.

Bentler, P.M. (1990). Comparative fit indexes in structural models. *Psychological Bulletin, 107,* 238-246.

Bentler, P.M. (2000). *EQS6 structural equations program manual*. Encino, CA: Multivariate Software.

Bentler, P.M., & Chou, C.-P. (1987). Practical issues in structural modeling. *Sociological Methods & Research, 16,* 78-117.

Bollen, K.A. (1989). *Structural equations with latent variables*. New York: Wiley.

Bollen, K.A. (1995). Structural equation models that are nonlinear in latent variables: A least-squares estimator. In P.M. Marsden (Ed.), *Sociological methodology* (pp. 223-252). Oxford, England: Blackwell.

Bollen, K.A., & Long, J.S. (1993). *Testing structural equation models*. Thousand Oaks, CA: Sage.

Boomsma, A. (1985). Nonconvergence, improper solutions, and starting values in LISREL maximum likelihood estimation. *Psychometrika, 52,* 345-370.

Boomsma, A. (2000). Reporting analyses of covariance structures. *Structural Equation Modeling, 7,* 461-483.

Boomsma, A., & Hoogland, J.J. (2001). The robustness of LISREL modeling revisited. In R. Cudeck, S. du Toit, & D. Sörbom (Eds.), *Structural equation modeling: A festschrift in honor of Karl Jöreskog* (pp. 139-168). Lincolnwood, IL: Scientific Software.

Breckler, S.J. (1990). Applications of covariance structure modeling in psychology: Cause for concern? *Psychological Bulletin, 107,* 260-273.

Brown, R.L. (1994). Efficacy of the indirect approach for estimating structural equation model with missing data: A comparison of five methods. *Structural Equation Modeling, 1,* 287-316.

Browne, M.W. (1984). Asymptotically distribution-free methods for the analysis of covariance structures. *British Journal of Mathematics and Statistical Psychology, 37,* 62-83.

Browne, M.W., Mels, G., & Cowan, M. (1994). *Path analysis: RAMONA: SYSTAT for DOS advanced application* (Version 6, pp. 167-244). Evanston, IL: SYSTAT.

Byrne, B.M. (2001). *Structural equation modeling with AMOS: Basic concepts, applications, and programming*. Hillsdale, NJ: Erlbaum.

Byrne, B.M., & Crombie, G. (2003). Modeling and testing change: An introduction to the latent growth curve model. *Understanding Statistics, 2,* 177-203.

Chou, C.-P., & Bentler, P. (1995). Estimates and tests in structural equation modeling. In R.H. Hoyle, (Ed.), *Structural equation modeling: Concepts, issues and applications* (pp. 37-55). Thousand Oaks, CA: Sage.

Cliff, N. (1983). Some cautions concerning the application of causal modeling methods. *Multivariate Behavioral Research, 18,* 115-126.

Clogg, C.C. (1995). Latent class models. In G. Arminger, C.C. Clogg, & M.E. Sobel (Eds.), *Handbook of statistical modeling for the social and behavioral sciences* (pp. 311-359), New York: Plenum Press.

Collins, L.M., Graham, J.W., Rousculp, S.S., & Hansen, W.B. (1997). Heavy caffeine use and the beginning of the substance use onset process: An illustration of latent transition analysis. In K. Bryant, K.M. Windle, & S. West (Eds.), *The science of prevention: Methodological advances from alcohol and substance use research* (pp. 79-99). Washington, DC: American Psychological Association.

Davis, W.R. (1993). The FC1 rule of identification for confirmatory factor analysis: A general sufficient condition. *Sociological Methods & Research, 21,* 403-437.

Diez-Roux, A.V. (2003). The examination of neighborhood effects on health: Conceptual and methodological issues related to the presence of multiple levels of organization. In I. Kawachi & L.F. Berkman (Eds.), *Neighborhoods and health* (pp. 45-64). New York: Oxford University.

Duncan, T.E., Duncan, S.C., Strycker, L.A., Li, F., & Alpert, A. (1999). *An introduction to latent variable growth curve model: Concepts, issues, and applications.* Hillsdale, NJ: Erlbaum.

Fishbein, M., & Ajzen, I. (1975). *Belief, attitude, intention, and behavior.* Reading, MA: Addison-Wesley.

Fisher, K.J., Li, F., Michael, Y., & Cleveland, M. (2004). Neighborhood-level influences on physical activity: A multilevel analysis. *Journal of Aging and Physical Activity, 11,* 49-67.

Fraser, C., & McDonald, R.P. (1988). COSAN: Covariance structure analysis. *Multivariate Behavioral Research, 23,* 263-265.

Gold, M.S., Bentler, P.M., & Kim, K.H. (2003). A comparison of maximum-likelihood and asymptotically distribution-free methods of treating incomplete non-normal data. *Structural Equation Modeling, 10,* 47-79.

Greenockle, K.M., Lee, A.M., & Lomax, R. (1990). The relation between selected student characteristics and activity patterns in required high school physical education class. *Research Quarterly for Exercise and Sport, 61,* 59-69.

Hartmann, W.M. (1992). *The CALIS procedure: Extended user's guide.* Cary, NC: SAS Institute.

Hayduk, L.A. (1987). *Structural equation modeling with LISREL: Essentials and advances.* Baltimore: Johns Hopkins University Press.

Hoyle, R.H., & Panter, A.T. (1995). Writing about structural equation models. In R.H. Hoyle (Ed.), *Structural equation modeling: Concepts, issues, and applications* (pp. 158-176). Thousand Oaks, CA: Sage.

Hu, L.T., & Bentler, P.M. (1995). Evaluating model fit. In R.H. Hoyle (Ed.), *Structural equation modeling: Concepts, issues, and applications* (pp. 76-117). Thousand Oaks, CA: Sage.

Hu, L.T., & Bentler, P.M. (1999). Cutoff criteria for fit indices in covariance structure analysis: Conventional criteria versus new alternatives. *Structural Equation Modeling, 6,* 1-55.

Jaccard, J., & Wan, C.K. (1995). Measurement error in the analysis interaction effects between continuous predictors using multiple regression: Multiple indicator and structural equation approaches. *Psychological Bulletin, 117,* 348-357.

Jaccard, J., & Wan, C.K. (1996). *LISREL approaches to interaction effects in multiple regression.* Thousand Oaks, CA: Sage.

Jackson, D.L. (2003). Revisiting sample size and number of parameter estimates: Some support for the N:q hypothesis. *Structural Equation Modeling, 10,* 128-141.

Jöreskog, K. (1969). A general approach to confirmatory maximum likelihood factor analysis. *Psychometrika, 34,* 183-202.

Jöreskog, K. (1971). Simultaneous factor analysis in several populations. *Psychometrika, 36,* 406-426.

Jöreskog, K.G., & Sörbom, D. (1996). *LISREL 8: User's reference guide.* Chicago: Scientific Software International.

Jöreskog, K., & Yang, F. (1996). Non-linear structural equation models: The Kenny-Judd model with interaction effects. In G.A. Marcoulides & R.E. Schumacker (Eds.), *Advanced structural equation modeling: Issues and techniques* (pp. 57-88). Hillsdale, NJ: Erlbaum.

Kenny, D., & Judd, C.M. (1984). Estimating the nonlinear and interaction effects of latent variables. *Psychological Bulletin, 96,* 201-210.

Kaplan, D. (2000). *Structural equation modeling: Foundations and extensions.* Thousand Oaks, CA: Sage.

Kline, R.B. (1998). *Principles and practice of structural equation modeling.* New York: Guilford Press.

Li, F. (1999). The Exercise Motivation Scale: Its multifaceted structure and construct validity. *Journal of Applied Sport Psychology, 11,* 97-115.

Li, F., Duncan, T.E., Duncan, S.C., & Acock, A. (2001). Latent growth modeling of longitudinal data: A finite growth mixture modeling approach. *Structural Equation Modeling, 8,* 493-530.

Li, F., Duncan, T.E., Duncan, S.C., McAuley, E., Caumeton, N.R., & Harmer, P. (2001). Enhancing the psychological well-being of elderly individuals through Tai Chi exercise: A latent growth curve analysis. *Structural Equation Modeling, 8,* 53-83.

Li, F., Duncan, T.E., McAuley, E., Harmer, P., & Smolkowski, K. (2000). A didactic example of latent curve analysis applicable to the study of aging. *Journal of Aging and Health, 12,* 388-425.

Li, F., & Fisher, K.J. (2004). A multilevel path analysis of the relationship between neighborhood physical activity and self-rated health status in older adults. *Journal of Physical Activity and Health, 1,* 398-412.

Li, F., Fisher, K.J., Bauman, A., Ory, M.G., Chodzko-Zaiko, W., Harmer, P., Bosworth, M., & Cleveland, M. (2005). Neighborhood influences on physical activity in older adults: A multilevel perspective. *Journal of Aging and Physical Activity, 13,* 32-58.

Li, F., Fisher, K.J., & Brownson, R. (2005). A multilevel analysis of change in neighborhood walking activity in older adults. *Journal of Aging and Physical Activity, 13,* 145-159.

Li, F., Fisher, J., Harmer, P., & McAuley, E. (2002). Delineating the impact of Tai Chi training on physical function among the elderly. *American Journal of Preventive Medicine, 23,* 92-97.

Li, F., & Harmer, P. (1996). Confirmatory factor analysis of the Group Environment questionnaire with an intercollegiate sample. *Journal of Sport & Exercise Psychology, 18,* 49-63.

Li, F., & Harmer, P. (1998). Modeling interaction effects: A two-stage least squares example. In R.E. Schumacker and G.A. Marcoulides (Eds.), *Interaction and nonlinear effects in structural equation modeling* (pp. 153-166). Hillsdale, NJ: Erlbaum.

Li, F., Harmer, P., Duncan, T.E., Duncan, S.C., Acock, A., & Boles, S. (1998). Approaches to testing interaction effects using structural equation modeling methodology. *Multivariate Behavioral Research, 33,* 1-39.

Li, F., Harmer, P., Fisher, K.J., McAuley, E., Chaumeton, N., Eckstrom, E., & Wilson, N.L. (2005). Tai Chi and fall reductions in older adults: A randomized controlled trial. *Journal of Gerontology: Medical Sciences, 60A,* 66-74.

Little, R.J.A., & Rubin, D.B. (1987). *Statistical analysis with missing data.* New York: Wiley.

Little, R.J.A., & Rubin, D.B. (2002). *Statistical analysis with missing data.* Hoboken, NJ: Wiley.

Loehlin, J.C. (1998). *Latent variable models: An instruction to factor, path, and structural analysis* (3rd ed.). Hillsdale, NJ: Erlbaum.

Long, J.S. (1983). *Confirmatory factor analysis: A preface to LISREL.* Beverly Hills, CA: Sage.

MacCallum, R.C. (1995). Model specification: Procedures, strategies and related issues. In R.H. Hoyle (Ed.), *Structural equation modeling: Concepts, issues, and applications* (pp. 17-36). Thousand Oaks, CA: Sage.

MacCallum, R C., & Austin, J.T. (2000). Applications of structural equation modeling in psychological research. *Annual Review of Psychology, 51,* 202-226.

MacCallum, R.C., Browne, M.W., & Sugawara, H.M. (1996). Power analysis and determination of sample size for covariance structural modeling. *Psychological Methods, 1,* 130-149.

Marsh, H. (1993). The multidimensional structure of physical fitness: Invariance over gender and age. *Research Quarterly for Exercise and Sport, 64,* 256-273.

Maruyama, G. (1998). *Basics of structural equation modeling.* Thousand Oaks, CA: Sage.

McArdle, J.J. (1994). Structural factor analysis experiments with incomplete data. *Multivariate Behavioral Research, 29,* 409-454.

McArdle, J. J., & Epstein, D. (1987). Latent growth curves within developmental structural equation models. *Child Development, 58,* 110-133.

McAuley, E. (1991). Efficacy, attributional, and affective responses to exercise participation. *Journal of Sport and Exercise Psychology, 13,* 382-393.

McDonald, R.P., & Ho, M.-H.R. (2002). Principles and practice in reporting structural equation analyses. *Psychological Methods, 7,* 64-82.

Meredith, W., & Tisak, J. (1990). Latent curve analysis. *Psychometrika, 55,* 107-122.

Micceri, T. (1989). The unicorn, the normal curve and other improbable creatures. *Psychological Bulletin, 105,* 156-165.

Mulaik, S.A. (1987). Toward a conception of causality acceptable to experimentation and causal modeling. *Child Development, 58,* 18-32.

Muthén, B.O. (1984). A general structural equation model with dichotomous, ordered categorical, and continuous latent variable indicators. *Psychometrika, 49,* 115-132.

Muthén, B.O. (1991). Analysis of longitudinal data using latent variable models with varying parameters. In L.C. Collins & J.L. Horn (Eds.), *Best methods for the analysis of change* (pp. 1-17). Washington, DC: American Psychological Association.

Muthén, B.O. (1994). Multilevel covariance structure analysis. *Sociological Methods and Research, 22,* 376-398.

Muthén, B.O. (1997). Latent variable modeling of longitudinal and multilevel data. In A. Raftery (Ed.), *Sociological methodology* (pp. 453-480). Oxford, England: Basil Blackwell.

Muthén, B.O. (2001). Latent variable mixture modeling. In G.A. Marcoulides & R.E. Schumacker (Eds.), *New developments and techniques in structural equation modeling* (pp. 1-33). Hillsdale, NJ: Erlbaum.

Muthén, B.O. (2003). Statistical and substantive checking in growth mixture modeling: Comment on Bauer and Curran (2003). *Psychological Methods, 8,* 369-377.

Muthén, B.O. (2004). Latent variable analysis: Growth mixture modeling and related techniques for longitudinal data. In D. Kaplan (ed.), *Handbook of quantitative methodology for the social sciences* (pp. 354-368). Newbury Park, CA : Sage.

Muthén, B., Kaplan, D., & Hollis, M. (1987). On structural equation modeling with data that are not missing completely at random. *Psychometrika, 52,* 431-462.

Muthén, B.O., & Shedden, K. (1999). Finite mixture modeling with mixture outcomes using the EM algorithm. *Biometrics, 55,* 463-469.

Muthén, L.K., & Muthén, B.O. (1998-2004). M*plus: User's guide.* Los Angeles: Muthén & Muthén.

Muthén, L.K., & Muthén, B.O. (2002). How to use a Monte Carlo study to decide on sample size and determine power. *Structural Equation Modeling, 4,* 599-620.

Nagin, D.S. (1999). Analyzing developmental trajectories: A semiparametric, group-based approach. *Psychological Methods, 4,* 139-157.

Neale, M.C., Boker, S.M., Xie, G., & Maes, H.H. (1999). *Mx: Statistical modeling* (5th ed.). Richmond, VA: Authors.

Pearl, J. (2000). *Causality: Models, reasoning and inference.* Cambridge, England: Cambridge University Press.

Raykov, T., Tomer, A., & Nesselroade, J.R. (1991). Reporting structural equation modeling results in Psychology and Aging: Some proposed guidelines. *Psychology and Aging, 6,* 499-503.

Rigdon, E.E. (1995). A necessary and sufficient identification rule for structural models estimated in practice. *Multivariate Behavioral Research, 30,* 359-383.

Rigdon, E.E., Schumacker, R.E., & Wothke, W. (1998). A comparative review of interaction and nonlinear modeling. In R.E. Schumacker & G.A. Marcoulides (Eds.), *Interaction and nonlinear effects in structural equation modeling* (pp. 1-16). Hillsdale, NJ: Erlbaum.

Satorra, A., & Bentler, P.M. (1994). Corrections to test statistic and standard errors in covariance structure analysis. In A. Von Eye & C.C. Clogg (Eds.), *Analysis of latent variables in developmental research* (pp. 399-419). Thousand Oaks, CA: Sage.

Schafer, J.L., & Graham, J.W. (2002). Missing data: Our view of the state of the art. *Psychological Methods, 7,* 147-177.

Schumacker, R.E., & Marcoulides, G.A. (1998). *Interaction and nonlinear effects in structural equation modeling.* Hillsdale, NJ: Erlbaum.

Schumacker, R.E., Randall, E. & Lomax, R.G. (1996). *A beginner's guide to structural equation modeling.* Hillsdale, NJ: Erlbaum.

Steiger, J.H. (1988). Aspects of person-machine communication in structural modeling of correlations and covariances. *Multivariate Behavioral Research, 23,* 281-290.

Steiger, J.H. (1990). Structural model evaluation and modification: An interval estimation approach. *Multivariate Behavioral Research, 25,* 173-180.

Steiger, J.H. (1995). *SEPATH: STATISTICA* (version 5). Tulsa: StatSoft.

Tucker, L.R., & Lewis, C. (1973). A reliability coefficient for maximum likelihood factor analysis. *Psychometrika, 38,* 1-10.

Weiner, B. (1985) An attributional theory of achievement motivation and emotion. *Psychological Review, 92,* 548-573.

West, S.G., Finch, J.F., & Curran, P.J. (1995). Structural equation models with non-normal variables In R.H. Hoyle (Ed.), *Structural equation modeling: Concepts, issues and applications* (pp. 56-75). Thousand Oaks, CA: Sage.

Willett, J.B., & Sayer, A.G. (1994). Using covariance structure analysis to detect correlates and predictors of individual change over time. *Psychological Bulletin, 116,* 363-381.

Wothke, W. (2000). Longitudinal and multigroup modeling with missing data. In T.D. Little, K.U. Schnabel, & J. Baumert (Eds.), *Modeling longitudinal and multilevel data: Practical issues, applied approaches, and specific examples* (pp. 219-240). Hillsdale, NJ: Erlbaum.

Young, Y.-F., & Bentler, P.M. (1996). Bootstrapping techniques in analysis of mean and covariance structures. In G.A. Marcoulides & R.E. Schumacker (Eds.), *Advanced structural equation modeling: Issues and techniques* (pp. 195-226). Hillsdale, NJ: Erlbaum.

Chapter 10

Bast, J., & Reitsma, P. (1997). Matthew effects in reading: A comparison of latent growth curve models and simplex models with structured means. *Multivariate Behavioral Research, 32,* 135-167.

Bollen, K.A. (1989). *Structural equations with latent variables.* New York: Wiley.

Bradley, D.R. (1988). *DATASIM.* Lewiston, ME: Desktop Press.

Bryk, A.S., & Raudenbush, S.W. (1987). Application of hierarchical linear models to assessing change. *Psychological Bulletin, 101,* 147-158.

Bryk, A.S., & Raudenbush, S.W. (1992). *Hierarchical linear models: Applications and data analysis methods.* Thousand Oaks, CA: Sage.

Cohen, J. (1988). *Statistical power analysis for the behavioral sciences* (3rd ed.). Hillsdale NJ: Erlbaum.

Collins, L.M., & Sayer, A.G. (Eds.). (2001). *New methods for the analysis of change.* Washington DC: American Psychological Association.

Curran, P.J., & Bollen, K.A. (2001). The best of both worlds: Combining autoregressive and latent curve models. In L.M. Collins & AG. Sayer (Eds.), *New methods for the analysis of change* (pp.107-135). Washington, DC: American Psychological Association.

Curran, P.J., West, S.G., & Finch, J.F. (1996). The robustness of test statistics to nonnormality and specification error in confirmatory factor analysis. *Psychological Methods, 1,* 16-29.

Davidson, M.L. (1972). Univariate versus multivariate tests in repeated-measures experiments. *Psychological Bulletin, 77,* 446-452.

Duncan, T.E., & Duncan, S.C. (1991). A latent growth curve approach to investigating developmental dynamics and correlates of change in children's perceptions of physical competence. *Research Quarterly for Exercise and Sport, 62,* 390-398.

Duncan, T.E., Duncan, S.C., Li, F., & Strycker, L.A. (2002). Multilevel modeling of longitudinal and functional data. In D.S. Moskowitz & S.L. Hershberger (Eds.), *Modeling intraindividual variability with repeated measures data: Method and application* (pp. 171-201). Hillsdale, NJ: Erlbaum.

Erdfelder, E., Faul, F., & Buchner, A. (1996). GPOWER: A general power analysis program. *Behavior Research Methods, Instruments & Computers, 28,* 1-11.

Goldstein, H.I. (1986). Multilevel mixed linear model analysis using iterative generalized least squares. *Biometrika, 73,* 43-56.

Goldstein, H.I. (1995). *Multilevel statistical models* (2nd ed.). New York: Halstead Press.

Gorman, B.S., Primavera, L.H., & Allison, D.B. (1995). POWPAL: A program for estimating effect sizes, statistical power, and sample sizes. *Educational & Psychological Measurement, 55,* 773-776.

Gottman, J.M. (Ed.). (1995). *The analysis of change.* Hillsdale, NJ: Erlbaum.

Graham, J.W., Taylor, B.J., & Cumsille, P.E. (2001). Planned missing-data designs in analysis of change. In L.M. Collins & A.G. Sayer (Eds.), *New methods for the analysis of change* (pp. 333-353). Washington, DC: American Psychological Association.

Hand, D.J., & Taylor, C.C. (1987). *Multivariate analysis of variance and repeated measures: A practical guide for behavioural scientists.* London: Chapman & Hall.

Heise, D.R. (1969). Separating reliability and stability in test-retest correlation. *American Sociological Review, 34,* 93-101.

Hertzog, C., & Rovine, M. (1985). Repeated-measures analysis of variance in developmental research: Selected issues. *Child Development, 56,* 787-809.

Hu, L.–T., & Bentler, P. (1999). Cut off criteria for fit indexes in covariance structure analysis: Conventional criteria versus new alternatives. *Structural Equation Modeling, 6,* 1-55.

Huberty, C.J., & Morris, J.D. (1989). Multivariate analysis versus multiple univariate analyses. *Psychological Bulletin, 105,* 302-308.

Jackson, D.L. (2001). Sample size and number of parameter estimates in maximum likelihood confirmatory factor analysis: A Monte Carlo investigation. *Structural Equation Modeling, 8,* 205-223

Jöreskog, K.G. (1970). Estimation and testing of simplex models. *British Journal of Mathematical and Statistical Psychology, 23,* 121-145.

Jöreskog, K.G., & Sörbom, D. (1996). *LISREL 8: User's reference guide.* Chicago: Scientific Software International.

Kenny, D.A., & Campbell, D.T. (1989). On the measurement of stability in over-time data. *Journal of Personality, 57,* 445-481.

Kenny, D.A., Bolger, N., & Kashy, D.A. (2002). Traditional methods for estimating multi-level models. In D.S. Moskowitz & S.L. Hershberger (Eds.), *Modeling intraindividual variability with repeated measures data* (pp. 1-24). Hillsdale, NJ: Erlbaum.

Keselman, H.J. (1982). Multiple comparisons for repeated measures means. *Multivariate Behavioral Research, 17,* 87-92.

Keselman, H.J. (1998). Testing treatment effects in repeated measures designs: An update for psychophysiological researchers. *Psychophysiology, 35,* 470-478.

Keselman, H.J., Algina, J., & Kowalchuk R.K. (2001). The analysis of repeated measures designs: A review. *British Journal of Mathematical and Statistical Psychology, 54,* 1-20.

Keselman, H.J., Keselman, J.C., & Shaffer, J.P. (1991). Multiple pairwise comparisons of repeated measures means under violation of multisample sphericity. *Psychological Bulletin, 110,* 162-170.

Laid, N.M., & Ware, H. (1982). Random-effects models for longitudinal data. *Biometrics, 38,* 963-974.

Lee, V., & Bryk, A.S. (1989). A multilevel model of the social distribution of high school achievement. *Sociology of Education, 62,* 172-192.

Marsh, H.W., & Grayson, D. (1994). Longitudinal stability of latent means and individual differences: A unified approach. *Structural Equation Modeling, 1,* 317-359.

McArdle, J.J. (1988). Dynamic but structural equation modeling of repeated measures data. In R.B. Cattel & J. Nesselroade (Eds.), *Handbook of multivariate experimental psychology* (2nd ed., pp. 561-614). New York: Plenum Press.

McArdle, J.J., & Hamagami, F. (1991). Modeling incomplete longitudinal and cross-sectional data using latent growth structural models. In L.M. Collins & J.C. Horn (Eds.), *Best methods for the analysis of change* (pp. 276-304). Washington, DC: American Psychological Association.

Meredith, W., & Tisak, J. (1984). *"Tuckerizing" curves.* Paper presented at the Psychometric Society Annual Meeting, Santa Barbara, CA.

Meredith, W., & Tisak, J. (1990). Latent curve analysis. *Psychometrika, 55,* 107-122.

Morris, C.N. (1983). Parametric empirical Bayes inference: Theory and applications. *Journal of the American Statistical Association, 78,* 47-65.

Moskowitz, D.S., & Hershberger, S.L. (Eds.) (2002). *Modeling intraindividual variability with repeated measures data: Method and application.* Hillsdale, NJ: Erlbaum.

Muller, K.E., LaVange, L.M., Ramey, S.L., & Ramey, C. (1992). Power calculations for general linear multivariate models including repeated measures applications. *Journal of the American Statistical Association, 87,* 1209-1224.

Muthén, B. (1994). Multilevel covariance structure analysis. *Sociological Methods & Research, 22,* 376-398.

Muthén, B. (1996). Growth modeling with binary responses. In A. von Eye & C. Clogg (Eds.), *Categorical variables in developmental research: Methods of analysis* (pp. 37-54). San Diego, CA: Academic Press.

Muthén, B. (2001). Second-generation structural equation modeling with a combination of categorical and continuous latent variables: New opportunities for latent class-latent growth modeling. In L.M. Collins & A.G. Sayer (Eds.), *New methods for the analysis of change* (pp. 289-322). Washington, DC: American Psychological Association.

Muthén, B. (2002). Beyond SEM: General latent variable modeling. *Behaviormetrika, 29,* 81-117.

Muthén, L.K., & Muthén, B. (1998). M*plus user's guide.* Los Angeles: Muthén and Muthén.

Number Cruncher Statistical Software. (1991). *PASS* (Power Analysis and Sample Size). Kaysville, UT: Author.

Park, I. (2001). *Latent growth models and reliability estimation of longitudinal physical performances.* Unpublished doctoral dissertation. Vancouver, BC: University of British Columbia.

Park, I., & Schutz, R.W. (1999). "Quick and easy" formulae for approximating statistical power in repeated measures ANOVA. *Measurement in Physical Education and Exercise Science, 3,* 249-270.

Potvin, J.P., & Schutz, R.W. (2000). Statistical power for the two-factor repeated measures ANOVA. *Behavior Research Methods, Instrumentation & Computers, 32,* 347-356.

Raudenbush, S.W. (2001). Toward a coherent framework for comparing trajectories of individual change. In L.M. Collins & A.G. Sayer (Eds.), *New methods for the analysis of change* (pp.35-64). Washington, DC: American Psychological Association.

Raudenbush, S.W., Bryk, A.S., Cheong, Y.F., & Congdon, R.T. (2001). *HLM 5: Hierarchical linear and nonlinear modeling.* Lincolnwood, IL: Scientific Software International.

Rodriguez, G., & Goldman, N. (1995). An assessment of estimation procedures for multilevel models with binary responses. *Journal of the Royal Statistical Society, A-158,* 73-90.

Rogosa, D., & Willet, J.B. (1985). Satisfying a simplex structure is simpler than it should be. *Journal of Educational Statistics, 10,* 99-107.

Schutz, R.W. (1989). Analyzing change. In J. Safrit, & T. Wood (Eds.), *Measurement concepts in physical education and exercise science* (pp. 206-228). Champaign, IL: Human Kinetics.

Schutz, R.W. (1998). Assessing the stability of psychological traits and measures. In J.L. Duda (Ed.), *Advances in sport and exercise psychology measurement* (pp. 393-408). Morgantown, WV: Fitness Information Technology.

Schutz, R.W., & Gessaroli, M.E. (1987). The analysis of repeated measures designs involving multiple dependent variables. *Research Quarterly for Exercise & Sport, 58,* 132-149.

StatSoft. (2001). *STATISTICA.* Tulsa, OK: Author.

Stoolmiller, M. (1995). Using latent growth curve models to study developmental processes. In J.M. Gottman (Ed.), *The analysis of change* (pp. 103-138). Hillsdale, NJ: Erlbaum.

Strenio, J.L. F., Weisberg, H. I., & Bryk, A.S. (1983). Empirical Bayes estimation of individual growth curve parameters and their relationship to covariates. *Biometrics, 39,* 71-86.

Tabachnick, B.G., & Fidell, L.S. (2000). *Using multivariate statistics* (4th ed.). Boston: Allyn & Bacon.

von Eye, A. (Ed.). (1990). *Statistical methods in longitudinal research* (Vols. I, II). San Diego, CA: Academic Press.

Werts, C.E., Linn, R.L., & Jöreskog, K.G. (1978). Reliability of college grades from longitudinal data. *Educational and Psychological Measurement, 38,* 89-95.

Wheaton, B., Muthén, B., Alwin, D.F., & Summers, G.F. (1977). Assessing reliability and stability in panel models. In D.R. Heise (Ed.), *Sociological methodology* (pp. 84-136). San Francisco: Jossey-Bass.

Winer, B.J., Brown, D.R., & Michels, K.M. (1991). *Statistical principles in experimental design* (3rd ed.). New York: McGraw-Hill.

Zhu, W. (1997). A multilevel analysis of school factors associated with health-related fitness. *Research Quarterly for Exercise and Sport, 68,* 125-135.

Zhu, W., & Erbaugh, S.J. (1997). Assessing change in swimming skills using the hierarchical linear model. *Measurement in Physical Education and Exercise Science, 1,* 179-201.

Chapter 11

Algina, J. (1994). Some alternative approximate tests for a split plot design. *Multivariate Behavioral Research, 29,* 365-384.

Bashein, B.J., & Markus, M.L. (2000). *Data warehouses: More than just mining.* Morristown, NJ: Financial Executive Research Foundation.

Berry, M.J.A., & Linoff, G. (1997). *Data mining techniques: For marketing, sales, and customer.* New York: Wiley.

Berson, A., Smith, S., & Thearling, K. (2000). *Building data mining applications for CRM.* New York: McGraw-Hill.

Berthold, M., & Hand, D.J. (Eds.). (1999). *Intelligent data analysis: An introduction.* New York: Springer.

Bhandari, I., Colet, E., Parker, J., Pines, Z., Pratap, R., & Ramanujam, K. (1997). Advanced scout: Data mining and knowledge discovery in NBA data. *Data Mining and Knowledge Discovery, 1,* 121-125.

Breiman, L., Friedman, J.H., Olshen, R.A., & Stone, C.J. (1984). *Classification and regression trees.* Monterey, CA: Wadsworth.

Cartwright, H. (Ed.) (2000). *Intelligent data analysis in science.* Oxford, England: Oxford University Press.

Chang, G., Healey, M.J., McHugh, J.A.M., & Wang, J.T.L. (2001). *Mining the World Wide Web: An information search approach.* Norwell, MA: Kluwer Academic.

Chapman, P., Clinton, J., Kerber, R., Khabaza, T., Reinartz, T., Shearer, C., & Wirth, R. (2000). *CRISP-DM 1.0: Step-by-step data mining guide.* Chicago: SPSS.

Delmater, R., & Hancock, M. (2001). *Data mining explained.* Woburn, MA: Butterworth-Heinemann.

Edgington, E.S. (1995). *Randomization tests* (3rd ed.). New York: Marcel Dekker.

Efron, B., & Tibshirani, R. (1993). *An introduction to the bootstrap.* New York: Chapman & Hall.

Elder, J.F., IV, & Abbott, D.W. (1998, August). *A comparison of leading data mining tools.* Paper presented at the Fourth International Conference on Knowledge Discovery and Data Mining, New York.

Fayyad, U.M., Piatetsky-Shapiro, G., Smyth, P., & Uthurusamy, R. (Eds.). (1996). *Advances in knowledge discovery and data mining.* Cambridge, MA: The MIT Press.

Franks, B.D., & Huck, S.W. (1986). Why does everyone use the .05 significance level? *Research Quarterly for Exercise and Sport, 57,* 245-249.

Hand, D.J. (1998). Data mining: Statistics and more? *The American Statistician, 52*(2), 112-118.

Hays, W.L. (1994). *Statistics* (5th ed.). Fort Worth: Harcourt Brace College.

Hearst, M.A. (1999). *Untangling text data mining.* The Proceedings of ACL'99: The 37th Annual Meeting of the Association for Computational Linguistics, University of Maryland [Online]. Available:www.sims.berkeley.edu/~hearst/papers/acl99/acl99-tdm.html.

Inmon, W.H. (1996). *Building the data warehouse* (2nd ed.). New York: Wiley.

Inmon, W.H., & Hackathorn, R.D. (1994). *Using the data warehouse.* New York: Wiley.

Kang, M., Zhu, W., & Ragan, B. (2002). Finding association rules among outside physical activities in elementary school children [Abstract]. *Research Quarterly for Exercise and Sport, 73*(Supp. 1), A-37 – A-38.

Kass, G.V. (1980). An exploratory technique for investigating large quantities of categorical data. *Applied Statistics, 29*(2), 119-127.

Kramer, S.H., & Rosenthal, R. (1999). Effect sizes and significance levels in small-sample research. In R.H. Hoyle (Ed.), *Statistical strategies for small sample research* (pp. 60-81), Thousand Oaks, CA: Sage.

Loton, T. (2002). *Web content mining with Java.* West Sussex, England: Wiley.

Maxwell, S.E. (1994). Optimal allocation of assessment time in randomized pretest-posttest designs. *Psychological Bulletin, 115,* 142-152.

Maxwell, S.E. (1998). Longitudinal designs in randomized group comparisons: When will intermediate observations increase statistical power? *Psychological Bulletin, 13,* 275-290.

Maxwell, S.E., & Arvey, R.D. (1982). Small sample profile analysis with many variables. *Psychological Bulletin, 92,* 778-785.

Maxwell, S.E., & Delaney, H.D. (1990). *Designing experiments and analyzing data: A model comparison perspective.* Belmont, CA: Wadsworth.

McClelland, G.H. (1997). Optimal design in psychological research. *Psychological Methods, 2,* 3-19.

McClelland, G.H. (2000). Increasing statistical power without increasing sample size. *American Psychologist, 55,* 963-964.

McCulloch, W.S., & Pitts, W.A. (1943). A logical calculus of the ideas immanent in neural nets. *Bulletin of Mathematical Biophysics, 5,* 115-133.

Piatetsky-Shapiro, G., & Frawley, W.J. (Eds.). (1991). *Knowledge discovery in databases.* Cambridge, MA: The MIT Press.

Riccia, G.D., Kruse, R., & Lenz, H.-J. (Eds.). (2000). *Computational intelligence in data mining.* New York: Springer-Verlag Wien.

Rogosa, D. (1980). A critique of cross-lagged correlation. *Psychological Bulletin, 88,* 245-258.

Shadish, W.R., Cook, T.D., & Campbell, D.T. (2002). *Experimental and quasi-experimental designs for generalized causal inference.* Boston, New York: Houghton Mifflin.

Shearer, C., & Jouve, O. (2002, March). *Text mining: Unlocking value in text.* Paper presented at the SPSS Online Seminar [Online]. Available: http://spssevents.webex.com/spssevents/onstage/mainframe.php?Rnd5530=.7309397263617905.

Thomas, J.R., Salazar, W., & Landers, D.M. (1991). What is missing in $p < .05$? Effect size. *Research Quarterly for Exercise and Sport, 62*, 344-348.

Two Crows Corporation. (1999). *Introduction to data mining and knowledge discovery* (3rd ed.). Potomac, MD: Author.

Venter, A., & Maxwell, S.E. (1999). Maximizing power in randomized designs when N is small. In R.H. Hoyle (Ed.), *Statistical strategies for small sample research* (pp. 33-59). Thousand Oaks, CA: Sage.

Venter, A., Maxwell, S.E., & Bolig, E. (2002). Power in randomized group comparisons: The value of adding a single intermediate timepoint to a traditional pretest-posttest design. *Psychological Methods, 7*(21), 194-209.

Welge, M., & Bushell, C. (2000). *Data mining in government.* Paper presented at ACCESS, Washington DC [Online]. Available: http://archive.ncsa.uiuc.edu/STI/ALG/dm_present_html/sld001.htm.

Winer, B.J., Brown, D.R., & Michels, K.M. (1991). *Statistical principles in experimental design* (3rd ed.). New York: McGraw-Hill.

Yung, Y.F., & Chan, W. (1999). Statistical analyses using bootstrapping: Concepts and implementation. In R.H. Hoyle (Ed.), *Statistical strategies for small sample research* (pp. 81-105), Thousand Oaks, CA: Sage.

Zhu, W. (1997). Making bootstrap statistical inferences: A tutorial. *Research Quarterly for Exercise and Sport, 68,* 44-55.

Zhu, W. (2002). Data mining and knowledge discovery in databases: An overview [Abstract]. *Research Quarterly for Exercise and Sport, 73*(Suppl. 1), A-37.

Zuckerman, M., Hodgins, H.S., Zuckerman, A., & Rosenthal, R. (1993). Contemporary issues in the analysis of data: A survey of 551 psychologists. *Psychological Science, 4*(1), 49-53.

Chapter 12

American Federation of Teachers, National Council on Measurement in Education, & National Education Association. (1990). *Standards for teacher competence in educational assessment of students.* Washington, DC: Authors.

Angoff, W.H. (1971). Scales, norms, and equivalent scores. In R.L. Thorndike (Ed.), *Educational measurement* (2nd ed., pp. 508-600). Washington, DC: American Council on Education.

Arter, J. (1991). Teaching about performance assessment. *Educational Measurement: Issues and practice, 18*(2), 30-44.

Arter, J., & McTighe, J. (2001). *Scoring rubrics in the classroom: Using performance criteria for assessing and improving student learning.* Thousand Oaks, CA: Corwin Press.

Baumgartner, T.A., & Chung, H. (2001). Confidence limits for intraclass reliability coefficients. *Measurement in Physical Education and Exercise Science, 5,* 179-188.

Berk, R.A. (Ed.). (1980). *Criterion-referenced measurement: The state of the art.* Baltimore: Johns Hopkins University Press.

Berk, R.A. (Ed). (1984). *A guide to criterion-referenced test construction.* Baltimore: Johns Hopkins University Press.

Bock, D.B. (1997). A brief history of item response theory. *Educational Measurement: Issues and Practice, 16*(4), 21-33.

Brennan, R.L. (2001). Some problems, pitfalls, and paradoxes in educational measurement. *Educational Measurement: Issues and Practice, 20*(4), 6-18..

Burger, S.E., & Burger, D.L. (1994). Determining the validity of performance-based assessment. *Educational Measurement: Issues and Practice, 13*(1), 9-15.

Chudowsky, N., & Behuniak, P. (1998). Using focus groups to examine the consequential aspect of validity. *Educational Measurement: Issues and Practice, 17*(4), 28-38.

Cizek, G.J. (Ed.). (2001). *Setting performance standards: Concepts, methods, and perspectives.* Hillsdale, NJ: Erlbaum.

Cooley, W.W. (1991). State-wide student assessment. *Educational Measurement: Issues and Practice, 10*(4), 3-6, 15.

Crocker, L. (1997). Assessing content representativeness of performance assessment exercises. *Applied Measurement in Education, 10,* 83-95.

Cutforth, N., & Parker, M. (1996). Promoting affective development in physical education. *Journal of Physical Education, Recreation & Dance, 67*(7), 19-23.

Danielson, C., & Abrutyn, L. (1997). *An introduction to using portfolios in the classroom.* Alexandria, VA: Association for Supervision and Curriculum Development.

Dunbar, S.B., Koretz, D.M., & Hoover, H.D. (1991). Quality control in the development and use of performance assessments. *Applied Measurement in Education, 4,* 289-303.

Frechtling, J.A. (1991). Performance assessment: Moonstruck or the real thing? *Educational Measurement: Issues and Practice, 10*(4), 23-25.

Glatthorn, A.A., Bragaw, D., Dawkins, K., & Parker, J. (1998). *Performance assessment and standards-based curricula: The achievement cycle.* Larchmont, NY: Eye on Education.

Guskey, T.R. (Ed.). (1996a). *Communicating student learning.* Alexandria, VA: Association for Supervision and Curriculum Development.

Guskey, T.R. (1996b). Reporting on student learning: Lessons from the past—Prescriptions for the future. In T.R. Guskey (Ed.), *Communicating student learning* (pp. 13-24). Alexandria, VA: Association for Supervision and Curriculum Development.

Guskey, T.R., & Bailey, J.M. (2001). *Developing grading and reporting systems for student learning.* Thousand Oaks, CA: Corwin Press.

Hensley, L.D., & East, W.B. (1989). Testing and grading in the psychomotor domain. In M.J. Safrit & T.M. Wood (Eds.), *Measurement concepts in physical education and exercise science* (pp. 297-321). Champaign, IL: Human Kinetics.

Herman, J.L., Klein, D.C.D., & Abedi, J. (2000). Assessing students' opportunity to learn: Teachers and student perspectives. *Educational Measurement: Issues and Practice, 19*(4), 16-24.

Jaeger, R.M. (1989). Certification of student competence. In R.L. Linn (Ed.), *Educational measurement* (3rd ed., pp. 485-514). New York: American Council on Education and Macmillan.

Joyner, A.B., & McManis, B.G. (1997). Quality control in alternative assessment. *Journal of Physical Education, Recreation & Dance, 68*(7), 38-40.

Kane, M.T. (1994). Validating the performance standards associated with passing scores. *Review of Educational Research, 64,* 425-461.

Kane, M.T. (2001). So much remains the same: Conception and status of validation in setting standards. In G.J. Cizek (Ed.), *Setting performance standards: Concepts, methods, and perspectives* (pp. 53-88). Hillsdale, NJ: Erlbaum.

Kane, M., Crooks, T., & Cohen, A. (1999). Validating measures of performance. *Educational Measurement: Issues and Practice, 18*(2), 5-17.

Kingston, N.M., Kahl, S.R., Sweeney, K.P., & Bay, L. (2001). Setting performance standards using the body of work method. In G.J. Cizek (Ed.), *Setting performance standards: Concepts, methods, and perspectives* (pp. 219-248). Hillsdale, NJ: Erlbaum.

Linn, R.L., Baker, E.L., & Dunbar, S.B. (1991). Complex, performance-based assessment: Expectations and validation criteria. *Educational Researcher, 20*(8), 15-21.

Linn, R.L., & Burton, E. (1994). Performance-based assessment: Implications of task specificity. *Educational Measurement: Issues and Practice, 13*(1), 5-8, 15.

Lund, J.L. (2000). *Creating rubrics for physical education.* Reston, VA: National Association for Sport and Physical Education.

Marzano, R.J. (2000). *Transforming classroom grading.* Alexandria, VA: Association for Supervision and Curriculum Development.

McMillan, J.H. (2001).Secondary teachers' classroom assessment and grading practices. *Educational Measurement: Issues and Practice, 20*(1), 20-32.

Mehrens, W.A. (1992). Using performance assessment for accountability purposes. *Educational Measurement: Issues and Practice, 11*(1), 3-9.

Melograno, V.J. (1994). Portfolio assessment: Documenting authentic student learning. *Journal of Physical Education, Recreation & Dance, 65*(8), 50-55, 58-61.

Melograno, V.J. (2000). Designing a portfolio system for K-12 physical education: A step-by-step process. *Measurement in Physical Education and Exercise Science, 4,* 97-115.

Messick, S. (1989). Validity. In R.L. Linn (Ed.), *Educational measurement* (3rd ed., pp. 13-103). New York: American Council on Education and Macmillan.

Messick, S. (1995). Standards of validity and the validity of standards in performance assessment. *Educational Measurement: Issues and practice, 14*(4), 5-8.

Messick, S. (1996). Validity of performance assessments. In G.W. Phillips (Ed.), *Technical issues in large-scale performance assessment* (pp. 1-18). Washington, DC: U.S. Department of Education, National Center for Education Statistics.

Meyer, C.A. (1992). What's the difference between authentic and performance assessment? *Educational Leadership, 49*(8), 39.

Miller, M.D., & Legg, S.M. (1993). Alternative assessment in a high-stakes environment. *Educational Measurement: Issues and practice, 12*(2), 9-15.

Mitzel, H.C., Lewis, D.M., Patz, R.J., & Green, D.R. (2001). The bookmark procedure: Psychological perspectives. In G.J. Cizek (Ed.), *Setting performance standards: Concepts, methods, and perspectives* (pp. 249-281). Hillsdale, NJ: Erlbaum.

Morrow, J.R., & Jackson, A.W. (1993). How "significant" is your reliability? *Research Quarterly for Exercise and Sport, 64,* 352-355.

Morrow, J.R., Jackson, A.W., Disch, J.G., & Mood, D.P. (2000). *Measurement and evaluation in human performance* (2nd ed.). Champaign, IL: Human Kinetics.

National Association for Sport and Physical Education. (1992). *Outcomes of quality physical education.* Reston, VA: Author.

National Association for Sport and Physical Education. (1995). *Moving into the future: National standards for physical education.* New York: McGraw-Hill.

National Association for Sport and Physical Education. (2004). *Moving into the future: National standards for physical education* (2nd ed.). New York: McGraw-Hill.

National Commission on Excellence in Education. (1983). *A nation at risk: The imperative for educational reform.* Washington, DC: U.S. Government Printing Office.

Oregon Department of Education. (2001). *Physical education common curriculum goals and content standards.* Salem, OR: Author.

Phillips, S.E. (1996). Legal defensibility of standards: Issues and policy perspectives. *Educational Measurement: Issues and Practice, 15*(2), 5-13, 19.

Phillips, S.E. (2001). Legal issues in standard setting for K-12 programs. In G.J. Cizek (Ed.), *Setting performance standards: Concepts, methods, and perspectives* (pp. 411-426). Hillsdale, NJ: Erlbaum.

Plake, B.S., & Hambleton, R.K. (2001). The analytic judgment method for setting standards on complex performance assessments. In G.J. Cizek (Ed.), *Setting performance standards: Concepts, methods, and perspectives* (pp. 283-312). Hillsdale, NJ: Erlbaum.

Quellmalz, E.S. (1991). Developing criteria for performance assessments: The missing link. *Applied Measurement in Education, 4,* 319-331.

Raymond, M.R., & Reid, J.B. (2001). Who made thee a judge: Selecting and training participants for standard setting. In G.J. Cizek (Ed.), *Setting performance standards: Concepts, methods, and perspectives* (pp. 119-157). Hillsdale, NJ: Erlbaum.

Safrit, M.J. (1989). Criterion-referenced measurement: Validity. In M.J. Safrit & T.M. Wood (Eds.), *Measurement concepts in physical education and exercise science* (pp. 119-135). Champaign, IL: Human Kinetics.

Safrit, M.J., & Wood, T.M. (Eds.). (1989). *Measurement concepts in physical education and exercise science.* Champaign, IL: Human Kinetics.

Safrit, M.J., & Wood, T.M. (1995). *Introduction to measurement in physical education and exercise science* (3rd ed.). St. Louis, MO: Mosby—Year Book.

Schafer, W.D. (1991). Essential assessment skills in professional education of teachers. *Educational Measurement: Issues and practice, 10*(1), 3-6, 12.

Sireci, S.G. (2001). Standard setting using cluster analysis. In G.J. Cizek (Ed.), *Setting performance standards: Concepts, methods, and perspectives* (pp. 339-354). Hillsdale, NJ: Erlbaum.

Smith, J.K., Smith, L.F., & De Lisi, R. (2001). *Natural classroom assessment: Designing seamless instruction & assessment.* Thousand Oaks, CA: Corwin Press.

Solano-Flores, G., & Shavelson, R.J. (1997). Development of performance assessments in science: Conceptual, practical, and logistical issues. *Educational Measurement: Issues and practice, 16*(3), 16-25.

Stiggins, R.J. (2001). The unfulfilled promise of classroom assessment. *Educational Measurement: Issues and practice, 20*(3), 5-15.

Taggart, G.L., Phifer, S.J., Nixon, J.A., & Wood, M. (1998). *Rubrics: A handbook for construction and use.* Lancaster, PA: Technomic.

Taleporos, E. (1998). Consequential validity: A practitioner's perspective. *Educational Measurement: Issues and Practice, 17*(2), 20-23, 34.

U.S. Department of Education. (2002). *No child left behind: What to know.* Retrieved August 2, 2002 from www.nclb.gov/next/overview/index.html.

Veal, M.L. (1992). School-based theories of pupil assessment: A case study. *Research Quarterly for Exercise and Sport, 63,* 48-59.

Whittington, D. (1999). Making room for values and fairness: Teaching reliability and validity in the classroom context. *Educational Measurement: Issues and practice, 18*(1), 14-22, 27.

Wichita Public Schools. (1999). *Physical education program standards: Curriculum alignment and assessment rubrics.* Wichita, KS: Author.

Wood, T.M. (1989). The changing nature of norm-referenced validity. In M.J. Safrit & T.M. Wood (Eds.), *Measurement concepts in physical education and exercise science* (pp. 23-44). Champaign, IL: Human Kinetics.

Wood, T.M. (1990). *Evaluation practices in public school physical education: A theoretician's perspective* (Cassette Recording No. 9015 Tape 16). Paper presented at the National Convention of the American Alliance for Health, Physical Education, Recreation and Dance, New Orleans, LA.

Wood, T.M. (1996). Evaluation and testing: The road less traveled. In S.J. Silverman & C.D. Ennis (Eds.), *Student learning in physical education: Applying research to enhance instruction* (pp. 199-219). Champaign, IL: Human Kinetics.

Wood, T.M. (2003). Assessment in physical education: The future is now! In S.J. Silverman & C.D. Ennis (Eds.), *Student learning in physical education: Applying research to enhance instruction* (2nd ed., pp. 187-203). Champaign, IL: Human Kinetics.

Chapter 13

Acheson, K.J., Campbell, I.T., Edholm, O.G., Miller, D.S., & Stock, M.J. (1980). The measurement of daily energy expenditure—An evaluation of some techniques. *American Journal of Clinical Nutrition, 33,* 1155-1164.

Ainsworth, B.E., Bassett, D.R., Strath, S.J., Swartz, A.M., O'Brien, W.L., Thompson, R.W., Jones, D.A., Macera, C.A., & Kimsey, C.D. (2000). Comparison of three methods for measuring the time spent in physical activity. *Medicine and Science in Sports and Exercise, 32*(9 Suppl.), S457-S464.

Ainsworth, B.E., Haskell, W.L., Leon, A.S., Jacobs, D.R., Montoye, H.J., Sallis, J.F., & Paffenbarger, R.S. (1993). Compendium of physical activities: Classification of energy costs of human physical activities. *Medicine and Science in Sports and Exercise, 25,* 71-80.

Ainsworth, B.E., Haskell, W.L., Whitt, M.C., Irwin, M.L., Swartz, A.M., Strath, S.J., O'Brien, W.L., Bassett, D.R., Schmitz, K.H., Emplaincourt, P.O., Jacobs, D.R., & Leon, A.S. (2000). Compendium of physical activities: an update of activity codes and MET intensities. *Medicine and Science in Sports and Exercise, 32*(Suppl.), S498-S516.

Ainsworth, B.E., Irwin, M.L., Addy, C.L., Whitt, M.C., & Stolarczyk, L.M. (1999). Moderate intensity physical activity patterns of minority women: The Cross-Cultural Activity Participation Study. *Journal of Women's Health, 8,* 805-813.

Ainsworth, B.E., LaMonte, M.J., Drowatzky, K.L., Cooper, R.S., Thompson, R.W., Irwin, M.L., Whitt, M.C., & Gilman, M. (2000). Evaluation of the CAPS Typical Week Physical Activity Survey among minority women. In *Proceedings of the Community Prevention Research in Women's Health Conference* (p.17). Bethesda, MD: National Institutes of Health.

Ainsworth, B.E., Leon, A.S., Richardson, M.T., Jacobs, D.R., & Paffenbarger, R.S. (1993). Accuracy of the College Alumnus physical activity questionnaire. *Journal of Clinical Epidemiology, 46,* 1403-1411.

Ainsworth, B.E., Montoye, H.J., & Leon, A.S. (1994). Methods of assessing physical activity during leisure and work. In C. Bouchard, R.J. Shephard, & T. Stephens (Eds.), *Physical activity, fitness, and health: International proceedings and consensus statement* (pp.146-159). Champaign, IL: Human Kinetics.

Ainsworth, B.E., Richardson, M.T., Jacobs, D.R., & Leon, A.S. (1992). Prediction of cardiorespiratory fitness using physical activity questionnaire data. *Medicine, Exercise, Nutrition, and Health, 1,* 75-82.

Ainsworth, B.E., Richardson, M.T., Jacobs, D.R., & Leon, A.S. (1993). Gender differences in physical activity. *Women in Sport and Physical Activity Journal, 2,* 3-16.

American College of Sports Medicine. (1995). *Guidelines for exercise testing and prescription* (5th ed.). Baltimore: Williams & Wilkins.

Arroll, B., & Beaglehole, R. (1991). Potential misclassification in studies of physical activity. *Medicine and Science in Sports and Exercise, 23,* 1176-1178.

Baecke, J.A.H., Burema, J., & Frijters, J.E.R. (1982). A short questionnaire for the measurement of habitual physical activity in epidemiological studies. *American Journal of Clinical Nutrition, 36,* 936-942.

Bassett, D.R., Ainsworth, B.E., Leggett, S.R., Mathien, C.A., Main, J.A., Hunter, D.C., & Duncan, G.E. (1996). Accuracy of five electronic pedometers for measuring distance walked. *Medicine and Science in Sports and Exercise, 28,* 1071-1077.

Bassett, D.R., Ainsworth, B.E., Swartz, A.M., Strath, S.J., O'Brien, W.L., & King, G.A. (2000). Validity of four motion sensors in measuring moderate intensity physical activity. *Medicine and Science in Sports and Exercise, 32*(9 Suppl.), S471-S480.

Belloc, N.B., & Breslow, L. (1972). Relationship of physical health status and health practices. *Preventive Medicine, 1,* 409-421.

Black, A.E., Coward, W.A., Cole, T.J., & Prentice, A.M. (1996). Human energy expenditure in affluent societies: An analysis of 574 doubly-labeled water measurements. *European Journal of Clinical Nutrition, 50,* 72-92.

Black, A.E., Prentice, A.M., & Coward, W.A. (1986). Use of food quotients to produce respiratory quotients for the doubly labeled water technique. *Human Nutrition: Clinical Nutrition, 40C,* 381-391.

Blair, S.N., Haskell, W.L., Ho, P., Paffenbarger, R.S., Vranizan, K.M., Farquhar, J.W., & Wood, P.D. (1985). Assessment of habitual physical activity by a seven-day recall in a community survey and controlled experiments. *American Journal of Epidemiology, 122,* 794-804.

Bouchard, C., Malina, R.M., & Perusse, L. (1997). *Genetics of fitness and physical performance.* Champaign, IL: Human Kinetics.

Bouchard, C., Tremblay, A., Leblanc, C., Lortie, G., Savard, R., & Theriault, G. (1983). A method to assess energy expenditure in children and adults. *American Journal of Clinical Nutrition, 37,* 461-467.

Bouten, C.V.C., Wilhelmine, P.H.G., De Venne, V., Westersterp, K.R., Verduin, M., & Janssen, J.D. (1996). Daily physical activity assessment: Comparison between movement registration and doubly labeled water. *Journal of Applied Physiology, 81,* 1019-1026.

Brooks, G.A., Fahey, T.D., & White, T.P. (1996). *Exercise physiology. Human bioenergetics and its applications* (2nd ed.). Mountain View, CA: Mayfield.

Campbell, K.L., Crocker, P.R.E., & McKenzie, D.C. (2002). Field evaluation of energy expenditure in women using Tritrac accelerometers. *Medicine and Science in Sports and Exercise, 34,* 1667-1674.

Caspersen, C.J., Powell, K.E., & Christenson, G.M. (1985). Physical activity, exercise, and physical fitness: Definitions and distinctions for health-related research. *Public Health Reports, 100,* 126-131.

Chasan-Taber, L., Erickson, J.B., Nasca, P.C., Chasan-Taber, S., & Freedson, P.S. (2002). Validity and reproducibility of a physical activity questionnaire in women. *Medicine and Science in Sports and Exercise, 34,* 987-992.

Ching, P.L.Y.H., Willett, W.C., Rimm, E.B., Colditz, G.A., Gortmaker, S.L., & Stampfer, M.J. (1996). Activity level and risk of overweight in male health professionals. *American Journal of Public Health, 86,* 25-30.

Conway, J.M., Seale, J.L., Jacobs, D.R., Irwin, M.L., & Ainsworth, B.E. (2002). Comparison of energy expenditure estimates from doubly labeled water, a physical activity questionnaire, and physical activity records. *American Journal of Clinical Nutrition, 75,* 519-525.

Cooper, K.H. (1968). A means of assessing maximal oxygen uptake. *Journal of the American Medical Association, 203,* 201-204.

Corbin, C.B., Pangrazi, R.P., & Franks, B.D. (2000). Definitions: Health, fitness, and physical activity. *President's Council on Physical Fitness and Sports Research Digest, Series 3,* No. 9, 1-8.

Coughlin, S.S. (1990). Recall bias in epidemiologic studies. *Journal of Clinical Epidemiology, 43,* 87-91.

Craig, C.L., Marshall, A.L., Sjostrom, M., Bauman, A.E., Booth, M.L., Ainsworth, B.E., Pratt, M., Ekelund, U., Yngve, A., Sallis, J.F., & Oja, P. (2003). International physical activity questionnaire: 12-country reliability and validity. *Medicine and Science in Sports and Exercise, 35,* 1381-1395.

Davis, J.A. (1996). Direct determination of aerobic power. In P.J. Maud & C. Foster, (Eds.), *Physiological assessment of human fitness* (pp. 9-17), Champaign, IL: Human Kinetics.

Durante, R., & Ainsworth, B.E. (1996). The recall of physical activity: Using a cognitive model of the question-answering process. *Medicine and Science in Sports and Exercise, 28,* 1282-1291.

Eason, K.E., Masse, L.C., Kelder, S.H., & Tortelero, S.R. (2002). Diary days needed to estimate activity among older African-American and Hispanic women. *Medicine and Science in Sports and Exercise, 34,* 1308-1315.

Elia, M., Fuller, N.J., & Murgatroyd, P.R. (1992). Measurement of bicarbonate turnover in humans: Applicability to estimation of energy expenditure. *American Journal of Physiology, 263,* E676-E687.

Elia, M., Jones, M.G., Jennings, G., Poppitt, S.D., Fuller, N.J., Murgatroyd, P.R., & Jebb, S.A. (1995). Estimating energy expenditure from specific activity of urine urea during lengthy subcutaneous $NaH_{14}CO_3$ infusion. *American Journal of Applied Physiology, 269,* E172-E182.

el-Khoury, A.E., Sanchex, M., Fukagawa, N.K., Gleason, R.E., & Young, V.R. (1994). Similar 24-h pattern and rate of carbon dioxide production, by indirect calorimetry vs. stable isotope dilution, in healthy adults under standardized metabolic conditions. *Journal of Nutrition, 124,* 1615-1627.

Ferrannini, E. (1988). The theoretical bases of indirect calorimetry. *Metabolism, 37,* 287-301.

Freedson, P.S., & Melanson, E.L. (1996). Measuring physical activity. In D. Docherty (Ed.), *Measurement in pediatric exercise science* (pp. 261-283). Champaign, IL: Human Kinetics.

Freedson, P.S., Melanson, E., & Sirard, S. (1998). Calibration of the Computer Science and Applications, Inc., accelerometer. *Medicine and Science and Sports and Exercise, 30,* 777-781.

Freedson, P.S., & Miller, K. (2000). Objective monitoring of physical activity using motion sensors and heart rate. *Research Quarterly for Exercise and Sport, 71,* 21-29.

Friedenreich, C.M., Courneya, K.S., & Bryant, H.E. (1998). The lifetime total physical activity questionnaire: Development and reliability. *Medicine and Science in Sports and Exercise, 30,* 266-274.

Fruin, M.L., & Rankin, J.W. (2004). Validity of a multi-sensor armband in estimating rest and exercise energy expenditure. *Medicine and Science in Sports and Exercise, 36,* 1063-1069.

Gretebeck, R.J., & Montoye, H.J. (1992). Variability of some objective measures of physical activity. *Medicine and Science in Sports and Exercise, 24,* 1167-1172.

Ham, S.A., Macera, C.A., Jones, D.A., Ainsworth, B.E., & Turczyn, K.M. (2004). Considerations for physical activity research: Variations on a theme. *Journal of Physical Activity and Health, 1,* 98-113.

Haskell, W.L., Yee, M.C., Evans, A., & Irby, P.J. (1993). Simultaneous measurement of heart rate and body motion to quantitate physical activity. *Medicine and Science in Sports and Exercise, 25,* 109-115.

Hastad, D.N., & Lacy, A.C. (1998). *Measurement and evaluation in physical education and exercise science* (3rd ed.). Boston: Allyn & Bacon.

Hayden-Wade, H.A., Coleman, K.J., Sallis, J.F., & Armstrong, C. (2003). Validation of the telephone and in-person interview versions of the 7-day PAR. *Medicine and Science in Sports and Exercise, 35,* 801-809.

Healey, J. (2000). Future possibilities in electronic monitoring of physical activity. *Research Quarterly for Exercise and Sport, 71*(2 Suppl.), 137-145.

Hendelman, D., Miller, K., Bagget, C., Debold, E., & Freedson, P. (2000). Validity of accelerometry for the assessment of moderate intensity physical activity in the field. *Medicine and Science in Sports and Exercise, 32*(9 Suppl.), S442-S449.

Henderson, K.A., Ainsworth, B.E., Stolarzcyk, L.M., Hootman, J.M., & Levin, S. (1999). Notes on linking qualitative and quantitative data: The Cross-Cultural Activity Participation Study. *Leisure Sciences, 21,* 247-255.

Hill, J.O., Melby, C., Johnson, S.L., & Peters, J.C. (1995). Physical activity and energy requirements. *American Journal of Clinical Nutrition, 62*(Suppl.), 1059s-1066s.

Horton E.S. (1983). An overview of the assessment and regulation of energy balance in humans. *American Journal of Clinical Nutrition, 38,* 972-977.

Howley, E.T. (2001). Type of activity: Resistance, aerobic and leisure versus occupational physical activity. *Medicine and Science in Sports and Exercise, 33,* S364-S369.

Jacobs, D.R., Ainsworth, B.E., Hartman, T.J., & Leon, A.S. (1993). A simultaneous evaluation of 10 commonly used physical activity questionnaires. *Medicine and Science in Sports and Exercise, 25,* 81-91.

Jacobs, D.R., Hahn, L.P., Haskell, W.L., Pirie, P., & Sidney, S. (1989). Reliability and validity of a short physical activity history: CARDIA and the Minnesota Heart Health Program. *Journal of Cardiopulmonary Rehabilitation, 9,* 448-459.

Jakicic, J.M., Marcus, M., Gallagher, K.I., Randall, C., Thomas, E., Goss, F.L., & Robertson, R.J. (2004). Evaluation of the SenseWear Pro Armband to assess energy expenditure during exercise. *Medicine and Science in Sports and Exercise, 36,* 897-904.

Jequier, E., Acheson, K., & Schutz, Y. (1987). Assessment of energy expenditure and fuel utilization in man. *Annual Review of Nutrition, 7,* 187-208.

Jequier, E., & Schutz, Y. (1983). Long-term measurements of energy expenditure in humans using a respiratory chamber. *American Journal of Clinical Nutrition, 38,* 989-998.

King, G.A., McLaughlin, J.E., Howley, E.T., Bassett, D.R., & Ainsworth, B.E. (1999). Validation of Aerosport KB1-C portable metabolic system. *International Journal of Sports Medicine, 20,* 304-308.

King, G.A., Torres, N., Potter, C., Brooks, T.J., & Coleman, K.J. (2004). Comparison of activity monitors to estimate energy cost of treadmill exercise. *Medicine and Science in Sports and Exercise, 36,* 1244-1251.

Kohl, H.W., Fulton, J.E., & Caspersen, C.J. (2000). Assessment of physical activity among children and adolescents: A review and synthesis. *Preventive Medicine, 31,* S54-S76.

Kriska, A.M., Sandler, R.B., Cauley, J.A., LaPorte, R.E., Hom, D.L., & Pambianco, G. (1988). The assessment of historical physical activity and its relation to adult bone parameters. *American Journal of Epidemiology, 127,* 1053-1063.

LaMonte, M.J., & Ainsworth, B.E. (2001). Quantifying energy expenditure and physical activity in the context of dose response. *Medicine and Science in Sports and Exercise, 33,* S370-S378.

LaMonte, M.J., Durstine, J.L., Addy, C.L., Irwin, M.L., & Ainsworth, B.E. (2001). Physical activity, physical fitness, and Framingham 10-year risk score: The Cross-Cultural Activity Participation Study. *Journal of Cardiopulmonary Rehabilitation, 21,* 63-70.

Leenders, N.Y., Nelson, T.E., & Sherman, W.M. (2003). Ability of different physical activity monitors to detect movement during treadmill walking. *International Journal of Sports Medicine, 24,* 43-50.

Leenders, N.Y., Sherman, W.M., & Nagarja, H.N. (2000). Comparisons of four methods of estimating physical activity in adult women. *Medicine and Science in Sports and Exercise, 32,* 1320-1326.

Le Masurier, G.C., & Tudor-Locke, C.E. (2003). Comparison of pedometer and accelerometer accuracy under controlled conditions. *Medicine and Science in Sports and Exercise, 35,* 867-871.

Levin, S., Jacobs, D.R., Ainsworth, B.E., Richardson, M.T., & Leon, A.S. (1999). Intra-individual variation and estimates of usual physical activity. *Annals of Epidemiology, 9,* 481-488.

Li, R., Deurenberg, P., & Hautvast, J.G.A.J. (1993). A critical evaluation of heart rate monitoring to assess energy expenditure in individuals *American Journal of Clinical Nutrition, 58,* 602-607.

Lifson, N., Gordon, G.B., & McClintock, R. (1955). Measurement of carbon dioxide production by means of D_2O^{18}. *Journal of Applied Physiology, 7,* 704-710.

Livingstone, M.B., Prentice, A.M., Coward, W.A., Ceesay, S.A., Strain, J.J., McKenna, P.G., Nevin, G.B., Barker, M.E., & Hickey, R.J. (1990). Simultaneous measurement of free-living energy expenditure by the doubly labeled water method and heart-rate monitoring. *American Journal of Clinical Nutrition, 52,* 59-65.

Macera, C.A., Ham, S.A., Jones, D.A., Kimsey, C.D., Ainsworth, B.E., & Neff, L.J. (2001). Limitations on the use of a single screening question to measure sedentary behavior. *American Journal of Public Health, 91,* 2010-2012.

Macera, C.A., & Pratt, M. (2000). Public health surveillance of physical activity. *Research Quarterly for Exercise and Sport, 71,* 97-103.

Malina, R.M., & Bouchard, C. (1991). *Growth, maturation, and physical activity.* Champaign, IL: Human Kinetics.

Masse, L.C. (2000). Reliability, validity, and methodological issues in assessing physical activity in a cross-cultural setting. *Research Quarterly for Exercise and Sport, 71*(2 Suppl.), S54-S58.

Matthews, C.E., DuBose, K.D., LaMonte, M.J., Tudor-Locke, C.E., & Ainsworth, B.E. (2002). Evaluation of a computerized 24-hour physical activity recall (24PAR). *Medicine and Science in Sports and Exercise, 34*(5 Suppl.): S41.

Matthews, C.E., Freedson, P.S., Hebert, J.R., Stanek, E.J., Merriam, P.A., & Ockene, I.S. (2000). Comparing physical activity assessment methods in the Seasonal Variation of Blood Cholesterol Study. *Medicine and Science in Sports and Exercise, 32,* 976-984.

Maud, P.J., & Foster, C. (Eds.). (1995). *Physiological assessment of human fitness.* Champaign, IL: Human Kinetics.

McCroy, M.A., Mole, P.A., Nommsen-Rivers, L.A., & Dewey, K.G. (1997). Between-day and within-day variability in the relation between heart rate and oxygen consumption: Effect on the estimation of energy expenditure by heart rate monitoring. *American Journal of Clinical Nutrition, 66,* 18-25.

Melanson, E., & Freedson, P. (1995). Validity of the Computer Science and Applications, Inc. (CSA) activity monitor. *Medicine and Science in Sports and Exercise, 27,* 934-940.

Montoye, H.J. (1971). Estimation of habitual physical activity by questionnaire and interview. *American Journal of Clinical Nutrition, 24,* 1113-1118.

Montoye, H.J., Kemper, H.C.G., Saris, W.H.M, & Washburn, R.A. (1996). *Measuring physical activity and energy expenditure.* Champaign, IL: Human Kinetics.

Montoye, H.J., Washburn, R., Servais, S., Ertl, A., Webster, J.G., & Nagle, F.J. (1983). Estimation of energy expenditure by a portable accelerometer. *Medicine and Science in Sports and Exercise, 15,* 403-407.

Nichols, J.F., Morgan, C.G., Sarkin, J.A., Sallis, J.F., & Calfas, K.J. (1999). Validity, reliability, and calibration of the Tritrac accelerometer as a measure of physical activity. *Medicine and Science in Sports and Exercise, 31,* 908-912.

Paffenbarger, R.S., Blair, S.N., Lee, I.M., & Hyde, R.T. (1993). Measurement of physical activity to assess health effects in free-living populations. *Medicine and Science in Sports and Exercise, 25,* 60-70.

Paffenbarger, R.S., Hyde, R.T., Wing, A.L., & Hsieh, C.C. (1986). Physical activity, all-cause mortality, and longevity of college alumni. *New England Journal of Medicine, 314,* 605-613.

Pate, R.R. (1988). The evolving definition of physical fitness. *Quest, 40,* 174-179.

Pate, R.R., Pratt, M., Blair, S.N., Haskell, W.L., Macera, C.A., Bouchard, C., Buchner, D., Ettinger, W., Heath, G.W., King, A.C., Kriska, A., Leon, A.S., Marcus, B.H., Morris, J., Paffenbarger, R.S., Patrick, K., Pollock, M.L., Rippe, J.M., Sallis, J.F., & Wilmore, J.H. (1995). Physical activity and public health. A recommendation from the Centers for Disease Control and Prevention and the American College of Sports Medicine. *Journal of the American Medical Association, 273,* 402-407.

Reis, J.P., DuBose, K.D., Ainsworth, B.E., Macera, C.A., & Yore, M. (2005). Reliability and validity of the Occupational Physical Activity questionnaire. *Medicine and Science in Sports and Exercise,* 37, 2075-2083.

Richardson, M.T., Ainsworth, B.E., Jacobs, D.R., & Leon, A.S. (2001). Validation of the Stanford 7-day recall to assess habitual physical activity. *Annals of Epidemiology, 11,* 145-153.

Richardson, M.T., Leon, A.S., Jacobs, D.R., Ainsworth, B.E., & Serfass, R. (1994). Comprehensive evaluation of the Minnesota Leisure Time Physical Activity questionnaire. *Journal of Clinical Epidemiology, 47,* 271-281.

Richardson, M.T., Leon, A.S., Jacobs, D.R., Ainsworth, B.E., & Serfass, R. (1995). Ability of the Caltrac accelerometer to assess daily physical activity levels. *Journal of Cardiopulmonary Rehabilitation, 15,* 107-113.

Riley, P.L. & Finnegan, L.P. (1997). The Prevention Research Centers Program: Collaboration in women's health. *Journal of Women's Health, 6,* 281-283.

Robertson, R.J., Caspersen, C.J., Allison, T.G., Skrinar, G.S., Abbott, R.A., & Metz, K.F. (1982). Differentiated perceptions of exertion and energy cost of young women while carrying loads. *European Journal of Applied Physiology, 49,* 69-78.

Roche, A.F., Heymsfield, S.B., & Lohman, T.G. (Eds.). (1996). *Human body composition.* Champaign, IL: Human Kinetics.

Saltin, B., & Rowell, L. (1980). Functional adaptations to physical activity and inactivity. *Federation Proceedings, 39,* 1506-1513.

Schneider, P.L., Crouter, S.E., & Bassett, D.L. (2004). Pedometer measures of free-living physical activity: Comparison of 13 models. *Medicine and Science in Sports and Exercise, 36,* 331-335.

Schulz, L.O., Nyomba, B.L., Alger, S., Anderson, T.E., & Ravussin, E. (1991). Effects of endurance training on sedentary energy expenditure measured in a respiratory chamber. *Journal of Applied Physiology, 260,* E257-E261.

Schulz, S., Westersterp, K.R., & Bruck, K. (1989). Comparison of energy expenditure by the doubly labeled water technique with energy intake, heart rate, and activity recording in man. *American Journal of Clinical Nutrition, 49,* 1146-1154.

Schutz, Y., & Herren, R. (1999). Assessment of speed of human locomotion using a differential satellite global positioning system. *Medicine and Science in Sports and Exercise, 32,* 642-646.

Seale, J.L., Rumpler, W.V., Conway, J.M., & Miles, C.W. (1990). Comparison of doubly labeled water, intake balance, and direct- and indirect-calorimetry methods for measuring energy expenditure in adult men. *American Journal of Clinical Nutrition, 52,* 66-71.

Slattery, M.L., & Jacobs, D.R. (1987). The inter-relationships of physical activity, physical fitness, and body measurements. *Medicine and Science in Sports and Exercise, 19,* 564-569.

Siscovick, D.S., Ekelund, L.G., Hyde, J.S., Johnson, J.L., Gordon, D.J., & LaRosa, J.C. (1988). Physical activity and coronary heart disease among asymptomatic hypercholesterolemic men. *American Journal of Public Health, 78,* 1428-1431.

Speakman, J.R. (1998). The history and theory of the doubly labeled water technique. *American Journal of Clinical Nutrition, 68,* 932S-938S.

Sternfeld, B., Cauley, J., Harlow, S., Liu, G., & Lee, M. 2000. Assessment of physical activity with a single global question in a large, multiethnic sample of midlife women. *American Journal of Epidemiology, 152,* 678-687.

Stevens, J., Cai, J., Pamuk, E.R., Williamson, D.F., Thun, M.J., & Wood, J.L. (1998). The effect of age on the association between body mass index and mortality. *New England Journal of Medicine, 338,* 1-7.

Stofan, J.R., DiPiettro, L., Davis, D., Kohl, H.W., & Blair, S.N. (1998). Physical activity patterns associated with cardiorespiratory fitness and reduced mortality: The Aerobics Center Longitudinal Study. *American Journal of Public Health, 88,* 1807-1813.

Strath, S.J., Bassett, D.R., & Swartz, A.M. (2003). Comparison of MTI accelerometer cut-points for predicting time spent in physical activity. *International Journal of Sports Medicine, 24,* 298-303.

Strath, S.J., Bassett, D.R., & Swartz, A.M. (2004). Comparison of the college alumnus questionnaire physical activity index with objective monitoring. *Annals of Epidemiology, 14,* 409-415.

Strath, S.J., Bassett, D.R., Thompson, D.L., & Swartz, A.M. (2002). Validity of the simultaneous heart rate-motion sensor technique for measuring energy expenditure. *Medicine and Science in Sports and Exercise, 34,* 888-894.

Strath, S.J., Swartz, A.M., Bassett, D.R., O'Brien, W.L., King, G.A., & Ainsworth, B.E. (2000). Evaluation of heart rate as a method for assessing moderate intensity physical activity. *Medicine and Science in Sports and Exercise, 32*(9 Suppl.), S465-S470.

Swartz, A.M., Strath, S.J., Bassett, D.R., O'Brien, W.L., King, G.A., & Ainsworth, B.E. (2000). Estimation of energy expenditure using CSA accelerometers at hip and waist sites. *Medicine and Science in Sports and Exercise, 32*(9 Suppl.), S450-S456.

Taylor, H.L., Jacobs, Jr. D.R., Schucker, B., Knudsen, J., Leon, A.S., & De Backer, G. (1978). A questionnaire for the assessment of leisure time physical activities. *Journal of Chronic Disease, 31,* 741-744.

Terrier, P., Ladetto, Q., Merminod, B., & Schutz, Y. (2000). High-precision satellite positioning system as a new tool to study the biomechanics of human locomotion. *Journal of Biomechanics, 33,* 1717-1722.

Terrier, P., & Schutz, Y. (2003). Variability of gait patterns during unconstrained walking assessed by satellite positioning (GPS). *European Journal of Applied Physiology, 90,* 554-561.

Tudor-Locke, C.E., & Myers, A.M. (2001). Methodological considerations for researchers and practitioners using pedometers to measure physical (ambulatory) activity. *Research Quarterly for Exercise and Sport, 71,* 1-12.

Tudor-Locke, C.E., Williams, J.E., Reis, J.P., & Pluto, D. (2002). Utility of pedometers for assessing physical activity: Convergent validity. *Sports Medicine, 32,* 795-808.

Tudor-Locke, C.E., Williams, J.E., Reis, J.P., & Pluto, D. (2004). Utility of pedometers for assessing physical activity: Construct validity. *Sports Medicine, 34,* 281-291.

Warnecke, R.B., Johnson, T.P., Chavez, N., Sudman, S., O'Rourke, D.P., Lacey, L., & Horm, J. (1997). Improving question wording in surveys of culturally diverse populations. *Annals of Epidemiology, 7,* 334-342.

Washburn, R.A. (2000). Assessment of physical activity in older adults. *Research Quarterly for Exercise and Sport, 71*(2 Suppl.), 79-88.

Washburn, R.A., & Montoye, H.L. (1986). Validity of heart rate as a measure of mean daily energy expenditure. *Exercise Physiology, 2,* 161-172.

Webb, P., Annis, J.F., & Troutman, S.J. (1980). Energy balance in man measured by direct and indirect calorimetry. *American Journal of Clinical Nutrition, 33,* 1287-1298.

Weir, J.B. (1949). New methods for calculating metabolic rate with special reference to protein metabolism. *Journal of Physiology, 109,* 1-9.

Welk, G.J. (Ed.). (2002). *Physical activity assessments for health-related research.* Champaign, IL: Human Kinetics.

Welk, G.J., Almeida, J., & Morss, G. (2003). Laboratory calibration and validation of the Biotrainer and Actitrac activity monitors. *Medicine and Science in Sports and Exercise, 35,* 1057-1064.

Welk, G.J., Blair, S.N., Wood, K., Jones, S., & Thompson, R.W. (2000). A comparative evaluation of three accelerometry-based physical activity monitors. *Medicine and Science in Sports and Exercise, 32*(9 Suppl.), S489-S497.

Welk, G.J., Corbin, CB., & Dale, D. (2000). Measurement issues in the assessment of physical activity in children. *Research Quarterly for Exercise and Sport, 71*(2 Suppl.), S59-S73.

Welk G.J., Differding, J.A., Thompson, R.W., Blair, S.N., Dziura, J., & Hart, P. (2000). The utility of the digi-walker step counter to assess physical activity patterns. *Medicine and Science in Sports and Exercise, 32*(9 Suppl.), S481-S488.

Westersterp, K.R., Brouns, F., Saris, W.H.M., & Ten Hoor, F. (1988). Comparison of doubly labeled water with respirometry at low- and high-activity levels. *Journal of Applied Physiology, 65,* 53-56.

Wilcox, S., Irwin, M.L., Addy, C.L., Ainsworth, B.E., Stolarczyk, L., Whitt, M.C., & Tudor-Locke, C.E. (2001). Agreement between participant-rated and compendium-coded intensity of daily activities in a triethnic sample of women ages 40 years and older. *Annals of Behavioral Medicine, 23,* 253-262.

Wilmore, J.H., & Haskell, W.L. (1971). Use of the heart rate-energy expenditure relationship in the individualized prescription of exercise. *American Journal of Clinical Nutrition, 24,* 1186-1192.

Wolf, A.M, Hunter, D.J., Colditz, G.A., Manson, J.E., Stampfer, M.J., Corsano, K.A., Rosner, B., Kriska, A., & Willett, W.C. (1994). Reproducibility and validity of a self-administered physical activity questionnaire. *International Journal of Epidemiology, 23,* 991-999.

Yore, M.M., Bowles, H.R., Ainsworth, B.E., Macera, C.A., & Kohl, H.W. III. (2006). Single-versus multiple-item questions on occupational physical activity. *Journal of Physical Activity and Health, 1,* 102-111.

Zhang, K., Pi-Sunyer, F.X., & Boozer, C.N. (2004). Improving energy expenditure estimation for physical activity. *Medicine and Science in Sports and Exercise, 36,* 883-889.

Zhang, K., Werner, P., Sun, M., Pi-Sunyer, F.X., & Boozer, C.N. (2003). Measurement of daily physical activity. *Obesity Research, 11,* 33-40.

Zhu, W. (1996). Should total scores from a rating scale be used directly? *Research Quarterly for Exercise and Sport, 67,* 363-372.

Zhu, W. (2000) Score equivalence is at the heart of international measures of physical activity. *Research Quarterly for Exercise and Sport, 71,* 121-128.

Zhu, W. (2002). A confirmatory study of Rasch-based optimal categorization of a rating scale. *Journal of Applied Measurement, 3,* 1-15.

Zhu, W., Timm, G., & Ainsworth, B.E. (2001). Rasch calibration and optimal categorization of an instrument measuring women's exercise perseverance and barriers. *Research Quarterly for Exercise and Sport, 72,* 104-116.

Chapter 14

Blair, S.N., Kohl, H.W., III., Paffenbarger, R.S., Jr., Clark, D.G., Cooper, K.H., & Gibbons, L.W. (1989). Physical fitness and all-cause mortality: A prospective study of healthy men and women. *Journal of the American Medical Association, 262*(17), 2395-2401.

Dolan, J.P., & Adams-Smith, W.N. (1978). *Health and society: A documentary history of medicine.* New York: The Seabury Press.

Fletcher, G.F., Blair, S.N., Blumenthal, J., Caspersen, C., Chaitman, B., Epstein, S., Falls, H., Froelicher, E.S.S., Froelicher, V., & Pina, I.L. (1992). Statement on exercise: Benefits and recommendations for physical activity programs for all Americans: A statement for health professionals by the Committee on Exercise and Cardiac Rehabilitation and the Council on Clinical Cardiology, American Heart Association. *Circulation, 86,* 340-344.

Hakim, A.A., Curb, J.D., Petrovitch, H., Rodriquez, B.L., Yano, K., Ross, G.W., White, L.R., & Abbott, R.D. (1999). Effects of walking on coronary heart disease in elderly men: The Honolulu Heart Program. *Circulation, 100,* 9-13.

Hu, F.B., Li, T.Y., Colditz, G.A., Willett, W.C., & Manson, J.E. (2003). Television watching and other sedentary behaviors in relation to risk of obesity and type 2 diabetes mellitus in women. *Journal of the American Medical Association, 289,* 1785-1791.

Jekel, J.F., Elmore, J.G., & Katz, D.L. (1996). *Epidemiology, biostatistics and preventive medicine.* Philadelphia: W.B. Saunders.

Kiely, D.K., Wolf, P.A., Cupples, L.A., Beiser, A.S., & Kannel, W.B. (1994). Physical activity and stroke risk: The Framingham study. *American Journal of Epidemiology, 140,* 608-620.

Manson, J.E., Nathan, D.M., Krolewksi, A.S., Stampfer, M.J., Willett, W.C., & Hennekens, C.H. (1992). A prospective study of exercise and incidence of diabetes among US male physicians. *Journal of the American Medical Association, 268*(1), 63-67.

Morris, J.N., Heady, J.A., Raffee, R.A.B., Roberts, C.G., & Parks, S.W. (1953). Coronary heart disease and physical activity of work. *Lancet, 2,* 1111-1120.

Paffenbarger, R.S., Jr., & Hale, W.D. (1975). Work activity and coronary heart disease mortality. *The New England Journal of Medicine, 292,* 545-550.

Paffenbarger, R.S., Jr., Wing, A.L., & Hyde, R.T. (1978). Physical activity as an index of heart attack risk in college alumni. *American Journal of Epidemiology, 108,* 165-175.

Pate, R.R., Blair, S.N., Haskell, W.L., Macera, C.A., Bouchard, C., Bruchner, D., Ettinger, W., Heath, G.W., King, A.C., Kriska, A., Leon, A.S., Marcus, B.H., Morris, J., Paffenbarger, R.S., Jr., Patrick, K., Pollock, M.L., Rippe, J.M., Sallis, J., & Wilmore, J. H. (1995).

Physical activity and public health: A recommendation from the Centers for Disease Control and Prevention and the American College of Sports Medicine. *Journal of the American Medical Association, 273,* 402-407.

Pomrehn, P.R., Wallace, R.B., & Burgmeister, L.F. (1982). Ischemic heart disease mortality in Iowa farmers: The influence of lifestyle. *Journal of the American Medical Association, 248,* 1073-1076.

Powell, K.E., & Paffenbarger, R.S., Jr. (1985). Workshop on epidemiologic and public health aspects of physical activity and exercise: A summary. *Public Health Reports, 100,* 118-126.

Sesso, H.D., Paffenbarger, R.S., Jr., & Lee, I.M. (2000). Physical activity and coronary heart disease in men: The Harvard Alumni Health Study. *Circulation, 102,* 975-980.

Stebbling, W. (1875). *Analysis of Mr. Mill's system of logic* (New edition). London: Longmans, Green.

Terris, M. (1975). Approaches to an epidemiology of health. *American Journal of Public Health, 65,* 1037-1045.

United States Department of Health and Human Services. (1996). *Physical activity and health: A report of the Surgeon General.* Atlanta, GA: United States Department of Health and Human Services, Centers for Disease Control and Prevention, National Center for Chronic Disease Prevention and Health Promotion.

Verloop, J., Rookus, R. A., van der Kooy, K., & van Leeuwen, F.E. (2000). Physical activity and breast cancer risk in women aged 20-54 years. *Journal of the National Cancer Institute, 92,* 128-135.

Chapter 15

Altman, D.G. (2000a). Diagnostic tests. In D.G. Altman, D. Machin, T.N., Bryant, & M.J. Gardner (Eds.), *Statistics with confidence: Confidence intervals and statistical guidelines* (2nd ed., pp. 105-119). Bristol, England: British Medical Journal Books.

Altman, D.G. (2000b). ROC curves and confidence intervals: Getting them right [Letter to the editor]. *Heart, 83,* 236.

Altman, D.G., & Bland, J.M. (1994a). Diagnostic tests 1: Sensitivity and specificity. *British Medical Journal, 308,* 1552.

Altman, D.G., & Bland, J.M. (1994b). Diagnostic tests 2: Predictive values. *British Medical Journal, 309,* 102.

Altman, D.G., & Bland, J.M. (1994c). Diagnostic tests 3: Receiver operating characteristic plots. *British Medical Journal, 309,* 188.

Ashley, E., Myers, J., & Froelicher, V. (2002). Exercise testing scores as an example of better decisions through science. *Medicine and Science in Sports and Exercise, 34,* 1391-1398.

Barr, W.B. (2001). Methodologic issues in neuropsychological testing. *Journal of Athletic Training, 36,* 297-302.

Basmajian, J.V., & Banerjee, S.N. (Eds.). (1996). *Clinical decision making in rehabilitation: Efficacy and outcomes.* New York: Churchill Livingstone.

Baumgartner, T.A., Jackson, A.S., Mahar, M.T., & Rowe, A. (2003). *Measurement for evaluation in physical education and exercise science* (7th ed.). Dubuque, IA: McGraw-Hill.

Begg, C.B. (1987). Biases in the assessment of diagnostic tests. *Statistics in Medicine, 6,* 411-423.

Begg, C.B., & Greenes, R.A. (1983). Assessment of diagnostic tests when disease verification is subject to selection bias. *Biometrics, 39,* 207-215.

Bernstein, J. (1997). Decision analysis. *Journal of Bone and Joint Surgery, 79,* 1404-1414.

Bossuyt, P.M., Reitsma, J.B., Bruns, D.E., Gatosonis, C.A., Glasziou, P.P., Irwig, L.M., et al. (2003). The STARD statement for reporting studies of diagnostic accuracy: Explanation and elaboration. *Clinical Chemistry, 49,* 7-18.

Brenner, H., & Gefeller, O. (1997). Variation of sensitivity, specificity, likelihood ratios, and predictive values with disease prevalence. *Statistics in Medicine, 16,* 981-991.

Cipriani, D., Fox, C., Khuder, S., & Boudreau, N. (2005). Comparing Rasch analyses probability estimates to sensitivity, specificity, and likelihood ratios when examining the utility of medical diagnostic tests. *Journal of Applied Measurement, 6,* 180-201.

Collins, M.W., Grindel, S.H., Lovell, M.R., Dede, D.E., Moser, D.J., Phalin, B.R., et al. (1999). Relationship between concussion and neuropsychological performance in college football players. *Journal of the American Medical Association, 282,* 964-970.

Dawson, B., & Trapp, R.G. (2001). *Basic and clinical biostatistics* (3rd ed.). St. Louis: Lange Medical Books/McGraw-Hill.

Eddy, D.M. (1996). *Clinical decision making from theory to practice: A collection of essays from the Journal of the American Medical Association.* Boston: Jones & Bartlett.

Fleiss, J.L., Levin, B., & Paik, M.C. (2003). *Statistical methods for rates and proportions* (3rd ed.). New York: Wiley.

Garrett, E.S., Eaton, W.W., & Zeger, S. (2002). Methods for evaluating the performance of diagnostic tests in the absence of a gold standard: A latent class model approach. *Statistics in Medicine, 21,* 1289-1307.

Glas, A.S., Pijnenburg, B.A.., Lijmer, J.G., Bogaard, K., deRoos, M.A.J., Keeman, J.N., et al. (2002). Comparison of diagnostic decision rules and structured data collection in assessment of acute ankle injury. *Canadian Medical Association Journal, 166,* 727-733.

Greenfield, M.L., Kuhn, M.V.H., & Wojtys, E.M. (1998). A statistics primer: Validity and reliability. *American Journal of Sports Medicine, 26,* 483-485. Retrieved January 15, 2002, from Periodical Abstracts Full Text database.

Guggenmoos-Holzmann, I., & van Houwelingen, H.C. (2000). The (in)validity of sensitivity and specificity. *Statistics in Medicine, 19,* 1783-1792.

Hadgu, A. (1996). This discrepancy in discrepant analysis. *Lancet, 348,* 592-593.

Hawkins, D.M., Garrett, J.A., & Stephenson, B. (2001). Some issues in resolution of diagnostic tests using an imperfect gold standard. *Statistics in Medicine, 20,* 1987-2001.

Hlatky, M.A., Pryor, D.B., Harrell, F.E., Jr., Califf, R.M., Mark, D.B., & Rosati, R.A. (1984). Factors affecting sensitivity and specificity of exercise electrocardiography. *American Journal of Medicine, 77,* 64-71.

Hui, S., & Walter, S. (1980). Estimating the error rates of diagnostic tests. *Biometrics, 36,* 167-171.

Jaeschke, R., Guyatt, G., & Sackett, D.L. (1994a). User's guides to the medical literature: III. How to use an article about a diagnostic test A. Are the results of the study valid? *Journal of the American Medical Association, 271,* 389-391.

Jaeschke, R., Guyatt, G., & Sackett, D.L. (1994b). User's guides to the medical literature: III. How to use an article about a diagnostic test B. What are the results and will they help me in caring for my patients? *Journal of the American Medical Association, 271,* 703-707.

Joseph, L., Gyorkos, T., & Coupal, L. (1995). Bayesian estimation of disease prevalence and the parameters of diagnostic tests in the absence of a gold standard. *American Journal of Epidemiology, 141,* 262-272.

Kent, D.L., & Larson, E.B. (1992). Disease, level of impact, and quality of research methods. Three dimensions of clinical efficacy assessment applied to magnetic resonance imaging. *Investigative Radiology, 27,* 245-254.

Kraemer, H.C. (1992). *Evaluating medical tests: Objective and quantitative guidelines.* Thousand Oaks, CA: Sage.

Kumbhare, D.A., & Basmajian, J.V. (2000). *Decision making and outcomes in sports rehabilitation.* New York: Churchill Livingstone.

Leddy, J.J., Kesari, A., & Smolinski, R.J. (2002). Implementation of the Ottawa ankle rule in a university sports medicine center. *Medicine and Science in Sport and Exercise, 34,* 57-62.

Lee, W. (1999). Selecting diagnostic tests for ruling out or ruling in disease: The use of the Kullback-Leibler distance. *International Journal of Epidemiology, 28,* 521-525.

Lijmer, J.G., Mol, B.W., Heisterkamp, S., Bonsel, G.J., Prins, M.H., Van Der Meulen, J.H.P., & Bossuyt, P.M. (1999). Empirical evidence of design-related bias in studies of diagnostic tests, *Journal of the American Medical Association, 282,* 1061-1066.

McCrea, M. (2001). Standardized mental status testing on the sideline after sport-related concussion. *Journal of Athletic Training, 36,* 274-279.

McCrea, M., Kelly, J.P., Kluge, J., Ackley, B., & Randolph, C. (1997). Standardized assessment of concussion in football players. *Neurology, 48,* 586-588.

McCrea, M., Kelly, J.P., Randolph, C., Kluge, J., Bartolic, E., Finn, G., et al. (1998). Standardized assessment of concussion (SAC) on-site mental status evaluation of the athlete. *Journal of Head Trauma Rehabilitation, 13,* 27-35.

McDowell, I., & Newell, C. (1996). The theoretical and technical foundations of health measurement (pp. 10-46). In *Measuring health: A guide to rating scales and questionnaires* (2nd ed.). New York: Oxford Press.

Miller, W.C. (1998). Bias in discrepant analysis: When two wrongs don't make a right. *Journal of Clinical Epidemiology, 51,* 219-231.

Newcombe, R.G. (1998). Two-sided confidence intervals for the single proportion: Comparison of seven methods. *Statistics in Medicine, 17,* 857-872.

Newcombe, R.G. (2004). Confidence intervals for proportions and related quantities. Retrieved July 20, 2004, from www.uwcm.ac.uk/study/medicine/epidemiology_statistics/research/statistics/explanation.htm.

Newcombe, R.G., & Altman, D.G. (2000). Proportions and their differences. In D.G. Altman, D. Machin, T.N., Bryant, & M.J. Gardner (Eds.), *Statistics with confidence: Confidence intervals and statistical guidelines* (2nd ed., pp. 45-56). Bristol, England: British Medical Journal Books.

Philbrick, J.T., Horwitz, R.I., Feinstein, A.R., Langou, R.A., & Chandler, J.P. (1982). The limited spectrum of patients studied in exercise test research. Analyzing the tip of the iceberg. *Journal of the American Medical Association, 248,* 2467-2470.

Qu, Y., & Hadgu, A. (1998). A model for evaluating sensitivity and specificity for correlated diagnostic tests in efficacy studies with an imperfect reference test. *Journal of the American Statistical Association, 93,* 920-928.

Randsohoff, D.F., & Feinstein, A.R. (1978). Problems of spectrum and bias in evaluating the efficacy of diagnostic tests. *New England Journal of Medicine, 299,* 226-930.

Reid, M.C., Lachs, M.S., & Feinstein, A.R. (1995). Use of methodological standards in diagnostic test research. *Journal of the American Medical Association, 274,* 645-651.

Sackett, D.L., Straus, S.E., Richardson, W.S., Rosenberg, W., & Haynes, R.B. (2000). *Evidence-based medicine: How to practice and teach EBM.* St. Louis: Churchill Livingstone.

Safrit, M.J., & Wood, T.M. (Eds.). (1989). *Measurement concepts in physical education and exercise science.* Champaign, IL: Human Kinetics.

Springer, B.A., Arciero, R.A., Tenuta, J.J., & Taylor, D.C. (2000). A prospective study of modified Ottawa ankle rules in a military population: Interobserver agreement between physical therapists and orthopedic surgeons. *American Journal of Sports Medicine, 28,* 864-868.

STARD Group. (n.d.). The STARD initiative: Towards complete and accurate reporting of studies on diagnostic accuracy. Retrieved from www.consort-statement.org/stardstatement.htm.

Valenstein, P. (1990). Evaluating diagnostic tests with imperfect standards. *American Journal of Clinical Pathology, 93,* 252-258.

Chapter 16

American Psychological Association. (1985). *Standards for educational and psychological testing.* Washington, DC: Author.

American Psychological Association. (1987). *Principles for the validation and use of personnel selection procedures.* Washington, DC: Division of Industrial-Organizational Psychology, American Psychological Association.

Arnold, J.D., Rauschenberger, J.M., Soubel, W.G., & Guion, R.M. (1982). Validation and utility of a strength test for selecting steelworkers. *Journal of Applied Psychology, 67,* 588-604.

Arvey, R.D., & Faley, R.H. (1988). *Fairness in selecting employees* (2nd ed.). Reading, MA: Addison-Wesley.

Arvey, R.D., Nutting, S.M., & Landon, T.E. (1992). Validation strategies for physical ability testing in police and fire settings. *Public Personnel Management, 21,* 301-312.

Åstrand, P.-O., & Rodahl, K. (1986). *Textbook of work physiology* (3rd ed.). New York: McGraw-Hill.

Ayoub, M. (1982a). Control of manual lifting hazards: II. Job redesign. *Journal of Occupational Medicine, 24,* 676-688.

Ayoub, M. (1982b). Control of manual lifting hazards: III. Preemployment screening. *Journal of Occupational Medicine, 24,* 751-761.

Barnard, R., & Duncan, H.W. (1975). Heart rate and ECG responses of firefighters. *Journal of Occupational Medicine, 17,* 247-250.

Baumgartner, T.A., Jackson, A.S., Mahar, M.T., & Rowe, D.A. (2003). *Measurement for evaluation in physical education and exercise science* (7th ed.). Dubuque: Brown.

Brooks, G., & Fahey, T. (1984). *Exercise physiology: Human bioenergetics and its applications*. New York: Wiley.

Brozek, J., & Keys, A. (1951). The evaluation of leanness-fatness in man: Norms and intercorrelations. *British Journal of Nutrition, 5*, 194-206.

Bruce, R.A., Kusumi, F., & Hosmer, D. (1973). Maximal oxygen intake and nomographic assessment of functional aerobic impairment in cardiovascular disease. *American Heart Journal, 85*, 546-562.

Cady, L.D., Bishoff, D.P., O'Connell, E.R., Thomas, P.C., & Allan, J.H. (1979). Back injuries in firefighters. *Journal of Occupational Medicine, 21*, 269-272.

Cady, L. J., Thomas, P., & Karwasky, R. (1985). Program for increasing health and physical fitness of fire fighters. *Journal of Occupational Medicine, 27*, 110-114.

Chaffin, D.B. (1974). Human strength capability and low-back pain. *Journal of Occupational Medicine, 16*, 248-254.

Chaffin, D.B., & Park, K.S. (1973). A longitudinal study of low-back pain as associated with occupational weight lifting factors. *American Industrial Hygiene Association Journal, 34*, 513-525.

Davis, P.O., Dotson, C.O., O'Connor, J.S., & Confessore, R.J. (1992). *The development of a job-related physical performance test for Saint Paul firefighters*. Saint Paul, MN: Saint Paul Fire Department.

Dehlin, O., Hendenrud, B., & Horal, J. (1976). Back symptoms in nursing aids in a geriatric hospital. *Scandinavian Journal of Rehabilitative Medicine, 8*, 47-53.

Doolittle, T.L., Spurlin, O., Kaiyala, K., & Sovern, D. (1988). Physical demands of lineworkers. *Proceedings of the Human Factors Society, 32nd Annual Meeting, 32*, 632-636.

Durnin, J.V.G.A., & Passmore, R. (1967). *Energy, work and leisure*. London: Heinemann Educational Books.

Durnin, J.V.G.A., & Wormsley, J. (1974). Body fat assessed from total body density and its estimation from skinfold thickness: Measurements on 481 men and women aged from 16 to 72 years. *British Journal of Nutrition, 32*, 77-92.

Equal Employment Opportunity Commission. (1978). *Uniform guidelines on employment selection procedures*. Federal Register, 43(38289-28309).

Foster, C., Jackson, A.S., & Pollock, M.L. (1984). Generalized equations for predicting functional capacity from treadmill performance. *American Heart Journal, 107*, 1229-1234.

Gael, S. (1988). *The job analysis handbook for business, industry, and government* (Vol. I). New York: Wiley.

Herrin, G.D., Jaraiedi, M., & Anderson, C.K. (1986). Prediction of overexertion injuries using biomechanical and psychophysical models. *American Industrial Hygiene Association Journal, 47*, 322-330.

Hodgdon, J.A. (1992). Body composition in the military services: Standards and methods. In B. M. Marriott & J. Grumstrup-Scott (Eds.), *Body composition and physical performance: Applications for the military services* (pp. 57-70). Washington, DC: National Academy Press.

Hodgdon, J.A., & Jackson, A.S. (2000). Chapter 5: Physical test validation for job selection. In S. P. Constable, B. (Ed.), *The process of physical fitness standards development* (pp. 139-177). Wright-Patterson Air Force Base, OH: Human Systems Information Analysis Center.

Hogan, J., & Quigley, A.M. (1986). Physical standard for employment and courts. *American Psychologist, 41,* 1193-1217.

Hogan, J.C. (1991). Chapter 11: Physical abilities. In M.D. Dunnette & L.M. Hough (Eds.), *Handbook of industrial and organizational psychology* (2nd ed., Vol. 2, pp. 743-831). Palo Alto: Consulting Psychologist Press.

Hosmer, D.W., & Lemeshow, S. (1989). *Applied logistic regression.* New York: Wiley .

Jackson, A.S. (1986). *Validity of isometric strength tests for predicting work performance in offshore drilling and producing environments.* Houston: Shell Oil Company.

Jackson, A.S. (2000). Chapter 4: Types of physical performance tests. In S. Constable & B. Palmer (Eds.), *The process of physical fitness standards development: SOAR* (pp. 101-137). Wright-Patterson Air Force Base, OH: Human Systems Information Analysis Center.

Jackson, A.S., Blair, S.N., Mahar, M.T., Wier, L.T., Ross, R.M., & Stuteville, J.E. (1990). Prediction of functional aerobic capacity without exercise testing. *Medicine and Science in Sports and Exercise, 22,* 863-870.

Jackson, A.S., Borg, G., Zhang, J.J., Laughery, K.R., & Chen, J. (1997). Role of physical work capacity and load weight on psychophysical lift ratings. *International Journal of Industrial Ergonomics, 20,* 181-190.

Jackson, A.S., & Osburn, H. (1983). *Preemployment physical test development for coal mining technicians. Technical report to Shell Oil Co.* Houston: Shell Oil Company.

Jackson, A.S., Osburn, H.G., & Laughery, K.R. (1984). Validity of isometric strength tests for predicting performance in physically demanding jobs. *Proceedings of the Human Factors Society 28th Annual Meeting, 28,* 452-454.

Jackson, A.S., Osburn, H.G., & Laughery, K.R. (1991). Validity of isometric strength tests for predicting endurance work tasks of coal miners. *Proceedings of the Human Factors Society 35th Annual Meeting, 1,* 763-767.

Jackson, A.S., Osburn, H.G., Laughery, K.R., & Sekula, B.K. (1998). *Revalidation of methods for pre-employment assessment of physical abilities at Shell Western Exploration and Production, Inc., and CalResources LLC.* Houston: University of Houston.

Jackson, A.S., Osburn, H.G., Laughery, K. R., Sr., & Vaubel, K. P. (1991). Strength demands of chemical plant work tasks. *Proceedings of the Human Factors Society 35th Annual Meeting, 1,* 758-762.

Jackson, A.S., Osburn, H.G., Laughery, K.R., & Vaubel, K.P. (1992). Validity of isometric strength tests for predicting the capacity to crack, open and close industrial valves. *Proceedings of the Human Factors Society 36th Annual Meeting, 1,* 688-691.

Jackson, A.S., Osburn, H.G., Laughery, K.R., & Young, S.L. (1993). *Validation of physical strength tests for the Federal Express Corporation.* Houston: Center of Applied Psychological Services, Rice University.

Jackson, A.S., Osburn, H.G., Laughery, K.R., Young, S.L., & Zhang, J.J. (1994). *Patient lifting tasks at Methodist Hospital.* Houston: Center of Applied Psychological Services, Rice University.

Jackson, A.S., & Pollock, M.L. (1978). Generalized equations for predicting body density of men. *British Journal of Nutrition, 40,* 497-504.

Jackson, A.S., Pollock, M.L., & Ward, A. (1980). Generalized equations for predicting body density of women. *Medicine and Science in Sports and Exercise, 12,* 175-182.

Jackson, A.S., & Sekula, B.K. (1999). The influence of strength and gender on defining psychophysical lift capacity. *Proceeding of the Human Factors and Ergonomics Society, 43*, 723-727.

Jackson, A.S., Zhang, J.J., Laughery, K.R., Osburn, H.G., & Young, S.L. (1993). *Final report: Patient lifting tasks at Methodist Hospital.* Houston: Center of Applied Psychological Services, Rice University.

Jeanneret & Associates (1999). *Evaluation of physical ability and written tests for entry-level firefighters.* St. Paul, MN: City of St. Paul Fire Department.

Keyserling, W.M., et al. (1980). Establishing an industrial strength testing program. *American Industrial Hygiene Association Journal, 41*, 730-736.

Keyserling, W.M., Herrin, G.D., & Chaffin, D.B. (1980). Isometric strength testing as a means of controlling medical incidents on strenuous jobs. *Journal of Occupational Medicine, 22*, 332-336.

Laughery, K.R., & Jackson, A.S. (1984). Pre-employment physical test development for roustabout jobs on offshore production facilities. Lafayette, LA: Kerr-McGee Corp.

Lemon, P., & Hermiston, R. T. (1977a). The human energy cost of firefighting. *Journal of Occupational Medicine, 19*, 558-562.

Lemon, P., & Hermiston, R. (1977b). Physiological profile of professional fire fighters. *Journal of Occupational Medicine, 19*, 337-340.

Liles, D.H., Deivanayagam, S., Ayoub, M.M., & Mahajan, P. (1984). A job severity index for the evaluation and control of lifting injury. *Human Factors, 26*, 683-693.

Magora, A. (1970). Investigation of the relation between low back pain and occupation. *Industrial Medicine and Surgery, 39*, 504-510.

Manning, J., & Griggs, T. (1983). Heart rates in fire fighters using light and heavy breathing equipment: similar near-maximal exertion in response to multiple workload conditions. *Journal of Occupational Medicine, 25*, 215-218.

McArdle, W.D., Katch, F.I., & Katch, V.L. (2002). *Exercise physiology: Energy, nutrition, and human performance* (5th ed.). Philadelphia: Lea & Febiger.

National Institute for Occupational Safety and Health. (1977). *Preemployment strength testing.* Washington, DC: U.S. Department of Health and Human Services.

National Institute for Occupational Safety and Health. (1981). *Work practices guide for manual lifting.* Washington, DC: U.S. Department of Health and Human Services.

O'Connell, E., Thomas, P., Caddy, L., & Karwasky, R. (1986). Energy costs of simulated stair climbing as a job-related task in fire fighting. *Journal of Occupational Medicine, 28*, 282-284.

Passmore, R., & Durnin, J.V.G.A. (1955). Human energy expenditure. *Physiological Review, 35*, 801-840.

Pedhazur, E.J. (1997). *Multiple regression in behavioral research: Explanation and prediction* (3rd ed.). New York: Harcourt Brace College.

Pollock, M.L., Bohannon, R.L., Cooper, K.H., Ayres, J.J., Ward, A., White, S.R., et al. (1976). A comparative analysis of four protocols for maximal treadmill stress testing. *American Heart Journal, 92*, 39-42.

Rayson, M. (2000). Chapter 3: Job analysis. In B.E.S.P. Constable (Ed.), *The process of physical fitness standards development* (pp. 67-98). Wright-Patterson Air Force Base, OH: Human Systems Information Analysis Center.

Reilly, R.R., Zedeck, S., & Tenopyr, M.L. (1979). Validity and fairness of physical ability tests for predicting craft jobs. *Journal of Applied Psychology, 64,* 267-274.

Rosier, B. (2000). *Fundamentals of biostatistics* (5th ed.). Pacific Grove, CA.: Duxbury.

Rothstein, M.A., Craver, C.B., Schroeder, E.P., & Shoben, E.W. (1999). *Employment law* (2nd ed.). St. Paul: West Group.

SAS. (2002). *JMP: User's Guide, Version 5.* Cary, NC: SAS Institute.

Snook, S.H., Campanelli, R.A., & Hart, J.W. (1978). A study of three preventive approaches to low back injury. *Journal of Occupational Medicine, 20,* 478-481.

Snook, S.H., & Ciriello, V.M. (1991). The design of manual handling tasks: Revised tables of maximum acceptable weights and forces. *Ergonomics, 34,* 1197-1213.

Sothmann, M.S., Saupe, K., Jasenor, D., & Blaney, J. (1992). Heart rate response of firefighters to actual emergencies. *Journal of Occupational Medicine, 34,* 797-800.

Sothmann, M.S., Saupe, K.W., Jasenor, D., Blaney, J., Donahue-Fuhrman, S., & Woulfe, T. (1990). Advancing age and the cardiorespiratory stress of fire suppression: Determining a minimum standard for aerobic fitness. *Human Performance, 3,* 217-236.

United State Department of Health and Human Services. (1996). *Physical activity and health: A report of the Surgeon General.* Atlanta, GA: U.S. Department of Health and Human Services, Centers for Disease Control and Prevention, National Center for Chronic Disease Prevention and Health Promotion.

Vecchio, T. (1986). Predictive value of a single diagnostic test in unselected populations. *New England Journal of Medicine, 274,* 1171-1177.

Vogel, J.A., & Friedl, K.E. (1992). Army data: Body composition and physical capacity. In B.M. Marriott & J. Grumstrup-Scott (Eds.), *Body composition and physical performance: Applications for the military services* (pp. 89-104). Washington, DC: National Academy Press.

Waters, T.R., Putz-Anderson, V., Garg, A., & Fine, L. J. (1993). Revised NIOSH equation for the design and evaluation of manual lifting tasks. *Ergonomics, 7,* 749-766.

Weiner, J. (1994). *Physical abilities test follow-up validation study.* Sacramento: California Commission on Peace Officer Standards and Training.

Wellens, R.I., Roche, A.F., Khamis, H.J., Jackson, A.S., Pollock, M.L., & Siervogel, R.M. (1996). Relationships between body mass index and body composition. *Obesity Research, 4,* 35-44.

Wilmore, J.H., & Costill, D.L. (1994). *Physiology of sport and exercise.* Champaign, IL: Human Kinetics.

Index

Note: The italicized *f* and *t* following page numbers refer to figures and tables, respectively.

About the Editors

Terry M. Wood has spent most of his professional life teaching and conducting research in measurement and evaluation. Best known for his contributions to measurement and evaluation in physical education, Terry co-authored the 1995 NASPE *National Content Standards for Physical Education* and was involved in developing and implementing physical education content standards and benchmarks in Oregon.

He has published numerous manuscripts, consulted, and presented scholarly papers at the state, national, and international levels; has served as measurement and evaluation section editor for the

Terry M. Wood

Research Quarterly for Exercise and Sport; and has coauthored two measurement and evaluation textbooks. Terry has served as chair of the American Alliance for Health, Physical Education, Recreation and Dance (AAHPERD) Measurement and Evaluation Council and President of the American Association for Active Lifestyles and Fitness (AAALF). In addition he served on the editorial boards of the *Journal of Physical Education, Recreation & Dance* (*JOPERD*) and *Measurement and Evaluation in Physical Education and Exercise Science* (*MPEES*).

Currently an Emeritus Associate Professor in the College of Health and Human Sciences at Oregon State University, Terry remains involved in teaching at the undergraduate and graduate levels while enjoying his family, playing tennis, and exploring the Buddhist path.

Weimo Zhu, PhD, is currently an associate professor in the Department of Kinesiology and Community Health at the University of Illinois at Urbana-Champaign and a visiting professor at the Guangzhou Institute of Physical Education and Shanghai Institute of Physical Education, both

Weimo Zhu

in China. His major area of research is in measurement and evaluation in kinesiology.

Dr. Zhu's primary research interests are in the study and application of new measurement theories (e.g., item response theory) and models to the field of kinesiology. His research works have earned him international recognition. He served as the measurement section editor of the *Research Quarterly for Exercise* and Sport from 1999 to 2005, and he is a fellow of the American Academy of Kinesiology and Physical Education, American College of Sports Medicine, and Research Consortium, AAHPERD. He is a member of the Science Board of the President's Council on Physical Fitness and Sports and the Fitnessgram/Activitygram Advisory Committee. He is also a member of the editorial board for three other journals and serves on the executive committees of several national and international professional organizations. Dr. Zhu was the chair of the Measurement and Evaluation Council, AAHPERD. Currently, Dr. Zhu is also examining the application of advanced measurement and statistical techniques to several measurement issues in the area of public health. A tangible practical application of Zhu's theoretical work has been his work in the assessment of physical activity, and he is exploring a new idea (physical activity space) and technologies (voice-recognition and automatic scoring) to solve the problems raised.